BEYOND NATURE'S HOUSEKEEPERS

Beyond Nature's Housekeepers

AMERICAN WOMEN IN ENVIRONMENTAL HISTORY

Nancy C. Unger

OXFORD
UNIVERSITY PRESS

OXFORD
UNIVERSITY PRESS

Oxford University Press is a department of the University of Oxford.
It furthers the University's objective of excellence in research,
scholarship, and education by publishing worldwide.

Oxford   New York
Auckland   Cape Town   Dar es Salaam   Hong Kong   Karachi
Kuala Lumpur   Madrid   Melbourne   Mexico City   Nairobi
New Delhi   Shanghai   Taipei   Toronto

With offices in
Argentina   Austria   Brazil   Chile   Czech Republic   France   Greece
Guatemala   Hungary   Italy   Japan   Poland   Portugal   Singapore
South Korea   Switzerland   Thailand   Turkey   Ukraine   Vietnam

Copyright © Oxford University Press 2012

Published in the United States of America by
Oxford University Press
198 Madison Avenue, New York, NY 10016

www.oup.com

Library of Congress Cataloging-in-Publication Data
Unger, Nancy C.
Beyond nature's housekeepers : American women in environmental history / Nancy C. Unger.
    p.   cm.
Includes bibliographical references and index.
ISBN 978-0-19-973506-8 (hardcover : acid-free paper)—ISBN 978-0-19-973507-5 (pbk. : acid-free paper)
1. Women and the environment—United States—History. 2. Sex role—United States—History.
3. Nature—Effect of human beings on—United States—History. 4. Human ecology—United States—History.
5. Conservation of natural resources—United States—History. 6. Environmentalism—United States—History.
7. United States—Environmental conditions—History. 8. United States—Social conditions. I. Title.
GF13.3.U6U54   2012
304.2082'0973—dc23   2012008759

9 8 7 6 5 4 3 2 1

Printed in the United States of America
on acid-free paper

To my children, Travis Whitebread and Megan Unger,
my joy and inspiration

Contents

Illustrations

Acknowledgments

THE BUSINESS HISTORIAN Edwin Perkins insists that no one writes two books. According to Ed, most authors write only one. People who write two books go on to write three or four. Since the completion of my first book, Ed has taken seriously his responsibility to remind me of the Perkins Maxim on a regular basis, so it is of considerable relief to us both to have this book completed at last. Thanks to Ed, for believing, and insisting, that I should not be a one-book wonder. Thanks, too, to John W. Wright, literary agent extraordinaire, who prodded me along as well. In addition, John shared his wealth of knowledge about the publishing industry and about good writing—and adds sparkle to my life.

My first book was *Fighting Bob La Follette: The Righteous Reformer* (University of North Carolina Press, 2000; rev. ed., Wisconsin Historical Society Press, 2008). Making the leap from political biography to women in environmental history required a lot of help from a large number of people without whom I could not have attempted to carry out this test of the Perkins Maxim.

As Robert Cherny of San Francisco State University has noted, you don't really know a subject until you've taught it. I offered my first environmental history class in 1994. Through funding provided by Santa Clara University's Institute on Environment, I was particularly fortunate to have Carolyn Merchant address that class and answer questions prompted by my students' reading of her textbook *Major Problems in American Environmental History* and some of her other works. Carolyn also took time to discuss with me some of the challenges of combining career and family and to encourage my scholarship. Over the years students in my environmental and gender history courses have made observations and raised questions that have challenged me to think deeply about a variety of issues and problems, causing me to reexamine, and sometimes jettison, what I assumed were immutable truths. Some students, including Katherine Bercovitz, alerted me to sources that proved valuable to this book.

Robert Carriker contributed significantly to this book by inviting me to present some of my preliminary thinking as the 2001 William L. Davis, S.J., Lecture at Gonzaga University. Preparing the talk "Beyond 'Nature's Housekeepers': American Women in Environmental History" gave me the first opportunity to express in a public forum my ideas on the relationship between gender and environment, and the response to that presentation was encouraging.

The following year, at the suggestion of Robert Cherny, Marie Bolton and I began the collaboration that would result in many papers and publications, all with environmental themes. Because Marie is Maître de conférences at University Blaise Pascal (Clermont-Ferrand II), almost all of our jointly authored works were composed, edited, and disseminated while we were on separate continents. A delightful exception was the 2003 Society for North America Studies conference, California: Periphery or Laboratory, in Montepellier, France, for which Annick Foucrier arranged that the American embassy provide my airfare. Despite the constraints on our ability to confer in person, Marie and I have enjoyed a productive partnership, and it was my work with her that gave me much of the experience and confidence necessary to pursue this book. A paper that we coauthored provided partial foundation for chapter 8. Moreover, despite her very demanding schedule, Marie read this entire manuscript on short notice, correcting errors and making many valuable suggestions.

Under the supervision of my department's office manager, Judy Gillette, and sometimes with her assistance as well, a number of students, including Maureen Babiarz, Michael Bates, Michael Easter, Brigid Eckhart, Ricardo Estrado, Daniel Felice, Kelly Greenwalt, Ryan Heal, Brian Hurd, Claire Ingram, Brian Kernan, Rachel Koch, Christina Lynch, Celina Mogan, Katherine Powers, Darby Riley, and Matthew Rinegar, aided in gathering materials, creating the bibliography, transcribing documents, and other related tasks. Blair Thedinger and Patti Adams conducted some of the research for one of the early essays coauthored by Marie Bolton and myself.

Students at Marquette University asked some thought-provoking questions after a guest lecture there, and my Santa Clara colleagues, especially Steven Gelber and Andrea Pappas, offered helpful suggestions following my presentations at various colloquia on our campus. My colleague Robert Senkewicz, an expert on early California history, answered my questions on the Spanish and Mexican periods and directed me to useful literature. Don Scherer was a visiting scholar at Santa Clara when he encouraged this project, supplying me with a number of articles concerning Ellen Swallow Richards. Janet Burns, President of the Cambridge Plant and Garden Club, kindly granted access to her organization's archives. In Wisconsin, Randall Davidson, Cassandra Dixon, Dorothy Lagerroos, Gail Lamberty, and the late Nan Cheney generously shared their expertise, as did Emily Greene and Barbara Stoll of Alapine Village, Alabama. My late colleague Catherine Bell forwarded to me the issue of *Smithsonian* magazine featuring the wonderful photograph by William T. Clarke of nature's housekeepers that appears on page 66 of this book. Elwood Mills, of Santa Clara's Media Services, cleaned up and reformatted the political cartoons that illustrate two of the chapters. Alan Lessoff, at Illinois State University, and my SCU colleague Naomi Andrews were particularly sympathetic sounding boards as I encountered many detours on the road to the book's completion. Their own commitments to excellence in historical research as well as their good humor inspired me to keep moving forward.

Organizations as well as individuals were generous. The Historical Society of Southern California/Haynes Foundation provided a research stipend that funded my work at the Department of Special Collections at the Davidson Library at the University of California, Santa Barbara. A Mary Lily Research Grant funded my work at the Sallie Bingham Center for Women's History and Culture at Duke University.

From the preliminary research to the finishing touches, Santa Clara University has been very liberal in its support. This project received research funding from the Center for Science, Technology, and Society as well as a Presidential Research Grant; an Arthur Vining Davis Grant; a Hackworth Grant for Faculty Research in Applied Ethics; and a Dean's Grant for document transcription. Additional internal grants paid for the illustration permissions and indexing.

The various grants funded the research; librarians and archivists helped to carry it out. Staff at a variety of research centers, including the Sallie Bingham Center for Women's History and Culture at Duke University, the Huntington Library, the Davidson Library at the University of California, Santa Barbara, the Bancroft Library, and the Schlesinger Library at the Radcliffe Institute all helped me find what I wanted and led me to sources I didn't even know I needed. The Wisconsin Historical Society continues to be one of the greatest research centers in the country. Staff at its various depositories throughout the state sent requested materials to Madison for my use. The librarians at Santa Clara University were also most helpful, especially Helene La France and Cynthia Bradley.

Photo librarians and archivists are a special breed and therefore deserve special thanks as they were sometimes forced to enter rather protracted correspondence to ensure that I received all the necessary paperwork as well as the requested image. Jessica Blesso, Larry Burtness, Wendy E. Chmielewski, Elizabeth Clemens, Ari Davidow,

William Donnelly, Sherrill Fouts, Madeline Kelty, Daniel Kosharek, Cathleen Miller, Merrideth Miller, William Offhaus, Vikki Schmidt, and Mary Beth Sigado all proved helpful and efficient. The Wisconsin Historical Society again comes in for special thanks, in particular Lisa Marine and especially John Nondorf. Photographers Harry Littell and Diana Mara Henry were exceptionally generous.

Some preliminary publications helped me to hone my writing and expand my thinking. Rachel Stein extensively and expertly edited my first essay on the topic, an overview called "Women, Sexuality, and Environmental Justice in American History," which appears in her edited collection, *New Perspectives on Environmental Justice: Gender, Sexuality, and Activism* (Rutgers University Press, 2004). In 2006, I presented "Gendered Approaches to Environmental Justice" at the 2006 meeting of the American Society for Environmental History in St. Paul, Minnesota. Comments on that presentation, especially those by Susan Schrepfer, proved most helpful in thinking about ways to approach this book.

Sylvia Washington also encouraged my development in the field. She included my essay "Gendered Approaches to Environmental Justice: An Historical Sampling" in the volume she edited with Paul C. Rosier and Heather Goodall, *Echoes from the Poisoned Well: Global Memories of Environmental Injustice* (Rowman and Littlefield/Lexington Books, 2006). It was Sylvia who invited me to join the board of editors of *Environmental Justice* and solicited my article "The Role of Gender in Environmental History" for the September 2008 issue of that journal.

Andrew (Drew) Isenberg invited me to write the chapter on gender for the forthcoming *Oxford Handbook of Environmental History*. The contributors met at Temple University in September 2007. I benefited greatly from all who contributed (including Walter Briggs, Emily Brock, Mark Carey, Kurk Dorsey, Jeff Featherstone, Petra Goedde, Angela Gugliotta, Nancy Langston, Thomas Lekan, Michael Lewis, Rob Mason, Sara Pritchard, Mark Stoll, Ellen Stroud, Paul Sutter, James Turner, Brett Walker, and Frank Zelko), particularly those who commented on my draft specifically, especially Connie Chiang and Beth Bailey. Drew's critique of a subsequent draft also substantially improved the final product, the chapter "Gender: A Useful Category of Analysis in Environmental History."

Two preliminary articles appeared in the *Wisconsin Magazine of History*: "The 'We Say What We Think' Club: Rural Wisconsin Women and the Development of Environmental Ethics" (Autumn 2006) and "Women for a Peaceful Christmas: Wisconsin Homemakers Seek to Remake American Culture" (Winter 2009–2010). A third article, "Wisconsin's League Against Nuclear Dangers: The Power of Informed Citizenship" is forthcoming. All articles benefited from the magazine's reviewers and fact-checkers, and most of all from the expertise of editor Jane De Broux.

A particularly delightful step along the journey was my participation in a model of cooperative scholarship, a workshop funded by the Social Sciences and Humanities

Research Council of Canada and hosted by organizers Catriona Mortimer-Sandilands and Bruce Erickson in Toronto in May 2007. I learned a great deal from my fellow participants Stacy Alaimo, David Bell, Dianne Chisholm, Giovanna Di Chiro, Andil Gosine, Katie Hogan, Gordon Brent Ingram, Ladelle McWhorter, Rachel Stein, and Noël Sturgeon, as well as Catriona and Bruce. The end result of our efforts, *Queer Ecologies: Sex, Nature, Biopolitics, and Desire* (University of Indiana Press, 2010), includes my essay "The Role of Nature in Lesbian Alternative Environments in the United States: From Jook Joints to Sisterspace," which benefited significantly from Catriona and Bruce's editing and the entire group's suggestions and encouragement.

In March 2010, it was my pleasure to chair and comment on the panel "Alternate Voices, Shared Visions: Women in Post–World War II Environmentalism," at the annual meeting of the American Society for Environmental History, in Portland, Oregon. Young scholars Carla Fisher, Brittany Fremion, and Megan Jones taught me a great deal about new trends in the study of the role of gender in environmental history, and I was heartened and inspired by their enthusiasm as well as their expertise.

Publications by my friend Mary Whisner, law librarian at the University of Washington, include *Practicing Reference: Thoughts for Librarians and Legal Researchers* (William S. Hein, 2006), a collection of the articles she contributes regularly to the *Law Library Journal*. In a 2011 contribution to the "Practicing Reference" series, an article called "Writing Buddies," Mary offers a wealth of good advice for writers as she explores the challenges and rewards of finding friends, colleagues, or relatives willing and able to offer meaningful feedback. I have been extremely fortunate to have several excellent writing buddies as this book has taken shape. High school student Sherry He volunteered her services, offering suggestions and scouring the notes and bibliography for errors. Alisa Chester took time from her master's program in women's studies at George Washington University to provide extensive commentary on several chapters. One of my most helpful writing buddies is Susan Goodier, who took time away from her own book project to provide many valuable suggestions on various drafts of mine. It has also been my good fortune to have had Mary Whisner herself as my writing buddy of longest and highest standing. Mary read carefully and thoughtfully the many drafts of this project and offered encouragement, suggestions, corrections, and helpful observations and comments. Knowing that Mary was going to give a chapter the "Whisner Treatment" inspired me to make my work the best it could be in order to take full advantage of her time and expertise. When she wasn't editing, she was sending me links to articles and news stories related to my topic. Mary's good humor, warmth, and ready wit complements her dedication to thorough research, proper sentence structure and word choice, and good, clear writing. All errors in this book are undoubtedly those I somehow managed to insert after Mary made her final meticulous edits to the page proofs.

Scholars lament that editors don't really edit anymore. Clearly, they have not worked with Susan Ferber, executive editor at Oxford University Press. Susan is a hands-on editor. Her historical expertise is wide ranging, and she is known to write almost as much on the manuscript as the author has. The expert reports she garnered for this manuscript were also enormously helpful. Following their recommendations and hers resulted in a leaner, cleaner manuscript, vastly improved in both content and mechanics. Susan also answers e-mail queries promptly, states expectations and processes clearly, and is all-around what every truly great editor should be. Others associated with the press match Susan in attention to detail and expertise. Stacey Hamilton meticulously applied her copy editing skills throughout the manuscript. Production manager Marc Schneider, in cooperation with Anitha Chellamuthu and the entire production team, saw to every detail of getting the pages between covers. Thanks too to Elyse Turr and many others at the press for their dedication and hard work.

Special thanks go to my mother-in-law, Virginia Remy Whitebread. The girl on horseback gracing the cover of this book is Virginia, aged sixteen, far from the confines of her native Chicago and glorying in the freedom of the Great American West. Her love for her adopted home, where she has lived for the last sixty years, still abides. Because of Virginia, I have a great book cover—and the world's best husband. Don Whitebread not only provided generous financial and endless emotional support, but also painstakingly read each draft of each chapter, offering useful feedback and invaluable encouragement. A prizewinning photographer, Don also provided technical support and advice concerning the book's illustrations.

Our children, Travis Whitebread and Megan Unger, didn't really contribute all that much to these pages. In fact, without them to provide endless distractions, I would have finished this book much sooner. I would not, however, have had nearly as much fun, for which I am truly and eternally grateful. They, and their dad, get me away from my desk, bring much-needed perspective, refresh my spirit, and bring me immeasurable joy.

BEYOND NATURE'S HOUSEKEEPERS

"A lack of cross-field knowledge"

Introduction

SEX, SEXUALITY, AND GENDER AS USEFUL CATEGORIES IN ENVIRONMENTAL HISTORY

"Nothing in Sight but Nature"

When John Jones became "carried away with the idea" of crossing the continent to live in California, his wife's first reaction was, "Oh, let us not go." But Mary A. Jones's objection "made no difference . . . & on the 4th day of May 1846 we joined the camp for California." Used to the privileges and relative ease of white middle-class life, Mary Jones was exhausted by the rigors of the overland journey. She was also pregnant. Upon their arrival she was occupied with the new baby and preferred not to travel any more than was necessary. She relented, however, when her husband, who had been scouting the countryside for a homesite, insisted that she make a preliminary trip to see his selection. "We camped that night," she recalled. "My husband stopped the team and said 'Mary, have you ever seen anything more beautiful?'" The young wife and mother was repulsed rather than impressed, noting with horror, "There was nothing in sight but nature. Nothing . . . except a little mud and stick hut." Mary Jones, notes historian Lillian Schlissel, "found nothing grand, nothing wonderful, nothing to suggest what her husband so clearly saw. She and other women did not find the new country a land of resplendent opportunities. They heard their children crying and longed for home."[1]

Why did so many men and women of the same race and class have such different and visceral reactions to the same landscape? Why did the majority of the white middle-class men on the overland trails in the mid-nineteenth century embark eagerly on their journeys, and why did so many of their wives accompany them only reluctantly, apprehensively, writing home to loved ones of their fear and dread?[2] Once in the West, some women eagerly threw off the social constraints of the East to embrace new opportunities, yet the reactions of men and women to the environment frequently remained polarized. Some sixty years later, many women were focused on protecting the natural resources (including trees, soil, birds, and animals) that so many men of their same race and class were determined, in the name of "progress," to exploit for profit. Lydia Adams-Williams represented the views of many women when she proclaimed in 1908, "Man has been too busy building railroads, constructing ships, engineering great projects, and exploiting vast commercial enterprises to take the time necessary to consider the problems which concern the welfare of the home and the future."[3] But her contemporary, George L. Knapp, called views like Adams-Williams's "unadulterated humbug" and dismissed conservationists' dire prophecies as "baseless vaporings." According to Knapp, men should be praised, not chastened, for turning "forests into villages, mines into ships and skyscrapers, scenery into work."[4]

This book is an effort to explain these kinds of extreme gendered divisions and to offer an enriched understanding of the powerful interplay between environment and sex, sexuality, and gender. The synergy produced by that interplay has been significant throughout American history, but it cannot be adequately understood and appreciated as long as those fields are discussed as discrete entities. The fields of gender and environment are growing, but scholars have seldom joined them together in analysis or heeded historian Carolyn Merchant's call that a gendered perspective be added to conceptual frameworks in environmental history.[5] They have not offered a unified analysis of the intersections that shaped gendered environmental concerns and activism and that framed as well the way the larger culture responded. Of the growing number of American environmental histories that feature women or gender, many remain narrowly focused on the modern environmental movement (environmentalism) or take a regional or otherwise limited approach.[6] Others offer fascinating global, gendered perspectives and profound philosophical insight but are not sufficiently focused for the reader specifically interested in American history.[7] Some of the existing scholarship concerning the role of gender in environmental history is even potentially damaging, such as the tendency to anthropomorphize and feminize nature through terms like "Mother Nature" and "Mother Earth," and calling environmental exploitation the "rape" of "virgin" land. Such tendencies devalue women and work against respecting nature as an agent in its own right: a partner, equal to humans in value and dignity.[8] Considerable work has been done, however, that is constructive and valuable. Many studies, for example, have been made of women naturalists and nature writers and of women in early environmental protection movements.[9]

In American environmental history surveys, women are most likely to appear in coverage of the Progressive Era (circa 1880–1917) as middle-class women claimed that their domestic expertise gave them a unique perspective on living in the new urban, industrialized society.[10] As families' production of their own food decreased, women became involved in activities to ensure that the store-bought foods were wholesome, free from impurities and harmful additives. Their campaigns for pure milk and better sanitation highlighted their new role as "municipal housekeepers." Other female reformers of the period applied their prescribed maternal role as caretakers to nonhuman nature and became active in the movements to create natural parks and nature preserves and to save the many species of birds that were being hunted into near extinction for their feathers. Journalist Wendy Kaminer didn't coin the term "nature's housekeepers" until 1992, but the housekeeping role for women that extended far beyond the confines of the home and local municipalities has long been important to environmental history.[11] It has, however, also overshadowed women's other contributions to the environment, such as opposition to nuclear war and support of soil conservation, activities not directly related to "housekeeping."

This study incorporates the better-known contributions made by women to environmental history but also moves beyond "Nature's Housekeepers" to provide that much needed overview of the role that sex, sexuality, and gender has played in the spectrum of American environmental history, from the pre-Columbian period to the present. In this view, gender is especially emphasized, particularly in the ways it has affected, and has been affected by, women.

What Is the Difference between Sex, Sexuality, and Gender?

Sex, sexuality, and gender are related terms, and sometimes it can be hard to know where one leaves off and another begins. Sex is determined by physiology, but it encompasses more than simply the differences between male and female genitalia. Sex-related functions (menstruation, pregnancy, childbirth, and breastfeeding) affected the way North American women interacted with the environment, starting with the farming practices of pre-Columbian Native Americans and continuing through to the present, as when female farm workers suffer the impact of pesticides on fetal development.[12] When faced with the challenge of domesticating the wilderness, slave buyers preferred the usually larger, stronger African males over the usually smaller, weaker African women, thereby skewing the sex ratio among slaves for several generations. Factors determined by sex also undoubtedly contributed to many women's reluctance to take the journey west. Some, like Mary Jones, were pregnant and worried by the prospect of the rigors of the journey and the knowledge that there could be no delays of more than a day or two while on the trail, and that there were no medical facilities along the way.[13] Women who were not pregnant faced another sex-based burden: enduring menstrual periods on the road in the pre-tampon, pre–disposable pad days of the nineteenth century. Sex-based bodily functions have played an important role in people's environmental attitudes and actions throughout American history.

Sexuality focuses on sexual practices and sexual identity. Both African and Native American women thwarted their own reproduction as a form of resistance to being used as forced laborers, retarding their masters' efforts to transform the land. Birth control played a role in Progressive-era efforts to improve conditions within urban environments, and lesbians used certain landscapes, both built and natural, to resist homophobia and to help foster within themselves a more positive identity.

Gender is perhaps the more complex of the three factors, encompassing the behavioral, cultural, or psychological traits typically associated with one sex. The ability to give birth and to breastfeed is determined by sex, as is getting ovarian or prostate cancer. Lung cancer, however, used to be a primarily male disease. Its increasing prevalence among women began once female smoking was no longer socially taboo and tobacco companies' advertisements began targeting women. The rise in women's lung cancer rates is the result of gender, of behavior resulting from changing ideas about what is considered acceptable for women. Gender, then, refers to culturally defined and/or acquired characteristics. Notions of gender affected the way people thought about themselves and others, and influenced the way they learned, lived, and wielded power.

Attributing to sex and anatomy the qualities and perceptions that resulted from gender and culture has resulted in false and damaging stereotypes throughout human history, but untying what sociologist Allan G. Johnson calls the "gender knot" is a daunting task.[14] With the rise of the second wave of feminism in the 1960s came an explosion in scholarship seeking to understand which qualities that have been described as "natural" to men and women were actually the result of the internalization of powerful gender prescriptions based in social, economic, and political factors. For example, the advent of the birth control pill and legalized abortion dramatically reduced fears of unplanned pregnancies, paving the way for the sexual revolution that belied the long-standing "truth" that women were "naturally" less sexual than men. "Certainties," such as one sex being inherently better at math or more emotional, also came under close scrutiny.

Gender scholars face a series of challenging and complicated questions: If differences previously chalked up to sex were actually prescribed and constructed as a result of gender, then how, when, and why were those differences constructed? Can they be deconstructed? Where do people get their ideas about what it means to be a man, and what it means to be a woman, and how and why have those ideas changed over time? How were the men and women who internalized the roles, values, and attitudes affected by the norms prescribed to them? What happened when gender roles perceived to be natural, and therefore fixed, changed?

This book focuses on how internalization of prescribed gender traits colored people's reactions to the world around them. It also reveals how significant and far-reaching the impact of sex, sexuality, and especially gender has been in women's responses to the environment and environmental issues throughout American history. Gender, sex, and sexuality shaped the social relationships between men and women, but they also influenced the way nonhuman nature was viewed, used, abused, protected, or preserved. Men and women frequently understood and responded to their environment and to environmental issues in decidedly different ways. Even as women uniquely contributed to how a particular environment developed, environments shaped the way women perceived themselves—as disempowered, as in the case of Mary A. Jones and other frontier women, or as strong and independent, as in the case of the lesbians who created alternative environments.

Environmental history is a relatively new discipline. In long-standing subfields of American history, the focus has traditionally been on the thoughts and actions of people, with the nonhuman world usually serving as a not-terribly-important stage upon which human actors perform. Studies of the colonial period, for example, tend to concentrate on the thoughts and actions of early settlers as they wrested civilization out of wilderness and laid the foundations for the American Revolution. Usually scant attention is paid to the changes that their practices brought to soils, water, and local flora and fauna. However, in the classic environmental history *Changes in the Land: Indians, Colonists, and the Ecology of New England*, William Cronon emphasizes the degree to which the early colonists transformed New England physically. Cronon exposes changes so far-reaching as to even affect climate and reveals how crucial such nonhuman factors were to the colonists' ultimate economic and political successes. By clearing land, farming, building fences, and introducing European domesticated animals and plants, colonists carried out an ecological revolution that was just as profound and far-reaching as the political revolution to come.

Despite the growing consensus on the importance of nonhuman nature, competing definitions of environmental history abound. One scholar went so far as to title his effort "A Death-Defying Attempt to Articulate a Coherent Definition of Environmental History."[15] Leading scholar J. R. McNeill issued one of the more inclusive and compelling definitions, asserting that there are three main strands of environmental history: material environmental history, focusing on changes in the biological and physical environment; cultural/intellectual environmental history, focusing on representations of the environment and what they reveal about society; and political environmental history, focused on government regulation, law, and official policy.[16] This book defines "environment" very broadly, incorporating all three of those strands. It includes, for example, ecology (the science focusing on the interrelationships of organisms) and ecosystems, as well as landscapes, both natural and constructed. It sees humans as rightful members of the natural world, not as inherently exploitative outsiders. It looks at the modern environmental movement dedicated to the conservation or preservation of wilderness and natural resources within the last century, but also examines how and why throughout American history landscapes and places have shaped people, their values and their cultural institutions—and how and why people shaped those landscapes and places in return.

Sex, sexuality, and gender are as much a part of men's lives as they are of women's. Throughout much of history, however, men have wielded disproportionate political and economic power and received the lion's share of the attention from historians. When the new women's history first emerged in the 1970s, many of the early studies emulated the male model of focusing on the "greats" of the past and provided accounts of female leadership in various traditionally male-dominated fields, such as politics, medicine, and literature. Environmental historians followed suit: Many of the environmental history studies of the 1970s and 1980s that focused on women examined the contributions of individual female scientists (Alice Hamilton, pioneer in occupational medicine and industrial toxicology), conservationists (naturalist Caroline Dormon), and nature writers (Mary Austin, *The Land of Little Rain*), with Rachel Carson, author of *Silent Spring*, becoming by far the most frequently cited and celebrated female environmentalist.[17] Because men have traditionally received greater attention from environmental historians, this book focuses primarily on women, but the uniqueness and importance of women's roles cannot be fully appreciated unless placed in an appropriately gendered context.[18]

In addition to sex, other factors, including race, ethnicity, and class, help construct gender roles, and the culture that results can change dramatically over time. These complexities must also be incorporated to appreciate fully the differences that gender, sex, and sexual identity have made in shaping men's and women's attitudes toward, and relationships with, the environment and each other. Race and class, for example, are at the heart of the environmental justice movements emerging in the 1970s.

Just as women's history rapidly developed from a rather pale imitation of men's history into a vibrant, rich, and important field in its own right, environmental history is broadening its focus to become a vast multidisciplinary field encompassing the entire globe. Women, not just individual female "greats," increasingly appear in its literature, and issues of masculinity as well as sexuality, including reproduction and homosexuality, are also recognized.[19] Studies of women in environmental history have also broadened from an emphasis on women activists who consciously worked to protect the environment to myriad gender-based environmental relationships across time and space.[20]

Beyond Nature's Housekeepers: American Women in Environmental History surveys a wide range of issues over a broad sweep of time and place from wilderness to cities to suburbia. Because gender played a key role throughout American environmental history, it (as well as sex and sexuality) is examined within the context of a variety of other factors. That is, explaining why and how women were thinking about the environment and interacting with it in different ways than men involves looking at free white women, enslaved women, white pioneer women, black pioneer women, poor urban immigrants, women of color living in areas formerly used by industry and often contaminated, white middle-class suburban women, and so on.[21] There is no single "woman's environmental experience" in any place and time—and yet across historical periods, age, sexuality, marital and maternal status, race, ethnicity, economic class, and gender consistently played a role in women's interactions with the environment. The topics examined in this book are exceptionally vivid and representative examples of a particular way in which gender affected significant environmental attitudes or actions, but they are by no means comprehensive.

The book proceeds chronologically for the most part, organized around key periods in American history with a greater emphasis on the twentieth century, given the abundance of research material and general interest. Occasionally a particular theme is studied over several time periods within a single chapter. Chapter 7, for example, emphasizes the gendering of sexuality, in this case homosexuality, and reveals the role of place in the evolution of lesbian identity over more than a century. As what it meant to be a lesbian changed over time, so did the kinds of environments in which lesbians felt the most at home and empowered.

Some women's organizations appear in more than one chapter. The Cambridge Plant and Garden Club (CPGC), founded in 1889, is one of the oldest garden clubs in the country, and its history reveals how the environmental consciousness of socially active white middle-class women developed over several generations from the care of houseplants to involvement in the nuclear freeze movement. Certain individuals are also highlighted because their actions or philosophies are representative of a group or trend rather than unique. For example, Adda Howie, a society matron turned dairy farmer, was initially ridiculed in the 1890s for keeping her barn scrupulously clean and even providing her cows with such feminine amenities as curtains. However, when her cows' butter and milk production broke all records, she was publicly praised for her skill in bringing traditional female values into barns and pastures, transforming her into a national, and then international, expert on dairy farming. Although Howie hailed from the rural Midwest, her kind of gendered thinking about the impact of the workplace environment on both quality of life and productivity formed the foundation for urban women's claim to authority as reform experts when they brought their "natural" expertise to the factories and other work environments of industrializing America.

No single book can reveal the totality of the intricate web of relationships between sex, sexuality, gender, and the environment throughout American history. This volume incorporates some of the leading scholarship documenting women's environmental contributions as nature writers (including poets and novelists, as well as authors of nonfiction) and scientists (such as biologists, botanists, and chemists). Its larger goal, however, is to show that people and movements that achieve the bulk of attention in more traditional environmental studies are not the sum total of the interplay between gender and environment and thus they may be relegated to the background here. For example, while the environmental movement that swept the nation in the 1970s merits significant notice, the largest share of attention is given to the aspects of that movement that feature distinct gender differences.

Like most surveys of American history, this one begins with Native Americans, but with a less conventional focus on prescribed gender roles within a variety of indigenous tribes, and more discussion of the entwined environmental repercussions of women's farming methods and their efforts to control their own reproduction. Chapter 2 features a topic familiar to any survey of American history: slavery. But it emphasizes the ways in which female slaves, using their environmental knowledge, subtly resisted their enslavement by limiting their masters' cotton production and accelerating soil exhaustion. The chapter dealing with the Great Depression and World War II de-emphasizes politics and economics to discuss how women employed conservation efforts in the wake of the Progressive Era to reassert some of their authority. *Beyond Nature's Housekeepers* will clearly not be the last word on the profound interrelationships between women's sex, sexuality, gender roles, race, power, and environment. It is intended instead to be a part of the larger conversation on the value of diversity and interdisciplinary approaches to understanding American history, particularly with regard to both gender and physical space.

Moreover, *Beyond Nature's Housekeepers* reveals the range of women's experiences as well as their contributions to American environmental history, and to the nation's political, economic, and social history as well. Not all women desired to stay as disengaged from the natural world as Mary Jones. From Native American women's ingenious practices to retard soil depletion, to enslaved women's efforts to subtly resist enriching their masters, to nineteenth-century women's claims that women were especially suited to the study of botany, to women's efforts to civilize the American West, the antebellum nation was profoundly shaped by women's engagement with their environments. Following the Civil War, women continued to wield significant influence as they served as midwives to the conservation movement, became empowered by the natural world through scouting and other organizations, contributed to the American victory in two world wars and combated the effects of the Great Depression by producing, preparing, and preserving their own food in order to avoid waste, and led efforts to save the planet while carrying out environmental justice. Combined, their stories reveal vibrant characters and shine a light on underappreciated aspects of the often inspiring and always complex history of American women.

"A new national script of gender"

1   Gendered Changes to the Land in Pre-Columbian and Colonial America

"The Women Now Went Willingly into the Field"

British efforts to colonize North America got off to such a horrific start that it is surprising that further plans were not quickly abandoned. Within a year and a half of the Virginia settlement's founding in 1607, of its original 608 settlers, 510 had died. Even after Jamestown had been established for more than a decade, the death toll remained staggering: between 1619 and 1622, roughly 3,000 of the colony's 3,600 settlers died. Despite the mythic tales of the bountiful first Thanksgiving in 1621, the Pilgrims of Plymouth, eking out a living on hilly, poor soil, fared only marginally better. It was an Indian man, Squanto, who explained that fertilizing the soil with the fish plentiful in the local brook was essential to a successful corn crop—which was ironic because it was Indian women, not men, who farmed virtually all indigenous crops except tobacco. Because of the intensive labor involved in the fertilization process, starvation might well have remained a constant Pilgrim fear, if not a likely companion. In the spring of 1622, however, Plymouth's governor, William Bradford, made a decision that brought his people's debilitating food shortages to an end: instead of farming communally, with each family sharing equally in the bounty (or shortfall), Bradford mandated that each household be assigned its own plot to cultivate, with each family keeping whatever it grew. The change in attitude—and productivity—was stunning. Whereas previously Pilgrim men had worked the fields while women tended the children and the home, "The women now went willingly into the field," according to Bradford, "and took their little ones with them to set corn." The embrace of private enterprise combined with the abandonment of traditional gender roles created an eager and expanded workforce in the fields that contributed significantly to the end of the starving times in Plymouth.[1]

The abandonment of traditional gender roles prescribed by the English allowed women to add field labor to the work they were already carrying out in the home and nearby outbuildings such as barns, chicken coops, and root cellars. Labor in this additional venue made them even more crucial contributors to the survival—and proliferation—of their people. The additional labor in the fields generated enough food to thrive, fueling the establishment of large families throughout New England. It also contributed to the transformation of the land.

In an area in which only about 20 percent of the land was suitable for farming, New England settlers averaged seven to eight children per family, children who grew up wanting farms of their own. By the late 1630s, local Indians, who also depended on agriculture, despaired over the ripple effect of English prosperity. The transformation of the environment was so profound that the Indians, once seen as the saviors of the Pilgrims, now feared the Pilgrims, Puritans, and other British invaders as a threat to their own traditional way of life and to their very survival. In 1642, when Narragansett leader Miantonomi realized that there were more Puritans in Massachusetts Bay than Native Americans in all of New England, he tried to persuade the Montauks to help the Narragansetts eliminate the English threat: "You know our fathers had plenty of deer and skins, our plains were full of deer, as also our woods, and of turkeys, and our coves full of fish and fowl. But these English having gotten our land, they with scythes cut down the grass and with axes fell the trees; their cows and horses eat the grass, and their hogs spoil our clam banks, and we shall all be starved."[2] Not only did English women help transform the land, so too did they affect the land practices of the indigenous men and women who had lived upon it in very different ways.

The First Americans

Understanding the role of sex and gender in early Native American environmental history proves difficult for several reasons. Scholars trained to glean history from written documents find a dearth of sources from the pre-Columbian period. Studying records of the descendants of the original peoples proves problematic, not only due to racial mixing, but also because in the centuries following the European invasions, virtually no Indian group was able to consistently practice traditional ways. Moreover, in what is now the United States there were hundreds of Indian nations. Their resource strategies, whether based in fishing, hunting, gathering, or farming, were as diverse as the landscapes they inhabited, making a definitive environmental history of the first Americans all the more challenging.[3]

Accounts written by the first European explorers and missionaries describe the practices and traditions of the native peoples they encountered before they were significantly affected by the interlopers' ways and attitudes. Nevertheless, using this evidence to reclaim the pre-Columbian environmental past paints a picture that is unreliable as well as incomplete. These earliest accounts of Native Americans were written by white observers whose own culture, especially their presumptions about gender, strongly colored their perspective on aboriginal ways. According to their views, women, especially in nonfarming tribes, remained within the local villages, caring for the children and doing the drudgery appropriate to their subordinate social status.[4] Early records portray women as producers of the finished goods (especially clothing and food that would not spoil) from raw materials provided by men. They therefore had little direct impact on the environment beyond gathering wild berries and roots. This old stereotype has been partially exchanged for the popular glorification of indigenous peoples as all leading nomadic lifestyles, living totally off the land, but having virtually no impact on it.[5]

Old and new stereotypes have gradually been replaced by more nuanced understandings of indigenous peoples' sex-based and gendered relationships with the environment across many cultures.[6] Archaeologists and anthropologists have used a variety of nonprint sources to glean much about pre-Columbian Americans' interactions with their environments.[7] In the Southwest, for example, Pueblo Indians including the Hohokam, Mimbres, Mongollon, and Anasazi developed complex societies and trade networks. By AD 700 southwestern women were developing corn hybrids adapted to desert soils and scarce water and creating complex tools with which to grind the dried kernels. As their farming methods became more sophisticated, they built increasingly elaborate dwellings near their fields. Some Puebloan women began making pottery in which to cook and store food, experimenting with different kinds of clay. Drawing upon pottery-making techniques, they built structures out of adobe, culminating in the complex multistoried houses of the Anasazi. Over time, the Pueblo exhausted the local soils, water, and trees and were forced to abandon their homes and move to fresh lands nearby. Long droughts in the fourteenth and fifteenth centuries, however, caused some thirty thousand people of the Colorado Plateau to abandon the entire area. They resettled hundreds of miles in different directions, where they mixed with the Indians already residing there.[8]

At about the same time as this major resettlement, the Athapaska, who lived as hunter-gatherers, migrated from Canada into what is now New Mexico. Some members learned agriculture from other local tribes, reforming as Navajo and Apache. Navajo women and men shared far more social and economic duties and responsibilities than the Apache, who practiced strict gendered divisions of labor, assigning hunting and raiding exclusively to men and a large variety of tasks, including the gathering and preserving of foods and preparation of skins, to women.[9]

Indians in the western Great Lakes region have been traced back to 11,000 to 9,000 BC. Their living habits were transformed by climate change, although hunting appears to have been a staple.[10] In the centuries preceding the arrival of Columbus, women of the Ho-Chunk, Mesquakie, and Sauk living in the southeastern area along the Mississippi mined lead, while women of the Okibwe and other, more nomadic native nations worked as traders.[11] In southeastern Wisconsin, where the climate became more conducive to agriculture, Potawatomi, Menominee, and Ho-Chunk men hunted, fished, and made weapons and tools while women farmed, dried fish and game, prepared hides, gathered local wild foods, and made the bags and baskets for food storage.[12] In the floodplains of the eastern United States, Indian women farmed, domesticating wild seed plants such as sunflowers, squash, and marsh elder.[13]

In Indian societies throughout what is now the United States, men frequently manipulated the environment by burning, hunting, and fishing. Women too manipulated the environment as they provided, via gathering or farming, much of their communities' total food. Sustenance, rather than maximum productivity for profit, was the goal. Where natural food supplies were generally abundant, agriculture was rarely practiced to any meaningful degree. In areas where tribes did farm, women were usually the primary distributors of the corn, beans, squash, and pumpkins they planted, weeded, and harvested.[14] In southeastern New England, for example, since about AD 1000, women's production of corn alone provided about 65 percent of their tribes' caloric input. By planting mixed crops, women shielded the soil from excessive sun, rain, and exhaustion and minimized weeding, which subsequent European farming methods would necessitate.[15] "We cared for our corn in those days, as we would care for a child," recalled one Dakota woman looking back on traditional ways, "for we Indians loved our fields as mothers love their children."[16]

Because the needs of the community and the skills of men and women frequently trumped any separation of tasks based on sex or gendered tradition, divisions of labor were rarely as rigid as those practiced by the Apache. While California Indian men, for example, were the primary hunters and fishers and women the primary gatherers and food preparers, men sometimes aided in gathering (such as knocking acorns off oak limbs) and women hunted, fished, and trapped small game.[17] In tribes that practiced agriculture, men frequently cleared the land and helped ready it for the planting, weeding, and harvesting carried out by women.[18]

FIGURE 1.1 In this 1591 engraving by Theodore de Bry, based on a 1564 watercolor by Jacques Le Moyne de Morgues, women of Florida's Timucua tribe plant seeds in soil prepared by men. (Library of Congress digital image LC-DIG-ppmscu-02937 of LC-USZ62-31869)

Despite this frequent overlap of activity, missionaries, soldiers, and officers in North America frequently perceived Indian women as working harder than Indian men.[19] In many tribes the work carried out by women year-round was more readily visible to an outsider than men's activities, like hunting, which took place only seasonally and away from the village. The French explorer Samuel de Champlain reported in 1616 that Huron women "have almost all the care of the house and the work; for they till the ground, plant the Indian corn, lay up wood for the winter, beat the hemp and spin it, make fishing-nets from the thread, catch fish . . . harvest their corn, store it, prepare it to eat . . . attend to their household affairs . . . [and] serve as mules [for their husbands]." The men, according to Champlain, "do nothing" but sporadically hunt, fish, go to war, feast, dance, and sleep.[20] French naturalist George-Louis de Buffon saw all aboriginal peoples as inferior to Europeans and asserted that in the New World, women's labor was exploited rather than respected. According to de Buffon, Indian men were savages who "look upon their wives only as servants for all work, or as beasts of burden" and whom they "compel without mercy, without gratitude, to perform tasks which are often beyond their strength."[21]

Much of the division of labor was determined by sex, as activities carried out communally within the village made it possible for women to combine productivity with frequent breast-feeding. Even when men and women were not doing the same tasks, there was often a more equitable division of labor than many outsiders realized, as both sexes contributed significantly to the tribe's survival and well-being.[22] In contrast to de Buffon, at least one government agent suggested that the traditional Indian division of labor was the product of a more egalitarian relationship between women and men than that of Europeans.[23] Jesuit missionary Paul Le Jeune, stationed along the St. Lawrence River, was disturbed by the gender relations he witnessed among the Algonquian in 1633, complaining to his superiors back in France, "The women have great power here." He worked to persuade the tribe's men that the women's husbands were the rightful masters, chiding them that "in France women do not rule their husbands."[24]

The ability of women to provide for their people, combined with their abilities and contributions as mothers, had gained them a place of respect and value within their societies. Arriving Europeans by no means shared this appreciation of Indian women's farming expertise. Colonists in New England, for example, were horrified rather than impressed by the practicality of the solution local Indian women brought to the problem of soil depletion. Native American crops, especially because they were so varied, did not leach nutrients from the soil at the same pace or to the same degree as did the Europeans' more monocultural, surplus-seeking methods of farming. Nonetheless, even those soils cultivated under native methods eventually tired, and crop yields lessened. Most Indian peoples, like the Pueblo, then moved onto new, untilled soils. Their small numbers, the simplicity of their dwellings, and their disinclination to collect large stockpiles of materials all allowed for ease of resettlement.[25] Europeans, having first learned from Indians the method of using fish to fertilize poor soils, were stunned at this flagrant "waste." They urged the fertilization of the land already under cultivation, especially where it could be done fairly easily due to the abundance of local fish. Indians in areas of seemingly endless stretches of accessible, untilled, rich soils rejected this solution as absurdly labor intensive. These contrasting approaches to the problem of soil depletion highlight the two cultures' dramatically different land values, ultimately based on issues of population.[26]

Although they did not seek to create a surplus beyond what could be bartered with other tribes to achieve subsistence, Indians did not live in total harmony with nature.[27] Like the Europeans to come, they altered and exploited the land. Native Americans occasionally hunted entire species into local extinction, farmed soil to exhaustion, and transformed vast lands through controlled burns. Because of their mobility, however, their lifestyle continued to be sustainable even as its individual elements changed over time. In some areas of North America, local conditions were sufficiently harsh to ensure a low population. Among more prosperous groups, the key to carrying out what historian William Cronon calls "living richly by wanting little" was that they controlled their numbers.[28]

Native American women's greatest environmental impact came not through their gathering, irrigation projects, horticulture, fishing, herding, or their ability to preserve foods. Instead, their greatest single impact came through their nearly universal practice of prolonged lactation. This effort to control their reproduction highlights the crucial role that population density plays in environmental issues. Breast-feeding was very common for the first three years after childbirth, but among some tribes it lasted for four years and sometimes even longer. Certainly breast-feeding in the first two years had enormous practical benefits, primarily convenience and mobility. It was also valued because frequent feedings (at least six a day) suppress ovulation, bringing decreased fertility. Because some Native American women actively sought to control their populations, they routinely nursed their babies past when children could easily thrive on solid foods. United States government agent Thomas Forsyth noted that Indian mothers continued this practice into the colonial period and beyond. They "allow their children to suck at least twice as long as white women do [and] they generally leave off child bearing at the age of thirty."[29] Along with prolonged lactation, Native American women, like their European counterparts, also practiced infanticide and abortion.[30] To guarantee population control, breast-feeding was sometimes combined, as in the case of the Huron and California's Ohlones, with sexual abstinence, a method also practiced by many indigenous peoples worldwide, including those who lived along the Amazon and within Africa's Congo basin.[31]

By carefully controlling their populations, to keep them below the land's "carrying capacity," Indian women made a crucial contribution to their peoples' ability to live easily sustainable lifestyles.[32] Indian populations were also periodically checked by other factors, including wars, droughts, and floods. In addition, some endured "lean" winters, during which the stores of food intentionally limited by the tribe ensured that the weakest were winnowed out. In his report "Manners and Customs of the Sauk and Fox Nations of Indians," Forsyth noted, "All Indians are very fond of their children and a sick Indian is loth [*sic*] to leave this world if his children are young, but if [the children are] grown up and married they know they are a burden to their children and don't care how soon they die."[33] Forsyth's position as an outsider may have colored his perspective, as respect and care for elders was nearly universal among Native Americans. The kind of indifference to their own individual fates observed by Forsyth, however, does reflect the Indians' recognition of the impact of their actions on future generations, as well as their belief that the soul is not extinguished in death, but lives on through reincarnation.[34] But the variety of external checks on the existing population cannot account for the remarkably stable (although larger than previously believed) numbers of Indians estimated to have populated what is now the United States.[35]

The traditional contributions to controlling tribal numbers made by Indian women, in cooperation with Indian men, were crucial to tribes' sustainability. Native American women's active and welcome role in limiting their people's population reflects Indian gender relations, in that most women shared more of a sense of control and partnership with their men, especially concerning reproduction, than did their European counterparts. It also reflects Indian perceptions of partnership with, rather than stewardship over, the land.

European Colonists

Europeans quickly changed the landscape of the places they came to dominate. Colonists and their African slaves cleared forests, drained swamps, built fences and permanent structures, plowed and planted fields, herded animals, hunted, fished, and in many other ways eliminated or disrupted pre-Columbian plant, animal, and human life. Colonial men did so partly based on their anthropocentric and patriarchal beliefs that the Bible commanded them to "subdue the earth." Even the relatively primitive homes of the early settlers reveal something of the environmental enormity of their ways of living compared to the far simpler and more transitory dwellings of the Indians. A modest seventeenth-century New England house required at least twelve tons of wood to build. Subsequently, it annually consumed around fifteen cords (1,920 cubic feet) of wood, leading a town of two hundred homes to deforest some seventy-five acres each year.[36] Members of the next generation, a substantially larger population, would also build—and heat—their homes.

Unlike their Native American counterparts, free white women in the early colonial period did not seek to limit their own reproduction or practice environmentally sustainable lifestyles. They bore many children and served as helpmates to their husbands, a role that was based in the harsh realities of their new homeland. Wives joined their husbands in the fields when the planting, hoeing, and harvesting necessary for survival required every pair of hands. The belief that woman's proper place as helpmate was within the home and its immediate environs remained strong, however. Based on women's lesser physical size and strength, the demands of menstruation, pregnancy, childbirth, and breast-feeding, and the patriarchal directives of the Bible, white colonial families yearned for the day when women could leave the heavy labor of the fields and return exclusively to their "natural" domestic tasks: feeding chickens and gathering eggs, milking cows, making butter and cheese, gardening, gathering berries, roots, nuts, and other wild edibles, preserving food, cooking, washing, making soap and candles, preparing flax and cleaning fleece, spinning, knitting, and weaving, as well as bearing and caring for children.[37] Yet even these more traditional women's activities frequently affected the environment significantly. The chickens and cows for which women were responsible, for example, required the building of fences and permanent outbuildings.

Even the first generations of settlers, both men and women, did more to subdue the earth than simply draw from it their subsistence. Unlike the Indians, they were determined to profit from the land, to increasingly meet the demands of distant markets for cattle, corn, fur, timber, and other goods.[38] Colonial women sold the fruit of their gender-prescribed labors as well, including surplus cloth, eggs, and butter. As early as 1653, the colonial historian Edward Johnson, when considering the New England ecosystems, marveled at the fact "that this Wilderness should turn [into] a mart for Merchants in so short a space [with] Holland, France, Spain, and Portugal coming hither for trade."[39] Both men and women contributed to the transformation of the land in a variety of ways, and the combined effect of their labor was overwhelming: European methods of hunting and farming and their concepts of ownership and progress quickly supplanted Indian traditional ways of life. Even more devastating to the Indians were the vast waves of death brought by exposure to European infectious diseases to which indigenous populations had no immunities. Although estimates vary, most historians agree that only 10 to 20 percent of the Indians across North America survived the first bouts of exposure.

Even as they were decimated by smallpox, measles, influenza, and venereal disease, native people's environmental knowledge and skills made them valuable as guides and as key contributors to the newcomers' ways of living on the land. The small percentage who survived the epidemics experienced great cultural confusion. Their traditional communal ways were suddenly rendered difficult, if not obsolete, by the changes not only in their numbers, but also by the transformation of the land wrought by the European concepts of property, particularly insistence on permanent structures and surplus commodities. If Native Americans were going to avoid extinction, they would need to adapt their traditional relationships with nature to the values of the invaders. Two experiences suggest the narrow spectrum and devastating environmental and cultural consequences of available options: Indians in the Great Lakes region turned to the fur trade as their means of survival, and California Indians supplied the labor for the Spanish missions.

The patriarchal traditions of the European invaders trumped the disdain some expressed for Native men's perceived laziness. More than their maleness brought native men greater esteem than native women.[40] Indian men's traditional hunting skills had been highly valued in the fur trading economy established at the very beginning of the colonial experience. Some settlers sold guns to New England Indians, who adapted to them quickly and became all the more adept at hunting. According to Governor Bradford, Indians brought more game and fur to English trading posts than did Europeans "by reason of their swiftness of foot and nimbleness of body, being also quick-sighted and by continual exercise well knowing the haunts of all sorts of game."[41] Supplies of game were quickly diminished in the original thirteen colonies by both settlement, which disrupted the natural habitat, and by hunting. European fur traders looked westward to the less settled, vast, fur-rich regions of the Great Lakes, and to Indian men for their superior hunting skills.

By contrast, European traders found Native American women's agricultural expertise far less lucrative, and even women's basket-weaving and pottery skills lost substantial value with the increasing availability of mass-manufactured goods. Indian women did not submit passively to the obliteration of their sustainable practices and other incursions into their peoples' traditions. Native women in the western Great Lakes region frequently married or were otherwise paired with French fur traders but strove to maintain their Indian identities largely through extensive kin networks.[42] These networks facilitated men's access to valuable pelts, fueling an industry so lucrative that women were empowered to negotiate positions of prominence. But as women of the Great Lakes and other areas rich in fur-bearing animals used variations of motherhood and kinship to enhance female autonomy, they contributed to the rapid depletion of game, including seals, otters, deer, buffalo, and beaver, all across North America.[43]

Indian women on southwestern missions frequently faced harsher conditions than did their sisters in the east. Immediately following the arrival of Columbus, missions were established in New Spain (the Caribbean, Mexico, and parts of the southwestern United States) in order to spread Catholicism and Spanish culture. Between 1769 and 1821 twenty-one missions were established in what would become the state of California. The acceptance by many Indian men of the missionaries' religious and secular teachings, both of which were highly patriarchal, contrasts with the resistance displayed by large numbers of Indian women.[44] European ways threatened to obliterate the ways of life and beliefs of all Indians, but Indian women, by virtue of Christian beliefs and traditions concerning gender, were forced to endure special devaluation. However, even as some Indian girls and women were converted to Christianity and inculcated with European gender norms as they were taught skills such as spinning and weaving, their traditional expertise as farmers and herders was valued and remained in demand.[45] While men performed the heavy field labor, women worked in orchards and kitchen gardens.

Even on missions where their skills were crucial to survival, women's greater agricultural expertise could not overcome the favored position men held in the European patriarchal value system. Nor could women's considerable expertise produce from a relatively hostile physical environment the surplus market crops so desired by the missionaries. Although individual missions produced some exports, they continued to depend on subsidies from Spain.[46] Despite these failures, Spanish colonization, much of it carried out by forced Indian labor, had profound ecological effects. Forests felled for buildings and fuel led to wood shortages. Erosion of the denuded landscape led to flooding. Large-scale irrigation systems redistributed local waters, while Spanish monoculture drained nutrients, eroded topsoils, and drove out many native species. Exhausted lands had to be abandoned and were usually reclaimed by the foreign weeds the Europeans had unintentionally introduced. Even more devastating to the native landscape than farming was ranching. Although it proved to be more successful in terms of short-term profit, livestock ranching wreaked environmental havoc far out of proportion to the relatively small number of colonists and Indians who supervised the grazing animals.[47] Dense herds of horses and especially cattle devoured vegetation, eroded hillsides, collapsed protective banks of streams, drove back the native deer and elk, and paved the way for other invading species. They also rendered indigenous ways of living on the land obsolete.[48] Considering the transformation of California over many generations, Kate Luckie, a member of the Wintu, deplored unsustainable white ways: "This world will stay [only] as long as Indians live. . . . White people never cared for the land. . . . We [Indians] don't chop down trees. We use only dead wood. White people plow up the ground, pull up the trees, kill everything. . . . The Indians never hurt anything, but the white people destroy all."[49]

Severe conditions on most missions left native men and women with few options but to carry out the invaders' unsustainable practices. Both sexes provided forced labor, but they were not completely powerless. Indian deaths outstripped births not only due to enforced separation of the sexes, disease, inadequate food supplies, and overwork, but also because in some circumstances women consciously limited their reproduction through sexual abstinence, abortion, and infanticide.[50] For a variety of reasons, native women chose not to be complicit in producing large supplies of future mission workers who, bereft of their native culture, would have perpetuated their own people as a permanent laboring class carrying out unsustainable environmental practices.

Across North America, European land use methods undercut traditional Indian gender and environmental relationships. As the colonial period came to a close, Native Americans were expected to adopt both the gender roles and environmental practices of whites, a program enthusiastically endorsed by Thomas Jefferson. The third president was confident that if Indian men gave up their traditional ways and became good American farmers, the lives of their wives would improve as well. In 1781 Jefferson asserted in *Notes on the State of Virginia* that in their traditional roles, Indian women were "subject to unfit drudgery," which he believed was "the case of every barbarous people."[51] If Indian men abandoned subsistence hunting in order to

raise cattle and crops, however, they would "acquire knowledge of the value of property" and bring prosperity to their families. Indian women would "spin and weave"
and, basking in the security of their husbands' reliable productivity, cease to practice
abortion. Family size would increase as the Indians prospered.[52] By 1803 Jefferson
was using this idyllic scenario to promote his plan for Indians to cede to the United
States their vast hunting grounds east of the Mississippi in favor of small tracts of
land to farm.[53]

Some Indian men became employees of white hunters and trappers. Most, however, rendered unable to enact their traditional roles and ways of living by the loss of
their homelands, were left with virtually no alternative except to farm, often on lands
ill suited to agriculture. The fact that farming was traditionally a female rather than
male enterprise was ignored by governmental mandates enforcing as national Indian
policy gender models based on European patriarchal traditions.[54] Forcing men into
traditional female roles in the years surrounding the American Revolution contributed to the several generations of neglect and decline suffered by Indians and Indian
culture.

Enslaved Africans

For centuries many of the dramatic changes made to the environment around the
globe were frequently carried out by the people the invaders brought with them as
forced labor to mine (in Brazil, for example) or to clear or drain and plant land.
African slavery in what is now the United States began in Virginia in 1619. As was
the case in many slave-owning societies, slave owners in early America depended on
trade rather than natural increase to maintain their supply of slaves. In view of
the heavy work to be done in the early years of turning wilderness into farms and
plantations, most owners valued size and brute strength, preferring to buy men
rather than women. Men composed roughly two-thirds of the seventeenth-century
African imports.[55] Only as the overseas trade was phased out did the numbers of
enslaved women in North America begin to approach the number of enslaved men,
with the sex ratio evening out around the 1740s.[56]

Most eighteenth-century slaves did not live in the primitive conditions suffered
by the first generations, nor did their daily routine consist exclusively of the same
backbreaking labor of draining swamps and clearing forests that helped speed their
predecessors to an early grave. As their material conditions improved throughout
the colonial period, slave life expectancy grew, although because of poor diet, inadequate medical care, and overwork, most slaves lived substantially shorter lives than
did their white owners.[57]

Although enslaved men and women were sometimes assigned different tasks, women were not exempt from heavy labor. More than sex or gender, a slate of interconnecting factors determined the type of labor to be performed, including individual skill level, size of the plantation or farm, season, and crop type.[58] The diary of slave-owner Mary Carr, for example, records that her husband, James, had directed their male slaves, Joe and Caesar, to plow and split rails, while Charity and Violet cleared brush and burned logs. There were days, however, when "James has all hands making fence," "all hands busy ploughing corn," "all hands busy ploughing cotton lands," or "all hands picking cotton."[59]

An individual slave's skills combined with the needs of slave owner trumped any desire to conform to gender stereotypes. However, detailed records of slaves' cotton-picking proficiencies are particularly abundant and reveal that sex was often the greatest predictor of productivity for certain tasks. Over one eleven-day period on Haller Nutt's Araby Plantation, for example, the top male pickers, Dick and Ike, averaged 259 and 255 pounds per day, respectively, substantially more than the top female pickers, Philis and Betsy, who averaged 185 and 182. Nevertheless, during a different, five-day period, the most productive picker, with a daily average of 299 pounds, was a man named Eoo, but Eliza came in second with 290 pounds, well ahead of previous winners Dick and Ike (282 and 272).[60] A scholar examining the picking quotas and records of a number of southern farms and plantations confirms that although slave women were not expected to pick as much cotton as men, some slave women were as strong as men and were able to pick as much, if not more, than most slave men.[61]

Although men were frequently (but not always) more productive at heavy labor in the fields than were women, enslaved men were more likely than women to be relieved of field work.[62] Because most slave owners shared the gendered perception that all males were smarter and more easily trained than women, they generally gave enslaved men and boys most available nonagricultural work, leaving a disproportionate amount of field labor to women and girls. Like the more elite enslaved men who were butlers and valets, enslaved women served as house servants (cooks, wet nurses, maids), but an additional variety of jobs remained almost exclusively within the male domain: stable worker, blacksmith, driver, groom, horse-breaker, cooper, carpenter. Figures for individual farms and plantations vary widely, but in his detailed study of rural South Carolina, historian Philip D. Morgan found that only 4 percent of enslaved females compared to 13 percent of males labored outside the fields in the 1740s.[63] By the 1790s, the percentage of female slaves laboring outside the fields had increased to 6 percent (and higher on large estates), while the figure for males was 26 percent.[64] The nation's slave population continued to increase in the decades preceding the Civil War, contributing to further changes in work assignments, but across the fields of the American South, female slaves consistently outnumbered males when it came to field work.[65]

With women serving disproportionately in the fields, their reproductive cycles proved a particular challenge, as owners balanced their own demands for strenuous labor from female slaves with the recognition that such labor could prohibit crucial human reproduction. Slave miscarriages "should never be the case on a well organized plantation," wrote owner Haller Nutt, and were a sign that "there is something wrong—[the female slave] has been badly managed and worked improperly." To avoid miscarriage, "women in the family way should avoid ploughing—and such heavy work as fit only for men."[66] In his "Rules for the Plantation," published in a South Carolina newspaper, John Billiller proclaimed that "no lifting or ploughing must be done by pregnant women" and noted that "sucking and pregnant women must be indulged as circumstances will allow, and never worked as much as others." He urged that nursing mothers "be allowed time to attend to their infants and if possible worked as near their homes as can be."[67] While women who successfully gave birth on Nutt's plantation were rewarded with exemption from field work for a month, women who miscarried received an even longer dispensation and were to be "nursed more carefully" to ensure successful subsequent pregnancies.[68] Some planters suspected their female slaves of exaggerating or inventing their various

FIGURE 1.2 Sketched from life near Fredericksburg, Virginia, on March 13, 1798, this watercolor by Benjamin Henry Latrobe depicts a white overseer supervising two slave women hoeing tobacco. (Image 1960.108.1.3.21, courtesy of the Maryland Historical Society)

maladies related to menstruation, pregnancy, and childbirth. A Virginia slave owner complained to journalist Frederick Law Olmsted that "the women on a plantation will hardly earn their salt after they come to the breeding age [claiming that they are] not fit to work. . . . You dare not set her to work [for fear of permanently damaging her health] . . . so she lays up till she feels like taking the air again, and plays the lady at your expense."[69] Many planters, however, willingly ran the risk of permanently damaging the health of their female slaves. Some assumed those who complained were malingerers; others deemed the granting of lighter loads to slaves, even those late in pregnancy or having just given birth, to be a luxury they could not afford.[70]

Male and female slaves used environmental knowledge gained in both Africa and North America and handed down not only to carry out (or to avoid) the work their owners required of them, but also to improve the quality of their own lives.[71] Slaves even manipulated plantation policies concerning reproduction for their own purposes. Methods used previously to control local populations were adapted in their new situations as forms of resistance to slavery. The demands of forced field labor precluded most enslaved women's ability to breast-feed with sufficient frequency to suppress ovulation. Instead, they limited reproduction by using the environmental knowledge brought from Africa and the Caribbean concerning the abortifacient qualities of a number of medicinal plants (especially cotton root) available as well in North America.[72] Such practices not only reduced their masters' supplies of new generations of forced laborers, but also served as a kind of strike, since reproduction was considered an important role for enslaved women, contributing to higher prices for women considered to be promising "breeders."[73] Enslaved women risked great harm when they intentionally terminated their own pregnancies. One owner advised, if "the woman is to blame herself [she] should be severely punished for it when she gets well."[74]

Enslaved women also used their environmental knowledge to raise their families' quality of life. They gathered naturally growing herbs, roots, and berries for dietary and medicinal purposes.[75] Women who were granted garden patches grew food that was used partially (sometimes nearly wholly) to provide for their families' diet and in some instances to sell or trade. The mixed crops women grew resulted in far less soil exhaustion than did the monocrops of their owners. In their search for additional food supplies, women sometimes joined men and boys in fishing and in trapping and hunting small game in surrounding areas.[76] Their ability to live off the land by fishing, hunting, and gathering in local swamps and woods allowed runaway slaves, male and female, to survive before making their way north to freedom or, as was more often the case, returning to their owners either through resignation or coercion.[77] With the rapid dispersal of the cotton gin, the South became increasingly dedicated to cotton production. With the growth of the Cotton Kingdom, enslaved women's environmental knowledge armed them with additional ways to thwart the institution that bound them.

. . .

Native Americans' transformation of the land was for the most part predicated on "living richly by wanting little" and sustainability. Both gender and sex played roles over many generations, as Indian women's farming methods and efforts to limit family size helped perpetuate native ways of living. European colonists held a different set of priorities and values. Once colonial women abandoned their prescribed gender sphere and joined men in the fields, the transformation of the American environment began in earnest. Both sex and gender contributed to the different ways in which native men and women fueled the colonists' transformation of California's environment as well as the fur trade that decimated many animal species across North America. With Indian numbers depleted and the environment so changed as to render their traditional ways of life obsolete, most survivors were eventually forced to take on gender roles to which they were unaccustomed. The lack of a male farming tradition, compounded by their being relegated to lands with

FIGURE 1.3 Slaves and their descendants incorporated their environmental knowledge into their daily lives. Here a girl in Seale, Alabama sweeps steps in 1915 using a homemade broom of locally growing sedge brush. (Wisconsin Historical Society 79984)

poor soil or inadequate water supplies, contributed to the continuing decline of Native American populations and culture. Africans were also forced to carry out European-mandated changes to the land. Their sex allowed women to limit to a degree the expansion of the slave labor force, but their gender did not protect them from the heavy demands of transforming pre-Columbian wilderness into farmlands dedicated to producing cash crops. In the interactions between colonists, Indians, and Africans, sex and gender as well as race clearly played significant roles in the shaping of the American environment in the seventeenth and eighteenth centuries.

"The greater household of earth"

2   The North and the South from Revolution to Civil War

"The Careless and Wasteful Nature of the Agriculture"

In 1834, the English actress Frances Anne Kemble retired from the stage in order to marry an American southerner, Pierce Butler. Four years later she traveled to the rice marshes and cotton fields that her husband had inherited in Georgia, where she recorded her observations in a series of letters and diaries. Her descriptions paint a vivid portrait of conditions from the perspective of a woman who was an outsider to the ways of the South. Her writings reveal the aesthetic (rather than economic) value of southern lands and climates. In trying to describe to a friend the natural southern beauty too dazzling to convey through words or works of art she exclaimed, "Italy and Claude Lorrain [renowned French painter of landscapes in the seventeenth century] may go hang themselves together!"[1] But overshadowing the joy she experienced from the local physical landscape was her reaction to another pervasive feature of the South: slavery. After writing of its evils, she went "out into the air to refresh my spirit." Although "the scene just beyond the house was beautiful," she added sadly, "I think I should die if I had to live here."[2]

Kemble was mortified by the institution of slavery, which she evaluated in terms of its devastating impact on the enslaved. "Scorn, derision, insult, menace—the handcuff, the lash—the tearing away of children from parents, of husbands from wives—the weary trudging in droves along the common highways, the labor of the body, the despair of the mind, the sickness of the heart—these are the realities which belong to the system, and form the rule, rather than the exception, in the slave's experience."[3] When it came to the white masters, Kemble emphasized moral degeneracy rather than the potential for financial profit as the sure consequence of slave ownership. When she saw how even her little daughter was instantly obeyed, it made her tremble as she considered the impact on a child of "learning to rule despotically . . . fellow creatures before the first lessons of self-government have been well spelled over!"[4] She found her husband "positively degraded in my eyes" when she heard him telling a work crew of pregnant slaves the necessity of their toil, and she "turned away in bitter disgust."[5]

During Kemble's residence in the United States her attentiveness to nature made her remarkably prescient in her predictions about the ultimate environmental repercussions of the South's refusal to diversify its crops. She observed, "Such a decrease as this in the value of one's crop, and the steady increase at the same time of a slave population . . . [with more] bodies to clothe and house, mouths to feed, while the land is being exhausted by the careless and wasteful nature of the agriculture itself, suggests a pretty serious prospect of declining prosperity." Indeed, she added, "unless these Georgia cotton planters can command more land, or lay abundant capital (which they have not, being almost all of them over head and ears in debt) upon that which has already spent its virgin vigor, it is a very obvious thing that they must all very soon be eaten up by their own property."[6] Kemble did not exaggerate. Southern planters' insistence on producing cotton exclusively led to soil exhaustion as well as a crop surplus. Once the Civil War began and England was unable to readily acquire southern cotton, it avoided a "cotton famine" by exhausting the surplus already on its docks and in its storehouses and factories before turning to new suppliers in Egypt, Brazil, and India. Kemble did not share the unquestioning devotion to King Cotton that was nearly universal among male planters. She foresaw that the expansion imperative fueled by a stubborn insistence that cotton was perpetually profitable—as well as the weakening soil and burgeoning slave population—would lead to a day of reckoning.

The evidence gathered by historian Philip D. Morgan cited in chapter 1 reveals that enslaved women and girls served disproportionately in the fields because men were more likely than women to be selected to carry out tasks requiring greater skill. Field slaves saw firsthand the effects of the soil exhaustion recognized by Kemble. It was their labor that brought about the environmental devastation, leading their owners to experiment with fertilizers and the terracing of hillsides deemed previously too steep to till. It was also their passive refusal to carry out these experiments successfully that forced the slave-owning forces in the South to demand new slave territory, hastening the coming of the Civil War. Enslaved women were even more economically and politically powerless than those white women, like Kemble, who opposed the "peculiar institution." However, like Kemble, they were shrewd observers of the natural world, and their environmental knowledge ultimately empowered them to play a role in bringing about the war that would eventually result in their freedom.

The Rise of Republican Motherhood

The harsh demands of a wilderness environment led women to "go willingly into the field" during the initial European settlement. As that wilderness was tamed and settled and women's field work became decreasingly vital to survival, white men and women reverted to more traditional divisions of labor. Wives in poor families, however, continued to work in the fields alongside their husbands and children. For the more affluent, the sharing of field work and food preservation gradually gave way to women working primarily in and around the home, with fields, paddocks, barns, and pastures increasingly becoming the exclusive domain of men. The amount of residual overlap between the two spheres was determined by economics: the richer the family, the more likely that its males and females inhabited very different worlds in their day-to-day activities. In times of war (including the French and Indian War, the American Revolution, and the War of 1812), middle-class wives and daughters who had worked increasingly indoors might once again be forced into the fields as husbands and sons served the war effort. Even among wealthier families, "The plan laid down for our [more genteel] education was entirely broken in upon by the War," noted Betsy Ambler Brent, who was eleven years old when the Revolution began in 1776. However, behavior inappropriate to elite women (like making rough homespun cloth rather than embroidering fancy British linens) that was encouraged in time of war was once again rejected as unacceptable in even the immediate postwar years. Expedient reversions were only temporary, and any gains in egalitarianism quickly faded when the men returned. From the vantage point of 1810 it was evident to Brent that "necessity taught us to use exertions [toward self-sufficiency independent of British finished goods] which our girls of the present day know nothing of."[7]

In the wake of their contributions to the war effort, upper-class women such as Brent considered themselves fully contributing patriots and hungered for new ways to use their talents to serve their country. Barred from voting, women were encouraged to make their contributions to the new nation in ways deemed appropriate to their sex. In 1976 historian Linda Kerber introduced the term *republican motherhood* to describe the key role assigned to women, especially women of the middle and upper classes, in the immediate postrevolutionary period.[8] Without a monarch to dictate rules and actions, the young republic was only as strong as the citizens and leadership it produced from within. Americans were educated to believe that women's political power lay not in the vote, but in the influence they held over their children through the values imparted to them within the home.[9] In the words of Benjamin Rush, one of the nation's founding fathers and a highly influential physician and educator, "The equal share that every citizen has in the liberty, and the possible share he may have in the government of our country, make it necessary that our ladies should be qualified to a certain degree by a peculiar and suitable education, to concur in instructing their sons in the principles of liberty and government."[10] If mothers raised their children to be self-sacrificing patriots—their sons to be good citizens and their daughters to be good wives and mothers—the nation would thrive. This civic prescription was endorsed and reinforced from pulpits all across the new republic, and many women took it to heart. Julia Cowles, reporting on a "well written and edifying" sermon delivered in her home parish of Connecticut in 1800, copied approvingly into her diary a long, unattributed quotation on the "respect and gratitude" owed to "dutiful mothers" who earned "their due rank and importance in society": "It is the ordinance of Providence that the heaviest and most important part of education should devolve upon the mother. . . . The fleeting period [a child] passes under the shadow of her wing is a season sacred to wisdom and piety."[11]

Women and Nonhuman Nature in the Early Republic

Women of the more prosperous classes increasingly spent their time within the home occupied by domestic concerns, which ideally included imbuing their children with the proper democratic values. Most women and girls were spending substantially less time outdoors than men and boys, and less time than had their mothers and grandmothers, with the greatest separation from nonhuman nature taking place at the highest economic levels. This separation of the male and female spheres was not complete or absolute in the decades following the Revolution, however, nor were women totally cut off from nature. Nevertheless, tanned skin was not accepted as the inevitable by-product of a woman's honest labor in the outdoors, but evidence of inferior, "common" status. Hannah Buchanan warned her traveling husband in 1809, "Do not be surprised to see me as black as Charlotte [one of their slaves]. I never go out in the sun, but am horribly tanned. I am told it is the water."[12] In 1820, Eleanor Hall Douglas, whose outdoor responsibilities included caring for forty chickens and nine cows, nonetheless wrote disapprovingly to her sister of a neighbor, Jane McKee, who was *shamefully tanned.*[13]

Even women like Hannah Buchanan, who claimed never to go out in the sun and who had slaves to carry out the more tedious and demanding aspects of "women's work," such as spinning thread, weaving fabric, washing clothes, and making soap, were nevertheless closely attuned to the details of farm production and the non-human nature upon which their families depended. They stuffed their families' mattresses and pillows with the best materials gathered close at hand: cotton, feathers, corn husks, straw, or leaves.[14] They wrote to absent husbands about weather, crops, and livestock and detailed the deadening impact of early frosts and long winters as well as the delights of spring, fruitful fields, fat pigs, and productive cows.[15]

Middle-class wives accompanied their husbands less frequently to the fields but still carried out chores such as feeding chickens and collecting eggs, and were routinely responsible for the kitchen garden. As reported in the *New-Hampshire Statesman and State Journal*, kitchen gardens did not generate cash, but "they are not the less valuable on that account, as every good housewife knows, who resorts to it, in order to procure a supply of wholesome and nutritious esculents for the table."[16] These gardens were a significant undertaking. Their size varied, but the minimum for a family of seven to eight people was usually considered to be a quarter of an acre of actual ground space, with additional room for the necessary access paths. Larger families required greater acreage.[17] The soil for the garden and raised beds was usually prepared by men who, the *Journal* noted disapprovingly, "do not duly appreciate, but disregard and neglect the garden, and throw the burden of its care and cultivation too much on the female branches of the family."[18] Women planted, weeded, and harvested their gardens, whose staples varied according to season and local climate but usually included beans, potatoes, peas, corn, cabbage, lettuce, beets, carrots, turnips, onions, leeks, asparagus, cauliflower, celery, broccoli, and herbs.[19] Women protected their crops from predators ranging from aphids to deer using a variety of methods, such as building scarecrows and picking off worms by hand.[20] They preserved in root cellars and through bottling (a precursor to canning) the food they grew, preparing it daily for their families throughout the year.

Kitchen gardens linked most women to the world of nonhuman nature on a daily basis. They ensured that women spent considerable time outdoors attuned to the weather and the changing seasons, and while they demanded women's labor, this exposure to the natural world provided a respite from the confines of their homes. Even ladies of the leisure class were urged by the popular magazine the *Lady's Book* "to devote a portion of every day, in favorable weather, in the open air, and in unfavorable weather, under a veranda, or in a green house to some of the lighter operations of gardening, for health's sake, and as a means of adding a zest to their ordinary in-door enjoyments."[21]

As late as 1890, fewer than 25 percent of American homes had running water.[22] Most irrigation projects were prohibitively expensive or labor intensive, so fields and kitchen gardens were planted in crops appropriate to the local area's natural rainfall.

Although women tending kitchen gardens were generally spared the rigors of extensive watering, like generations before them much of their labor nevertheless revolved around this precious natural resource. Men, women, and children spent a considerable amount of energy drawing and carrying water for drinking and use in cleaning and cooking. Sources included wells, springs, creeks, rain barrels, and, for the more well-to-do, pumps. Menstruating women and those recovering from childbirth were particularly inconvenienced by the difficulties in acquiring water for personal hygiene.

Standards of cleanliness and hygiene increased during the first half of the nineteenth century, and women, the guardians and protectors of the home, were responsible for the cleanliness of their homes and the cleanliness of all who dwelled within them.[23] This expectation put women at odds with their local environments. Window screens were not widely available until after the Civil War, which meant that especially in hot weather women bent on keeping their homes clean sought in vain to ward off infestations of biting pests like bedbugs, mosquitoes, and ticks as well as the general nuisance caused by flies, cockroaches, and ants. Winter offered some relief from these intruders, but women's responsibility for keeping their families' clothes clean was a burden the year round. Monday was washday for most women and involved hauling water, heating it, scrubbing clothes, hauling and heating more water for rinsing, then hanging the clothes to dry. During the winter months in cold climates, clothing would freeze on the clothesline and have to be brought indoors and dried before the fire. An 1862 Department of Agriculture report documented the backbreaking labor involved in keeping a family in clean clothes, contributing to its conclusion that on three out of four farms the wife was the hardest-working and most long-suffering member of the family.[24]

Even in cities indoor plumbing was scarce, leaving urban women with their own struggles to find and transport water. Prior to the Civil War, apartment dwellers hauled upstairs the water they drew from wells or pumps in central courtyards, areas that also housed the privies shared by all residents. Like their sisters on farms, urban women then heated the water for personal hygiene, laundry, cooking, and cleaning. They also had to haul back down the leftover dirty water, as well as the human waste collected in chamber pots during the night when the long trek to the privy was too dark and too cold.

Despite her battles with dirt and the elements, a good wife and mother was also attuned to the healing powers found within nature. Well before Elizabeth Blackwell became the nation's first woman physician in 1849, women had been responsible for the day-to-day health of their families. Recipes for healing salves, teas, and other natural remedies made from roots, leaves, flowers, nuts, and all manner of fruits, vegetables, and animal products were handed down from mother to daughter. Cloves, for example, were routinely used to ease toothaches, willow bark tea to bring down fevers, and a syrup of onion juice and honey to cure a cough.[25] Such home remedies were staples in women's magazines and in the many "how-to" pamphlets and books published for women concerning proper household management. One of the most popular such books, Lydia Marie Child's *The American Frugal Housewife*, went through thirty-two printings between 1832 and 1845.[26] Knowledge of natural remedies and the ability to identify and procure their ingredients were part of virtually every woman's education, regardless of class or race.[27] Parisian immigrant Marie d'Autremont was so resolutely proper that, even after sixteen years "without getting anything but the most urgent necessities," she dressed for dinner when living in a log cabin in rural New York in 1807.[28] Such determinedly refined living was not deemed inconsistent with her seeking out the local red flowers that, when brewed as a tea, were "very good for colds and coughs." So confident was she of their curative powers that she even sent some to her son in Paris.[29]

As the generation of girls criticized by Betsy Ambler Brent for knowing nothing of her generation's sacrifices matured into women and became mothers themselves, the behavior prescribed to middle-class men and women as socially appropriate and politically necessary changed further, widening the divide between the sexes.[30] Material conditions continued to improve, especially in the more settled East. Even on farms men were increasingly attuned to modern markets, yet rural women continued to focus almost exclusively on the household economy.[31] The divide between men and women was still more pronounced for the growing number of Americans living in towns and cities, where both sexes spent less time in the natural world.

In "Spring in the Woodlands," published in the July 1845 issue of the *Lady's Book*, Mary Clavers mourned the diminished capacity of city dwellers to appreciate the beauties and complex mysteries of nature, made all the more vivid with the coming of spring. She noted with a mixture of pity and contempt the young woman who exclaimed, "Oh, I do love the spring so dearly! The windows on Broadway are full of new goods and all so light and lively!" According to Clavers, "There spoke nature, but it was a nature in bodice and stays. That bright-eyed girl, with all her joyous impulses, would be ennuyée [bored] to the last degree if you should introduce her to the wild woods, with their endless store of beauty and music."[32] Ten years later the best-selling magazine *Godey's Lady's Book* ran a feature on kitchen gardens, noting approvingly, "Our lady readers in rural districts are now enjoying a greater pleasure than all the 'spring shopping' in the world can give."[33] On the eve of the Civil War, such urbanites who were increasingly divorced from nature constituted less than a fifth of the national population, yet their gender relations led the way to the future. Republican motherhood had given way to a different definition of women's proper sphere, a concept dubbed the "Cult of True Womanhood."[34]

FIGURE 2.1 ˙Although they are standing in a wooded area, these three women are lavishly dressed, complete with gloves and elaborate hats, and are clearly out of their "natural" element. (Wisconsin Historical Society 68015)

True Womanhood as the Foundation for Women as Nature's Housekeepers

What did it mean to be a "natural" woman or "natural" man in the eyes of most Americans in the decades prior to the Civil War? For white Americans, economic status continued to widen the gulf between the prescribed spheres. Men, who in the right to vote held exclusive political power, were to be strong, practical, and capable and to shoulder all financial responsibilities. Qualities admired in men included strength, leadership, aggression, and business acumen as measured by wealth. The home was the man's to provide, the woman's to serve.[35]

By the mid-nineteenth century, woman's proper sphere was the idealized domestic environment of the middle-class home, upheld by four pillars: piety, purity, submissiveness, and domesticity. Within this home, women were described as innately dependent, affectionate, gentle, nurturing, benevolent, and sacrificing. Morally and spiritually superior to men, women (mothers, ideally) within this sphere maintained a high level of purity in all things, including immaculate furnishings, clean language, noble motives, reverent beliefs, wholesome entertainment, and modest dress. Women were to embody, by virtue of their sex, each of these qualities on an individual level, while also bearing total responsibility for inspiring and cultivating purity within all inhabitants of the home.[36]

According to this view, if bereft of women's soothing, civilizing efforts to elevate life to a higher plane, men would remain mired in a constant competition with each other, dedicated to creating surpluses for profit rather than to beauty, truth, or concerns about future generations. Or as one young woman put it, bitter at having to leave "one of the most beautiful of all the beautiful little villages" for the wilds of Wisconsin in 1855, "What will not men do for money?" She "would rather be very poor and live where I could see the bright streams and green hills of old Pennsylvania," but her father was bent on improving the family's material status.[37]

In order for women to carry out their "true natures," and to retain the control they were obliged to exercise over their domestic world, all outside influences or experiences were considered suspect. Men, and the most important aspects of "their" world, posed a constant threat. Women should not be distracted or diverted from their domestic work by politics or business. However, according to the prescriptive rhetoric of the day, this surrender to domesticity did not doom women to a life of powerless drudgery. Although the concept of True Womanhood tied women more closely to their preindustrial daily routines, it ostensibly delivered to them a greater, more powerful, and frequently autonomous role within their own homes, as middle-class men were increasingly tied to the more industrialized world of politics, power, business, professions, higher education, and money.[38] The possibility of living out these prescriptions was limited primarily to the urban middle class, yet those values ultimately spread across geography, class, and even color lines.[39]

Significantly, women's letters and diaries in the decades prior to the Civil War abound with detailed efforts to live up to the prescribed ideals of True Womanhood, and they urged other women to do the same. Some women, however, resented the role assigned to them and rejected it as much as was in their power to do so. As a child in the 1850s, Lucinda Lee Dalton, for example, "longed to be a boy, because boys were so highly privileged and free." She particularly resented the advice from her teacher that it would be a waste for her to study algebra, because she "had already had more learning than was necessary for a good housekeeper, wife and mother, which was woman's only proper place on earth."[40] Dalton often "winced under the unconcealed contempt for 'females' expressed by [men] of all grades." She railed at the unfairness to the once beautiful girl who, upon marriage, was quickly worn out from "drudging from morning till night to keep [her husband's] house in order . . . only to be looked on by him as an inferior being, designed by nature to serve him. . . . He will talk about supporting her as if she did not perform more actual work . . . in twenty-four hours than her lord and master [did] in a week." Dalton believed that the separate spheres were contrived, not natural: "There is nothing in Nature to prevent woman from sharing all the great things of this world."[41]

Lucinda Dalton was one of the few women to blame society rather than personal failings for her feelings of disappointment and frustration. Most women's diaries, letters, and journals didn't criticize the prescribed sphere for being unrealistic in its demands and expectations. Nor, however, did they celebrate carrying out its rigid rules to the letter. Instead, women bemoaned the personal inadequacies and weaknesses that caused them to resist living out the prescribed behavior to the fullest, preventing them from being true or natural women. Lorena Hays wrote in her journal in 1852, "I know my own incapabilities and feel the need for greater proficiency in many things. . . . Why don't I like work better? I am afraid I am very selfish and like enjoyment too well, and am not desirous as I should be of being useful, yet I do sometimes think I want to be, but perhaps I do not improve every opportunity that I have."[42]

Compared to the powerful worlds of business and politics, and in view of women's self-deprecating attitudes, it seems curious to consider the limited world of domestic life as potentially positive or liberating for women. Even such a restricted sphere of influence, however, is better than no influence at all. Previously, routine household labor garnered little praise or even attention.[43] By the nineteenth century, all aspects contributing to domestic tranquility were subjects of careful study and promotion because the elevated, spiritual nature of such a state demanded greater esteem, as did the women who made it their life's work.[44] Excluded from the conventional worlds of prestige and power, some middle-class women like Lucinda Dalton enjoyed a new consciousness and value of themselves as unique contributors to society. That consciousness would come to have significant environmental repercussions.

Women as the Mothers of Civilization

The widespread acceptance of women as the nation's natural civilizers constituted a potential advantage for women.[45] Even Lucinda Dalton, who had longed to be a boy and who criticized the artificiality and unfairness of the women's prescribed sphere, ultimately internalized the more positive aspects of the values and behaviors assigned to her as a woman, and celebrated them. She came to be grateful that she had not been born a boy, reporting that she was "now thankful that I belong to a more respectable class of society. Not for all [men's] boasted 'supremacy,' 'superiority,'—and extensive advantages would I have women come down to their low moral level. [Male] intellectual acquirements of fame, power, [and] wealth . . . are as feathers in the scale against [female] moral purity."[46]

Women who were raised to take their rightful place within their prescribed sphere and who truly internalized its values found themselves on the horns of a dilemma. In theory, their world was wholly divorced from that of politics, business, and money. Yet as the nation became more urban and industrial, in order to keep their domestic, feminine world safe from masculine evils, women often had no recourse but to immerse themselves in those evils. By granting women expertise as especially attuned to purity and nonremunerative values due to their inherent moral superiority and civilizing influence, the prescribed female sphere laid the groundwork that would ultimately force women to claim and exert environmental authority.

By the mid-1800s, women's participation in efforts to provide social uplift beyond the home was by no means universal, even among the middle class. Such activities were deemed a luxury that many women, exhausted from trying to meet ever-increasing standards of cleanliness, godliness, household efficiency, and child rearing, simply did not have the desire or energy to pursue. Others, however, found that their charge as natural civilizers and uplifters left them no choice but to remedy the huge societal wrongs brought on by immoral, impure, and impious men. From the early 1800s on, middle-class women banded together in voluntary associations bent on uplifting those they deemed spiritually and materially deprived.

Women quickly moved from participation in long-standing prayer circles and missionary societies to addressing a variety of specific issues that threatened the sanctity of their homes and aspects of the outside world that men could not be counted on to protect. These evils included alcohol abuse, urban slums, and crime (especially prostitution), and, for a few northern women, the greatest immorality of them all, slavery.[47] Protection of natural resources was not yet on most groups' agendas, but women's obligation to improve urban areas in particular marked early assertions of female environmental expertise.

True Womanhood Expands Women's Sphere

It can be argued that the concept of True Womanhood contained the seeds of its own destruction. Compelled by society to bear the responsibility of imparting the prescribed female virtues of piety, purity, and moral superiority to young, impressionable minds, women were deemed natural teachers in the classroom as well as in the home. Women paid to work outside the home were decidedly antithetical to the notion of women as wholly domestic. However, many safeguards were put into place to ensure that women's alleged natural suitability to teach in the lower grades could not put them on equal professional footing with men, who retained control and direction of the schools as administrators, policymakers, politicians, and members of the school board.[48] A female teacher was paid also substantially less than a male teacher who likely had a wife and children to support. In more rural areas she often boarded with students' families, and in urban areas minimal pay frequently forced even the most frugal teachers into low-grade boardinghouses.[49]

Teaching was not thought of as a lifelong career for women, but rather as an interim stage between girlhood and marriage that would provide training, discipline, and exposure to children, experiences that would help prepare a young woman to be a better wife and mother.[50] A female teacher was usually expected to resign upon marriage.[51] Nevertheless, single women flocked to one of the few careers open to them. By 1860, roughly one-quarter of the nation's teachers were women, and they would soon come to dominate that profession, which previously had been occupied almost exclusively by men.[52] Female teachers incorporated into their lessons the importance of caring for the environment, a value that was increasingly part of the prescribed woman's sphere.

True Womanhood and the "Greater Household of Earth"

Many women shared with their husbands the belief that land should be manipulated in whatever way made it most profitable. Although in 1851 Mary Irving noted that plowing the prairie was "much to the disadvantage of its picturesqueness," she nonetheless wrote approvingly that "young trees spring up and flourish most luxuriantly" and "that which has been redeemed by much toil from the [newly planted] forest is, in a few years, sufficient . . . to pay for [the prairie's] 'clearing.'"[53] Mary Clavers agreed: "The hardy pioneer [is] actually preparing the way, with his single arm [plow], for future comfort and civilization. Well may we say the 'sacred plough,' and consider its furrows as blest of Heaven."[54]

For many women, however, internalization of their prescribed sphere led them to look at nonhuman nature differently than did men. Although *Godey's Lady's Book* gave a passing nod to trees' "utility to man," it measured the worth of trees not in terms of board feet, but aesthetics: "A tree in itself is, indeed, the noblest of inanimate nature; it combines every species of beauty, from its sublime effect as a whole, to the individual beauty of its leaves; it exhibits that majestic uniformity and infinite variety which constitute the essence of relative beauty; and the natural expressions of individual species are as various as are their forms and magnitude."[55] An increasing number of women felt obligated by their prescribed sphere to concern themselves with men's exploitation of nonhuman nature, especially as it posed a danger to future generations. Accordingly, they immersed themselves in nature study.

Botanist Almira Phelps wrote in 1829 that "the study of Botany seems peculiarly adapted to females" and counseled that mothers and female teachers must become deeply familiar with specific aspects of the environment if they wanted to inspire an understanding of nature in their children and students.[56] Nineteenth-century women produced nature drawings and sketches. Although most art academies did not admit women, a few female artists, including Mary Blood Mellen, Susie M. Barstow, and Laura Woodward, joined other painters of the Hudson River School to produce landscapes depicting the nation's natural beauty.[57]

Women were particularly drawn to nature writing. Susan Fenimore Cooper's seasonal journal *Rural Hours*, published in 1850, helped to popularize the nature essay.[58] Within its pages, Cooper in no way challenged the prescribed rightful place for women: "Home, we may rest assured, will always be, as a rule, the best place for a woman; her labors and interests, should all be centre [*sic*] there, whatever be her sphere of life."[59] Moreover, according to Cooper, women had special obligations concerning nonhuman nature. She charged women with both domesticating the wilderness and preserving aspects of it.[60] Nature, she argued, functioned like the home. Tended crops balanced by a region's native plants and animals created a place of harmony and happiness. She urged her readers to think of their land as "really one room in the greater household of earth," which was "the common home of all."[61]

Cooper wrote extensively about how the study of nature revealed proper gender roles, and how those roles in turn empowered people to better understand the true value of nonhuman nature.[62] She urged her primarily female readers to closely observe flowering plants because they reflected the virtues of modesty, constancy, and sisterhood.[63] She praised trees using the same kind of language found in *Godey's Lady's Book*, but went further. She condemned the wanton cutting of trees because it demonstrated a "careless indifference to any good gift of our gracious Maker . . . [and] betrays a reckless spirit of evil."[64] She demanded that Americans change their confrontational attitude toward this larger "home" and called for laws to protect certain animals endangered by hunting and settlement.[65] Women, she believed, bore a special responsibility for the preservation and conservation of natural resources from the avarice of men. Their protection of nature would benefit themselves as well as future generations. If they shirked their responsibility, however, not only would natural resources be squandered, but women would also forfeit their moral authority. Her widely read work piqued many women's interest in nature study and in the protection of plants and animals.[66]

A pioneer of literary realism, novelist Rebecca Harding Davis emphasized the dangers that the race to develop natural resources posed to people as well as to the nonhuman environment. In her story "Life in the Iron Mills," published in the *Atlantic Monthly* in 1861, she describes the scene in Wheeling, West Virginia:

"Smoke everywhere! A dirty canary chirps desolately in a cage beside me. Its dream of green fields and sunshine is a very old dream—almost worn out, I think." It was not only the canary that was trapped in "a reality of soul starvation" but the mill workers as well: "Masses of men, with dull, besotted faces bent to the ground, sharpened here or there by pain or cunning; skin and muscle and flesh begrimed with smoke and ashes; stooping all night over boiling cauldrons of metal, laired by day in dens of drunkenness and infamy; breathing from infancy to death an air saturated with fog and grease and soot, vileness for soul and body."[67] Through a variety of literary forms—essays, poems, and stories—women expressed their appreciation of nature and their fears about the forces that threatened human and nonhuman nature alike. Teachers increasingly incorporated these writings into their lesson plans as part of their obligation to properly educate America's youth.

Enslaved Women's Role in the Land Expansion Imperative

White women were not alone in recognizing the environmental impact of men's dedication to profiting from domination of the earth. Agricultural experience and wisdom combined with gender roles to empower enslaved women.[68] In response to the development of southern agriculture, the actions of enslaved field workers, disproportionately female, hastened the necessity for the geographic expansion of slavery.

In 1855, David Christy published a defense of slavery and cotton with the unambiguous title *Cotton Is King*.[69] Christy spoke for many southern planters and small farmers when he asserted that the key to the South's economic success was to keep southern lands dedicated to the production of cotton. "Here [in Georgia] nothing signifies except the cotton crop," noted Frances Kemble.[70] Long before Christy insisted that cotton was the key to wealth, however, Kemble wrote at length of the enormous decrease in both quality and quantity of the famous Sea Island, long-staple cotton that her husband's family had been growing for generations. When her husband's grandfather first sent his plantation's cotton to England, "it was of so fine a quality that it used to be quoted by itself in the Liverpool cotton market."[71] As the area's cotton quality and quantity fell victim to soil erosion, Kemble appreciated the wisdom of the minority of local farmers who grew rice instead, and she noted that the increased profits meant improved conditions: "All the slaves' huts on [cotton-dedicated] St. Simons are far less solid, comfortable, and habitable than those at the rice[-growing] island."[72] She could not understand most southern planters' stubborn refusal to diversify despite the clear potential for profit: "Riding home I passed some beautiful woodland, with charming pink and white blossoming peach and plum trees, which seemed to belong to some orchard that had been attempted, and afterward delivered over to wilderness. . . . What a pity it seems! For in this warm, delicious winter climate any and every species of fruit might be cultivated with little pains and to great perfection."[73]

Although southern agricultural periodicals and state geological surveys repeatedly stressed the need for crop rotation and manuring, most men dismissed the idea of replacing cotton with a variety of crops as the kind of fanciful thinking only a woman could produce.[74] Even long after the Civil War, when cotton prices were at an all-time low, a male speaker at a convention of cotton farmers responded to the notion that "we cotton planters ought to learn to plant lettuce, cabbage, onions, etc., in order to make a profit" by thundering, "Now, every cotton planter here present knows how absolutely foolish this is."[75] White planters saw cotton production as necessary to the existing infrastructure. More important, because of the nature of the labor required, it served as a guarantee of the existing racial hierarchy. For that reason, observed Mississippian David Cohn, even if a planter could "sharply increase by growing asparagus," he would refuse because "asparagus is not cotton. Arguments for diversification left [white] men's hearts untouched."[76]

Much of the work of slaves had profound environmental impact, especially in the field, where productivity accelerated soil exhaustion. Planter Alfred Holt Stone "saw his own [Mississippi] plantation as a kind of monument to both man's power over nature and to white supremacy." But even as he credited white men for the money and ideas that led to the environmental transformation of the Delta, he acknowledged, "The work itself is the negro's."[77] Slaves, male and female, were quick to assess the agricultural potential of their new environment.[78] While slave owners may have considered the field work carried out by women to be unskilled labor left to them by default, they nevertheless benefited from the gendered expertise of female field hands.[79]

Women were confident in their knowledge of agriculture and soils. Sukey, a slave owned by Virginia planter Langdon Carter, told her master in no uncertain terms where he should plant tobacco, for she "knew the ground." When he disputed her assessment, she repeated that she "knew the ground, knew how it was dunged [fertilized]" and insisted that it would produce good tobacco."[80] Such expertise was not always used to the benefit of slave owners.[81]

Most slave owners had little capital due to the high cost of purchasing and maintaining their workers. Top quality fertilizer and the equipment and draft animals needed to bury it by deep plowing were beyond the financial reach of all but the wealthiest plantation owners. The vast acreage of the plantations of the few who had ready cash rendered fertilization virtually impossible. Certainly these factors contributed to the South's failure to revitalize its soils, but the passive refusal of field workers to fertilize increasingly depleted cotton fields or to terrace untilled hillsides also played a significant role. Maximum supervision was required to obtain even minimum results in reclaiming exhausted soils, for fertilizer required all the time, care, and attention that slaves either would not or could not provide.[82] A few of the less prosperous planters with more manageable fields either sold some of their slaves to buy fertilizer, or made do with cheaper, lower quality fertilizers. To optimize cotton yields, fertilizers are not just worked into the field overall, but incorporated specifically into the rows, a process that requires considerable care.[83] While field workers did not refuse outright to increase their masters' crop production, they consistently applied the costly fertilizers improperly, which was not only wasteful but, depending on the fertilizer, could actually damage the land.[84] The tools required were ill used, forever breaking or disappearing mysteriously. Overseers and owners had neither the time nor inclination to watch over the slaves with the unrelenting vigilance that was needed.[85] In view of the cost of fertilizer and the careless and wasteful way in which the slaves worked, planter James S. Peacocke of Redwood, Louisiana, noted, "In respect to our worn out lands, it is almost useless for anyone to waste paper and ink to write the Southern planter telling him to manure."[86] Slave failures at fertilization were so widespread that owners preferred to view them as further proof of their slaves' laziness and stupidity rather than recognize such behaviors as calculated forms of resistance, and many quickly abandoned expensive terracing and fertilizing efforts.[87] As the soils became exhausted and cotton yields shrank, expansion onto fresh lands became imperative if King Cotton was to thrive, or even to survive.[88]

Frances Kemble's prediction of declining production due to soil exhaustion was quickly realized. The passive resistance of field workers contributed to the dwindling production from exhausted soils, threatening the wealth and position of their white owners. Frederick Law Olmsted traveled through the South in 1861 on assignment for the *New-York Daily Times*, noting abandoned plantations with their "stables and negro quarters all . . . given up to decay."[89] Olmsted saw terracing of the hillsides as a solution to making the land once again productive. However, in view of the labor involved in terracing, he noted that "with negroes at fourteen hundred dollars a head and fresh land in Texas at half a dollar an acre, nothing of this sort can be thought of."[90]

The actions of field workers, disproportionately female, hastened the necessity for the geographic expansion of slavery. A series of political compromises opened some new territories to the institution, delaying but ultimately not preventing the day of reckoning: a civil war that freed all slaves, including those field workers who had contributed to the urgency of expansion. In other words, enslaved women's environmental knowledge empowered them to indirectly play a role in facilitating their own freedom. Like the Native American women who were considered powerless due to their gender, race, and status, enslaved women asserted considerable control over the production of lives and of crops, effectively resisting an oppressive institution. Not surprisingly, once the same people who were so inept as slaves attained their freedom, they managed tools and fertilizers quite effectively.[91]

The Civil War also resulted in a new awareness of the importance of cleanliness to human health. The U.S. Sanitary Commission, a forerunner of the American Red Cross, was signed into being by President Abraham Lincoln in 1861 and dedicated to improving conditions in Union army camps. Commission officers coordinated Union women's volunteer and fundraising activities. The commission's directives promoted a healthy diet for soldiers and, at a time when more soldiers died from illness than war wounds, emphasized the importance of cleanliness in the fight against disease. The new standards in Union military hospitals included more frequent washing of bed linens and patients. These tasks were carried out almost exclusively by female volunteers and significantly reduced the number of deaths caused by disease. Word of the powerful role of cleanliness in preventing and curing disease was spread by the commission before it disbanded in 1866, its message reinforced by the returning volunteers who preached the gospel of cleanliness within their communities, stepping up women's war against dirt and disorder as a safeguard to their families' health.[92]

FIGURE 2.2 After the Civil War, African-American women and children continued to labor in the fields. (Wisconsin Historical Society 947015)

. . .

Following the Revolutionary War, women's interactions with the environment were both empowering and disempowering. Although white women generally spent less time performing heavy labor in the fields than did previous generations, their work in kitchen gardens required considerable effort and expertise. At the same time, women's prescriptive sphere bound them more tightly to the confines of their homes and demanded increasing standards of cleanliness. However, it also ultimately granted women nascent authority as keepers of the greater household of earth. Enslaved women continued to carry out the heavy demands of field work, but they too experienced mixed results, as their environmental knowledge contributed to their passive refusal to remedy soil exhaustion, thereby exacerbating the land expansion imperative. As the young nation took shape, Americans' relationships with the environment continued to be affected by race and class as well as prescribed gender roles.

"From a well-stored and comfortable home in Connecticut to this wretched den in the wilderness"

3   The Frontier Environment as Test of Prescribed Gender Spheres

"Singularly Fitted to Be a Useful Pioneer in a New Country"

In mid-nineteenth-century women's magazines, the ideal woman selflessly and even cheerfully bore up under the harshest frontier conditions. Pioneer Jessie May, for example, recounted in *Godey's Lady's Book* the story of a newcomer, Effie Howard, "reared in the lap of luxury" who had never before been long away from her "tender mother and loving brothers and sisters." May "trembled for the domestic happiness" of Howard when she thought of the "unpretending and somewhat rude little home" that awaited the new bride. But Howard, "ever finding a bright side," exclaimed, "It will be so nice to live on an uncarpeted floor; no dust raised in sweeping, and it will be so cool for the summer, to rinse it off with clear cold water. Oh it will be so healthy, so simple, and altogether pleasant." "She was a true woman," noted May, "in the highest sense of the word. She had married the man of her choice; and, knowing that he had yet to make his way in the world, she stationed herself by his side, to cheer and encourage, to help and sustain him when needful." Howard turned her "humble abode" into "a paradise" by virtue of "the hand of [a] woman . . . of taste and refinement." Nicknamed "Sunbeam" for her sterling virtues and optimistic outlook, the inexperienced and naive "household pet" became an excellent cook, nurse, and friend. "Her example was better to me," mused May, "than all the books of 'Advice to Women,' that were ever written; than all the sermons on faith, hope, and contentment that were ever preached. She inspired me to emulate her in looking on the bright side of life, she learned me patience, and firmness and self reliance, she filled my heart with trust, and peace, and satisfied happiness, and was far more the teacher than the taught." Because of her friend's sterling example, May did not long for society, but was "happy in the bosom of [her] own little family . . . never pined for the privileges and enjoyments . . . left behind" in the East, and remained "well contented." Moreover, concluded May, the frontier was home to "many such women as Effie Howard: lovely in their heroic simplicity, their noble disinterestedness, their brave gentleness; whose lives are psalms of beauty, set to lofty music, all the more worthy for being unwritten and unsung by historian or bard."[1]

Throughout the nineteenth century, publications aimed at middle-class women printed endless firsthand accounts from women like May on frontiers stretching from Michigan to California. In one, a woman with the initials M. C. P., writing from "the Minnesota side of the St. Louis Bay," acknowledged cheerfully, "Yes, we live in a rough board shanty." As much as M. C. P. had "dreaded to come" west, she quickly adapted to her new surroundings, noting that "the charm of novelty is effectual in causing a hearty laugh over little discomforts which, at home, would have been unendurable" and concluding, "I am very happy in my new home on the frontiers."[2] Another settler, Mrs. E. F. Ellett, celebrated her neighbor, Mary Ann Rumsey, one of the "Pioneer Mothers of Michigan," as "singularly fitted to be a useful pioneer in a new country, where difficulties and discouragements must be met with unblinding courage, fortitude, and patient perseverance. Her commanding aspect, whether natural, or the result of a habit of being foremost in enterprise was well suited to her qualities of determination and strength of purpose." Such qualities made Rumsey the ideal wife: "By such aid and encouragement it is that woman, a true helpmeet, can hold up man's hands and strengthen his heart when disquieted by care and vexation."[3]

Not every description of women's frontier experience presented it as ennobling, however. Carolyn Kirkland wrote a rare satirical firsthand account of the frontier experience in *A New Home—Who'll Follow?* Kirkland makes clear her elite status by beginning the chapters of her account of life in 1830s Michigan with epigraphs from various literary classics. She writes, with condescending wit, of the huge mud holes dotting the untamed countryside and the primitive dress, diet, housing, and manners of the local population. When she and her husband arrived at a crude log cabin, for example, the woman who lived there tried to awaken her husband, who was drunk and had fallen asleep at the table. As Kirkland tells it, "Thus conjured, the master of the mansion tried to overcome the still potent effects of his evening potations, enough to understand what was the matter, but in vain. He could only exclaim, 'What the devil's got into the woman?' and down went the head again."[4] Kirkland sincerely pities one woman's descent "from a well-stored and comfortable home in Connecticut to this wretched den in the wilderness" but is openly critical of the "self-gratulation" of Michigan settlers who have sacrificed quality of life in exchange for quantities of land.[5] Kirkland emphasized that it was not only elite urban women who suffered from the move to the wilds of Michigan. Any woman who had previously lived on a farm found herself stripped "of the old familiar means and appliances," while her husband "goes to his work with the same axe or hoe which fitted his hand in his old woods and fields, he tills the same soil or perhaps a far richer and more hopeful one—he gazes on the same book of nature which he has read from his infancy, and sees only a fresher and more glowing page."[6] Kirkland's book was widely read, causing a "whirlwind of indignation" among her neighbors, who heartily resented her unflattering portrayals.[7] Kirkland learned the hard way the wisdom of Mary Irving, defender of life on the prairie frontier: "'Home is home, all the world over,' and those who have fixed their home on the Prairie, or in the 'backwoods,' do not like to have its homeliness thrown into relief by any descriptive allusions to Eastern scenery or customs."[8]

As these various accounts attest, there was no single "woman's frontier experience." It is undeniable, however, that gender did play a significant role in how most Americans anticipated and experienced crossing the frontier and settling in the West. The prescriptive literature made it clear that women's nurturing qualities were inherent and therefore portable. A greater test of the theories of the evolution of republican motherhood, the cult of true womanhood, and women as the mothers of civilization could not have been set than that provided by western expansion.[9]

"Go West, Young Man, Go West"

The move west for many Americans was not a single, dramatic journey from coast to coast, but rather was taken in a series of smaller segments spread out over time, often over generations.[10] Charles Ingalls, for example, immortalized in the *Little House* books by his daughter, Laura Ingalls Wilder, moved his family repeatedly. They left the big woods of Wisconsin because it was attracting "too many people" and "he liked a country where the wild animal lived without being afraid."[11] They moved first to Missouri and then on to the Indian Territory (present-day Oklahoma), Minnesota, and Iowa before finally settling in South Dakota. The Hays family of Erie, Pennsylvania, moved to Illinois in 1839. Thirteen years later the family, inspired by tales of prosperity in the Far West, decided to move to California.

The way husbands and wives of the same race, class, and ethnicity responded to the idea of making the journey west highlights gender differences. The trip was nearly always planned at the instigation of a man or men. Men were the targets of political, social, and economic appeals presenting westward expansion as not only good for the country, but also golden opportunities for the individual men who carried it out. In 1837, newspaper editor Horace Greeley, concerned about the threat that overcrowded cities posed to the nation's already unstable economy, urged any worker "unable to get employment at a fair price," and whose financial circumstances would allow it, to "go to the Great West. . . . The West is the true destination." As the economic situation worsened the following year, Greeley stepped up his exhortations: "If any young man is about to commence the world, we say to him publicly and privately, 'Go to the West. There your capacities are surely to be appreciated, and your industry rewarded'" for "land and freedom lay to the West."[12]

Would-be migrants were not only being pushed west by hostile economic conditions in the East but were actively courted by western boosters seeking to hasten development by enlarging their state or territory's population. In an 1857 open letter "To the Unemployed in Our Eastern Cities," Iowa's Edwin H. Grant boasted of the "unrivalled beauty and fertility" of the Des Moines valley, where "lands can be procured for reasonable prices" and "every necessity of life can be obtained at prices far below those paid in the crowded east." Grant described the "liberal-hearted, generous people" who lived in this land where labor is "appreciated and amply rewarded."[13] That same year the Belleville, Illinois, *Advocate* advertized its state more succinctly, proclaiming simply, "Free Soil for Free Men."[14] According to the fanciful prose of Peter Burnett, the future governor of California, "Out in Oregon the pigs are running about under the great acorn trees, round and fat, and already cooked, with knives and forks sticking in them so that you can cut off a slice whenever you are hungry."[15]

While Greeley and Grant presented westward migration as a source of individual opportunity, journalist John O'Sullivan presented it as the salvation of the world. In 1839, O'Sullivan described the "divine destiny" of the United States to spread its freedoms and democratic principles. In 1845, his vision of America living up to its true potential became more specific. He claimed that, as a nation, it is "our manifest destiny to overspread and to possess the whole of the continent which Providence has given us for the development of the great experiment of liberty and federated self-government entrusted to us."[16] The term "manifest destiny" was swiftly embraced by advocates of the annexation of Texas and used to justify the Mexican-American War. By the 1850s the directive, "Go West, young man, go West," was a call to patriotism, an opportunity to provide a vital service to the world while serving God and improving one's own circumstances.

Life on the frontier provided men with the true test of their mettle. Looking back from the vantage point of 1893, only three years after the U.S. Census declared the frontier to be closed (that is, populated), Frederick Jackson Turner described the American frontier to his fellow historians as the cradle of democracy. In Turner's view, only in their efforts to tame the uncultivated and uncivilized lands of the West did American men gain true strength and individuality and, in the process, give the United States its unique democratic identity. The notion that the western frontier represented freedom, opportunity, and independence resonated strongly with generations of Americans, especially men.[17] By the middle of the nineteenth century, vast expanses of uninhabited western lands were becoming increasingly scarce. Men who wanted to serve their country while improving their fortunes by relocating had to act fast.

The Overland Journey

Many who traveled to the Far West in the mid-nineteenth century followed the Oregon Trail, departing St. Louis, Missouri, in April or May in order to complete before winter the journey that, if all went well, could be achieved in four to six months, but with complications sometimes stretched into as many as eight. A few married men joined traveling parties on their own, leaving wives and children with members of their extended families. John Wilburn, for example, traveled with his sister and her nephew, seeking gold in Montana and leaving behind "a wife and two small children with his widowed mother, to watch, and wait, and pray for his success and safe return home."[18] Others planned to send for their families once they were established in the West. It was impractical, however, for all but the wealthiest families to host a venturing husband's wife and children for an indefinite period. Most married men seeking to resettle permanently chose to move west with their families intact. The overwhelming majority of women who made the journey west were married.[19] Some scholars insist that these wives were considerably more assertive than has been recognized when it came to the decision to move west and were able to exert veto power.[20] Fathers and husbands were the decision makers in most cases, however, leaving women no choice because man's role as provider and decision maker prevailed over woman's moral or material objections.[21]

Certainly children, both boys and girls, tended to embrace the move with enthusiasm.[22] Dora Sanford was "in extacy" in anticipation of her family's journey westward from Indiana, and her eight-year-old brother clicked his heels over the news and shouted, "Hurrah! For Nebraska!"[23] Children often welcomed the change and the adventure, and many girls especially looked forward to the journey.[24] For girls who spent their time almost exclusively within the home under the sometimes stifling supervision of their mothers and other women, the journey west offered a welcome immersion into the natural world and the chance to enjoy the company of other people, especially other young people. The diaries and letters of girls and young, single women demonstrate their excitement at the knowledge that the rigors of the trip would allow them to throw off with impunity at least some of the repressive "ladylike" behaviors already being thrust upon them.[25]

Age as well as marital and maternal status figured significantly into the eagerness, or reluctance, of many females who faced the prospect of being uprooted. An emphasis on these factors, however, runs the risk of ignoring the very real physiological differences between the sexes that may provide some of the most compelling reasons that so many women's diaries and journals reveal that they were not as eager as men to make the westward journey. More than just cultural constraints shaped women's trepidation about leaving their middle-class homes and joining the some 350,000 Americans who made the crossing of two thousand miles or more between 1840 and 1870.[26]

The journey west was a young person's undertaking. The elderly rarely attempted to negotiate the physical demands of the terrain. Traversing the Great Plains promised monotony, but also the threat of hostile Indians. Most routes westward included rivers to be forded, requiring especially skillful handling of ungainly wagons and frightened livestock. Desert crossings had to be done efficiently or parties ran the risk of exhausting their supply of water. On some routes wagons had to be hoisted over mountain passes. Men tackled this undertaking while in their physical prime, as did women. But for women this meant walking (or riding in a wagon) across the country and facing its many obstacles while experiencing menstrual cycles or pregnancy, or having just given birth. Some women experienced all three conditions at various points in a single journey.

Until the first mass production of the disposable sanitary pad in 1919, menstruation significantly hindered women's freedom of movement. The clean rags used to stanch a woman's menstrual flow were quickly saturated, making leaks such a concern that women of the upper classes frequently took to their beds during the heaviest days. Women's monthly "sickness" confirmed perceptions of women as the weaker, inferior sex and proved a particular burden for women travelers. At home, saturated rags were left to soak before being laundered, then dried on a clothesline. In preparing for the journey west, all a woman could do was pack a supply of rags and hope for conditions conducive for their reuse. Days spent crossing the desert or plains without access to rivers or other water sources precluded the rinsing or washing of saturated rags, and heavy rains left no opportunity for clean rags to dry.[27] When caught short, a woman lucky enough to have female traveling companions might borrow clean rags.

Pregnancy carried its own unique burdens. For a class of women mortified by having to carry out bodily functions, the routine lack of privacy on the trail was a particular humiliation.[28] Frequent urination common in the final trimester proved a special embarrassment on the plains and deserts that afforded no trees or bushes. Even in the best of circumstances, infection and other complications led to high rates of maternal and/or infant death. On the trail, childbirth could take place at the most inconvenient and dangerous times and places—halfway up a mountain pass or during a hailstorm—and traveling parties rarely stopped for more than a day to accommodate a woman's labor and recovery. Fears of Indian attacks, the need to reach grazing land and gather fuel supplies, and the specter of being trapped by an early snowfall trumped any inclination toward a longer period of rest. Mothers of newborns faced immediate travel in a wagon with no springs.[29] Even over relatively smooth trails, travelers would attach a churn to the front of a wagon and fill it with cream. The jarring motion of the wagon over the course of the day was sufficient to produce butter, but also was extremely uncomfortable for any woman who had just given birth.[30]

Another problem that affected women disproportionately along the trail was uncertain access to water for drinking, bathing, or washing clothes.[31] When it was not possible to maintain a steady supply of clean, dry cloth diapers, babies suffered from diaper rash. Breast-feeding was valued for its convenience as well as its ability to inhibit ovulation (and thereby prevent menstruation) but was not always easy when drinking water was scarce (nursing women need to be well hydrated) or when weather and trail conditions exacerbated the exhaustion of caring around the clock for a newborn. That responsibility was added to the immediate resumption of such chores as cooking, gathering fuel, and caring for any older children. Men and women not only saw the trail through different eyes, but they also experienced it in very different bodies.

More than physiology, however, was at work in determining women's attitudes and experiences on the trail. Lorena Hays, a young, single, struggling schoolteacher with ambitions to be a writer, kept a journal that revealed many internalized qualities of True Womanhood. Contemplating the journey, she was primarily concerned with being "useful in some way to others." Like many a single young woman as yet unburdened by husband or children, she saw the West as a land of opportunity for her family and herself, but worried that "the society is so bad [that] I am afraid we will want to come back very much . . . sometimes I almost wish I was not going."[32] As the journey began she felt "some feelings of regret," for "it is with some misgivings that I think of starting across the plains. I expect to see some trying times."[33] Eighteen-year-old Mollie Dorsey Sanford, also a schoolteacher, had enjoyed "the benefits of schools, churches, and the best of society" in Indiana. She knew she would "be deprived of all these in the 'wild unsettled country'" of the West but deemed it her responsibility to "not shrink from unpleasant and untried realities." She nevertheless spoke of the "sadness of my heart," for "all of the bright anticipations of the probable interest of our trip does [*sic*] not compensate me for the sorrow at parting from friends." She and her best friend "wept as only loving friends can weep, to think that we must part—perhaps forever."[34] Most adolescent boys would also miss their friends, but friends were less integral to their identities. Writing "in the shade of our prairie schooner" on their first day of travel, Sarah Raymond Herndon observed, "When people who are comfortably and pleasantly situated pull up stakes and leave all, or nearly all, that makes life worth living, start on a long, tedious, and perhaps dangerous journey, to seek a home in a strange land among strangers, with no other motive than that of bettering their circumstances, by gaining wealth, and heaping together riches that perish with the using . . . the motive does not seem to justify the inconvenience, the anxiety, the suspense that must be endured." She lamented, "Why have we left home, friends, relatives, and loved ones, who have made so large a part of our lives and added so much to our happiness?"[35]

Some girls and teenagers anticipated the journey with genuine joy and excitement. When Sarah Herndon's friend Cash Kerfoot stated, "It is so jolly to be going across the continent," Herndon replied, "I wish I felt that way," before asking, "Aren't you sorry to leave your friends?" "Of course I am," Kerfoot acknowledged cheerfully, noting that she and her friends could always exchange letters, adding, "and I will make new friends wherever I go, and somehow I am glad that I am going."[36] Many teenaged girls and young women spent their honeymoons on the trail. Lucena and George Parsons married in Illinois on March 18, 1850, and left for California the next day.[37] Harriet Fish waited until her fiancé had acquired a mining job in Colorado before traveling to meet him. The two married the day she arrived. She and other young brides wrote with great hope and enthusiasm of starting their married lives in a new land ripe with opportunities.[38]

Although most wives of longer standing did not share the enthusiasm of the new brides, there were exceptions. Margaret and Ledyard Frink married in Kentucky in 1839. They moved first to Ohio, then to Indiana, where they "succeeded very well," but "were not yet satisfied." When they heard the "exciting news" of California's "delightful climate" and "abundance of gold," they resolved jointly to cross the plains. They remained undaunted by the discouragement of their neighbors.[39] A common characteristic of wives who shared their husbands' zeal for the West was a lack of children, but even many childless wives expressed reluctance to leave the comforts of home and family, to forsake culture, prosperity, and peace for crude living and adversity.[40] Although women believed men to be the rightful leaders of the family, many mothers sought to dissuade their husbands from exchanging the stability of the East for the uncertainties of the West. Even the dutiful Caroline Ingalls, the long-suffering mother in the *Little House* series, responded to her husband's decision to once again uproot their family, "comfortable" in their "snug house," with, "Oh, Charles, must we go now?"[41] Sarah Jones was more unequivocal still in her plea to her husband: "*O let us not go.*"[42] As her family prepared to leave Indiana, fourteen-year-old Sallie Hester reported, "My mother is heartbroken over this separation of relatives and friends. Giving up old associations for what?" When they suffered an accident a few days out, her mother "thought it a bad omen and wanted to return and give up the trip."[43] Such objections were overruled by husbands, as drawn to the promise of wealth, adventure, and rugged individualism as their wives were to the safety, comfort, and female companionship of their existing homes. The same promotional literature extolling the virtues of Kansas that W. B. Caton found compelling and convincing spelled to his wife "destruction, desperadoes, and cyclones": "I could not agree with my husband that any good would come out of such a country, but the characteristic disposition of the male prevailed, and [we were] bound for the 'Promised Land.' To say I wept bitterly would but faintly express the ocean of tears I shed upon leaving my beloved home and state to take up residence in the 'wild and wooly West.'"[44]

The degree to which most nineteenth-century men and women internalized their prescribed spheres quickly became evident in their reactions to the call of the wild, although factors including race, class, age, marital status, and ethnicity all played roles as well. At midcentury, men far outnumbered women on the trail and in the West. Only two days into her party's journey from Indiana to California in 1850, Margaret Frink heard the locals in a town comment on her presence, saying, "Surely there's no man going to take a woman on such a journey as that across the plains."[45] Frink observed that on the trail "there are but few women; among these thousands of men, we have not seen more than ten or twelve."[46] Clarence Kellogg remembered the California gold rush as "a human stampede" that "originally consisted entirely of men."[47] The non-Indian population of California in 1852 was more than 90 percent male, and only in the late 1860s did it begin to approach parity.[48]

Many an unmarried young man, unencumbered by a wife or children, undertook the journey, but rarely on his own. It was too long, too dangerous, and too complicated a trip. It was also rarely a poor person's endeavor. Argonauts (gold seekers) who planned to make a quick fortune and then return east did not attempt to bring such bulky items as furniture, but they did need foodstuffs and equipment, both to survive the trip and to provide for basic necessities once in the West. The cost of outfitting the overland journey, including wagon, draft animals, foodstuffs, and tools, was far beyond the reach of the working class, although hardy individuals, women as well as men, did attempt the trip on foot, supplied only with what they could carry on their backs. Margaret Frink noted, "Among the crowds on foot [in August 1850], a negro woman . . . tramping along through the heat and dust, carrying a cast-iron bake oven on her head, with her provisions and blanket piled on top—all she possessed in the world—bravely pushing on for California."[49] Some came to see the advantages of setting one's own pace, unencumbered by the needs and desires of others. A member of Frink's own traveling party concluded that "he could walk to California sooner than we could get there, at the rate we were traveling. We gave him all the provisions he could carry and he started, with blankets, clothing, and provisions strapped on his back, to walk fifteen hundred miles to California."[50] Others came to this mode of travel involuntarily when their oxen died or their wagons broke down; some were reduced to begging from other travelers they met on the trail.[51] Frink described these emigrants as a "woe-begone, sorry-looking crowd" made up of "men with long hair and matted beards, in soiled and ragged clothes, covered with alkali dust, [who] have a half-savage appearance."[52]

Responding to Trail Environments

Many factors influenced the overland experience: preparation, destination, departure year and date, size and health of the traveling party, route, weather, interactions with Indians, and individual temperaments. Some travelers experienced a relatively trouble-free crossing, while others suffered devastating losses due to disease, dehydration, starvation, and accidents. Men tended to record their observations of the country, distance traveled, the status of their animals and provisions, and their hopes for what lay ahead. By no means were they cavalier, however. Even men unencumbered by wives and children wrote movingly of their sorrow at the deaths of traveling companions, and as the journey progressed, many recorded a variety of fears and concerns.[53] For men with wives and children in tow, the weight of their responsibilities as the providers and protectors of their families was plain. Even men unaffiliated with the women in their party accepted, however reluctantly, their responsibility to look out for them. Margaret Frink observed, when there were worries about a possible Indian attack, "I had a very strong feeling . . . that these men would have felt more at ease if there had not been a woman in the party, to be taken care of in case of danger."[54]

In all-male parties, there could be no division of labor along traditional gender lines. Where women were present, men usually performed the most demanding physical labor. Women carried out domestic chores: the daily packing and unpacking of the wagons, preparing meals, washing clothes, and caring for children. But as in earlier frontier times, divisions of labor were rarely rigid when there was so much to be done, and the life or death of the party hinged on cooperation in the prompt and efficient completion of tasks. Men joined women in foraging for food and fuel. Women drove oxen and herded cattle.

Although some wealthier families brought their servants with them, most did not. At the start of her party's journey to California, Sarah Herndon noted that two women "have been [struggling while] getting a supper for a family of twelve, no small undertaking for them, as they have been used to servants and know very little about cooking," and were forced to learn under the most primitive conditions.[55] While some urban men rejoiced in the opportunity to reconnect with the outdoors, few of their wives relished performing the manual labor they had delegated to others at home. Despite some initial misgivings, Sarah Herndon discovered that she and her mother enjoyed life on the trail, and she found herself "feeling sorry for the people that have to stay at home, and cannot travel and camp out." She observed, however, that their traveling partner, Mrs. Kerfoot, "is homesick, blue and despondent" under difficult conditions, for "she has always had such an easy life that anything disagreeable discourages her."[56]

Sarah Herndon recorded in her overland diary that the man's word "is law in this camp."[57] Husbands on the trail made the crucial decisions about route, pace, and the breaking up of larger parties into smaller ones, even over the strenuous objections of wives who had been carrying out "men's work." Sharing men's chores, but gaining no compensatory powers, most wives clung to the social roles that had once guaranteed them their larger sphere of influence. Whereas many men's letters record their hopes for the new country, their wives' letters are filled with longing for the homes and friends left behind. As much as was possible in trail conditions, women remained "ladies," making clear their intention to maintain that identity and to reclaim their rightful position once established in the West. All but a few persisted in wearing long dresses and aprons, despite the impracticality of these garments.[58] In addition, white women were mortified if they tanned and strove to protect their skin from the elements.

Mothers worried especially about the dangers posed to their daughters by prolonged exposure to the trail environment. Adrietta Hixon's recollection was typical: "While traveling, mother was always particular about Louvina and me wearing sunbonnets and long mitts in order to protect our complexions, hair, and hands. . . . Mother was always reminding Louvina and me to be ladies." The girls in Sarah Herndon's traveling party wore not only sunbonnets, but also "riding-habits, made of dark-brown denim" that covered their clothes completely as protection against mud and dust. Nevertheless, because they wore "thick shoes" and "short dresses," that is, ones without trailing skirts, some of the girls suffered "mortification" at the thought of "town" people seeing them so attired.[59] When Mollie Dorsey Sanford donned her father's suit to entertain her companions, "It was very funny to all but Mother, who feared I am losing all the dignity I ever possessed."[60] When nine-year-old Mary Ellen Todd learned to crack the whip while driving the oxen, she initially felt "a secret joy in being able to have the power to set things going." She was gratified by her father's pride in her new skill, but then she heard her mother confide in her father that she feared a girl cracking the whip wasn't very ladylike. After that, "There was also a sense of shame over this new accomplishment."[61]

Mothers were preoccupied by more than fears that their daughters would lose the qualities that elevated women within their prescribed sphere. They were almost solely responsible for infants who required round-the-clock care and small children who proved particularly exhausting. Removed from the routine and level of supervision that characterized their lives in the East, children fell out of wagons, wandered off, drowned, and suffered all manner of accidents. They also came down with the usual childhood diseases, as well as cholera, an epidemic that was particularly virulent on the trail between 1848 and 1854. In the absence of women, men nursed themselves through illness and injury, but since women's prescribed role included nurturance, traveling parties with women looked to them to care for all the injured and ill. More women's overland diaries and journals than men's reveal a preoccupation with death, as they tallied each day the number of graves and animal carcasses

FIGURE 3.1 Travelers camped on their journey west. The two girls on the left wear the "short" skirts deemed more practical for travel. A woman stands before the wagon in the center. Next to her is a seated woman with a small child. (Ben Wittick, courtesy Palace of the Governors Photo Archive NMHM/DCA 003083)

they passed along the trail. Those travelers who survived the journey faced new challenges.

Making a Frontier Life

Anna Shaw remembered well her mother's response to the log house that would become their home on the Michigan frontier: "Something within her seemed to give way, and she sank upon the ground . . . she buried her face in her hands, and in that way she sat for hours without moving or speaking. . . . Never before had we seen our mother give way to despair." Other women broke into tears when they saw the crude structures that were the culmination of their long trek away from family and the familiar comforts of the East. A seventeen-year-old bride took in her new home's mud roof and dirt floor, then proclaimed with indignation, "My father had a much better house for his hogs!"[62] When Debby Allan, newly arrived on the frontier, learned that a panther was menacing the area, she cried to her husband, "This is a terrible region. Do, Peter, let us go back home."[63] Not surprisingly, depression plagued frontier women.[64]

Western climates and frontier conditions made unhappy women miserable. According to land speculator and journalist Charles Dana Wilbur in 1881, the act of plowing and planting crops on the Great Plains would change atmospheric conditions, transforming the prairie into a garden.[65] Wilbur's thesis boiled down to the succinct phrase, "Rain follows the plow." Pioneers settled on the vast prairie, building houses out of sod because of the dearth of trees. Women raised on the gospel of cleanliness despaired in homes where the walls as well as the floor were made of earth. Farmers planted crops and waited for rain that never came. What they experienced instead were blistering summers, often punctuated by droughts that left native grasses brown and parched, destroying grazing lands and feeding the fast-moving prairie fires sparked by lightning strikes. Swarms of locusts appeared without warning, devouring everything in their path, including crops and even wooden plow handles. Blizzards could come on so early (starting sometimes in mid-October) and violently that potatoes yet to be harvested froze in the ground almost instantly. Families who had yet to take their crops to be milled were reduced to grinding wheat in coffee mills, a procedure so time consuming it took nearly constant effort to produce the flour necessary for survival.[66]

A scarcity of fuel sources plagued settlers in climates where temperatures would remain below zero for as long as two months at a time. Tasks such as gathering and preparing fuel often fell to women and were relentless. In the days before coal and oil were brought in by the railroads, stoves had to be fed a steady supply of dried buffalo or cattle dung, corn stocks and cobs, or hay twisted into hard sticks. Weather extremes were unpredictable and could be deadly. Mothers who prized education above all else nevertheless feared sending their children to school. On January 13, 1888, a cold front swept across a wide swath of the nation with amazing speed, resulting in an immense blizzard. In southern Dakota Territory, the temperature, with wind chill, dropped from five degrees above zero in the mid-morning to forty below zero by 9:00 p.m. More than a hundred children froze to death on the Dakota-Nebraska prairie, blinded by snow and unable to make their way home from school in the freezing, gale-force winds.[67] The Nebraska environment was so harsh that 43 percent of homesteaders failed, necessitating either a retreat to the East or, more often, a new move farther west.[68]

Like their female counterparts, men suffered from extreme climates and unpredictable weather, crude living conditions, a limited and monotonous diet, and ceaseless labor, but debilitating loneliness beleaguered women far more than men.[69] Annie Green insisted on accompanying her husband to Colorado when he decided to participate in a homesteading experiment sponsored by Horace Greeley. She wrote of the "terrible gloomy days" that she spent in her "lonely tent" with her two small children in 1870. When she learned of another wife who had "wisely stayed in her happy home" in the East, she "reproached . . . [herself] for not doing likewise," but tried to make the best of her "mistake." She "resolved to cultivate a cheerful disposition" but confessed that she had to feign appreciation of her new surroundings "in order to coincide" with her husband, who clearly loved Colorado and "displayed much anxiety" that his wife share his enthusiasm.[70] Alice Neal, an officer's wife, "could not bear" her husband to reproach himself for taking her from her "delightful home" only "to encounter one privation after another." Like Green she resolved to be cheerful, "but there are still many times when I find my spirits sinking for the lack of the companionship of my own sex." She confided in her diary, "Oh! It is so very, very desolate and lonely here!"[71] In an 1852 tribute to the "pioneer mothers of Michigan," *Godey's Lady's Book* saluted the women who had to "struggle with that feeling of isolation and loneliness which presses heavily on those who have severed all the endearing times of home, where cluster those fond attachments only formed in youth. Many a sad hour was passed in remembrance and regret by the young wife . . . when she had no sympathizing friend in whose bosom she could pour her griefs."[72]

Women's experiences, and their reactions to them, varied widely. Some wives were eager to partner with their husbands in the business of homesteading. Others sought homesteads in their own right, but as a result of a number of legal barriers, women counted for less than 10 percent of Americans filing claims before 1860. The Homestead Act of 1862, however, was not gender specific, and women constituted a larger portion of the population filing for land under its provisions. Up to one-third of claimants in the Dakotas were women.[73] Two days before Ada Colvin filed for her Nebraska timber claim of 60 acres, she spent the evening "setting with a bed quilt around my chest, hat on my head, playing dice with Fred. Such is western life." When she saw her land she felt "pretty well satisfied with it, ... better than I expected. [The men] all seemed to think I am hoggish but I took the same as the rest did."[74] Louisa Clapp, writing under the pen name "Dame Shirley," presented her life as the childless wife of a California miner as one great adventure. She concluded the series of letters she wrote to her friend in the East: "I like this wild and barbarous life. . . . I took kindly to this existence, which to you seems so sordid and mean."[75]

Like Colvin and Clapp, some frontier women were neither lonely nor unhappy. Susan Shelby Magoffin, for example, recorded positive impressions of her new life in Santa Fe in 1846.[76] Unlike Green, who lived in a tent that was constantly blowing over and who "would have given almost everything [she] possessed" to have a friend with her, Magoffin, the bride of a successful merchant, lived in a house staffed with servants and enjoyed the company of her mother-in-law and a constant stream of visitors.[77] The companionship of at least one other woman of the same race and class seems to have been enough to recreate crucial aspects of the traditional sphere of comfort and control, and served as a key indicator of many women's level of satisfaction upon settling in the West.

According to the prescribed ideals, men, the natural protectors and the material providers for their families, gloried on the frontier where they were made (or broken) by their ability to wrest a fortune, or at least a livelihood, from the land in the form of precious metals, crops, or livestock. Men did not see the transformation of the land wrought by such endeavors as exploitative, destructive, or shortsighted, but rather as their right and their familial and even religious obligation. In the words of Reverend Thomas Starr King, in an 1862 address before California's San Joaquin Valley Agricultural Society, "The true farmer is an artist. He brings out into fact an idea of God." Men in environments that did not naturally lend themselves to carrying out God's idea were not thereby exempt from their obligation. Even the desertlike floor of the San Joaquin Valley must be transformed because "the *earth is not yet finished*. . . . It was made for grains, for orchards, for the vine, for the comfort and luxuries of thrifty homes." How must such a transformation take place? According to King, "Through the educated, organized, and moral labor of men."[78]

In fact, the labor of women as well as men was required to settle the West and remedy its crude conditions. Whatever their objections, women had no choice but to stay and do their part in civilizing the wild country.[79] In the early years of settlement, the lines that formerly separated women's work from men's remained blurred.[80] Nontraditional labor, such as plowing, harvesting, or hunting, was the new norm for many formerly middle-class eastern women, but, as on the trail, this did not create an egalitarian society. Some women found life in the West liberating from the restrictions of their previously tightly bound sphere, yet aspects of these gendered ideas and values remained stubbornly in place. As the outdoor work carried out by women and necessary to their families' survival gradually gave way to more traditional divisions of labor, women's indoor activities resurged. In Clarence Kellogg's recollection of family life in the California mining camps in the late 1840s and early 1850s, the "pioneer women were restricted to the confines of their homes, never paraded in public, and [their] contact with the male element was extremely limited. Distance served to enhance sweet memories of home and female loved ones, hence the female was placed upon a high pinnacle." Women and girls, according to Kellogg, "never went 'downtown'" or "passed in front of a saloon." When a woman did go out, "she never went without her corsets" and "never went outdoors bareheaded for fear of the dreaded freckle," but always with "bonnet, veil, and parasol."

Within their homes, however crude, women held all occupants to middle-class standards of cleanliness. Kellogg remembered the thoroughness of his mother's spring cleaning, requiring "wash-boiler after wash-boiler-full o' hot water [and a] big pot o' soft-soap. Everything was moved out of the house." The floors were "soused and slopped with soap-suds-water," swabbed with broom and mop, then rinsed with clean hot water. Furniture was polished using rags saturated with kerosene. The stove was polished, and "knives and forks [received] a good scourin' with wet ashes." Mattresses were stuffed with new straw. War was declared on pests as "bedbugs hidin' places received especial attention," new flycatchers were hung, and mosquito-bars (netting usually fashioned to fit over beds or cots) received "a good-goin'-over and darnin'." Blankets and quilts "got a good boilin' 'n' airin'," and the kitchen table was covered with new black oilcloth. When the house was finally clean to her exacting standards, Kellogg's mother would add the crowning touch of civilization by having her children "go out and pick a nice bouquet of sailor-caps and mountain stars, and put them in the pickle-bottle." Only then, "flushed, and tired, but radiant," would she declare with satisfaction, "thank goodness this house is clean again," before reminding her family, "Now please don't track-in any mud with you, *please* wipe your feet."[81]

Civilizing the Frontier

In Kellogg's memory, women and girls remained ensconced within the home, appearing regularly on the streets only after farmers had replaced miners as the new immigrants, bringing wives and children. He noted approvingly that those wives strove to recreate in the West the civilizing institutions of the East: the schoolhouse and the church.[82] Gender values prescribed to men the responsibility for providing for their families materially, while most pioneer women saw themselves as crucial to the civilizing of the wilderness, a task they took on with some relish. Women insisted upon the establishment of schools and churches rather than saloons and turned crude dwellings into homes. Lorena Hays observed proudly as her party passed through the thriving farm lands of Utah in 1852 that it was "only a few years since this great and beautiful valley was waste wilderness, on which roamed only the indolent savage . . . and now the busy hum of civilized life is heard everywhere."[83] Upon arrival in California, however, she worried about the very qualities that made the state so appealing to Louisa Clapp. This "wild, romantic, and almost barbarous country" struck Hays as dangerous rather than appealing. She observed disapprovingly that "all restraint were laid aside by everyone when they arrive here," as "sober-minded young men of good habits become wild and reckless—husbands almost forget the anxieties of a confiding wife, and all go helplessly rushing, hurrying on in wild strife for gold." Hays identified the cause of the widespread drinking and gambling by the men around her: "no profitable places of recreation and . . . no female society."[84] Her concerns were eased, however, by her observation that "females who are to be a purifying influence upon society are cordially welcomed and politely treated by all." Faced with "habits [in others] we cannot approve of," Hays sought to "set a good example which may have an influence upon others," emphasizing that setting a favorable example was woman's "privilege."[85] She contrasted California's natural beauty with the moral depravity of its male population: "Sin and darkness have darkened and cast a moral gloom over its hills and its ravines and valleys, but we trust that the star of reform will ere long shine upon us, that the sun of righteousness will shed its bright rays upon us and sweep immorality and crime from our beautiful land."[86] In her view, California's physical and moral environments both required transformation: California "needs the hand of cultivation and art to render it still more beautiful, to strip off its barren and somewhat unclothed appearance, to give its valleys and mountains a finishing stroke of beauty, a homelike charm and attractiveness and then what may not be the ideal for the future of such a land with wise laws and institutions."[87]

Lorena Hays was not alone in her certainty that it was the province of women to improve the culture of the land. Transplanted Virginia native Louise Palmer noted approvingly that "the entire religious and social life of Nevada is conducted by ladies."[88] Not all women in the West were equally committed to turning the physical and moral environments of the West into shining new versions of the East, however. Slave-holding women, for example, brought their slaves with them when they migrated to Texas.[89] Other women worked in saloons and gambling establishments,

FIGURE 3.2 Nature's housekeepers. As these families in a small logging camp in north central Pennsylvania "civilize" the wilderness circa 1896, two of the women hold brooms and a third pours coffee from a silver pot into a china cup. (William T. Clarke photograph, Lois Barden Collection, originally published in *William T. Clarke, Photographer: The Epic Transformation of North Central Pennsylvania from "Black Forest to Bleak Desert," Circa 1878–1917*, by Ronald E. Ostman and Harry Littell)

and some were prostitutes and madams.[90] Many more were simply too busy cooking, cleaning, mining, farming, and running boardinghouses, restaurants, laundries, stores, and other businesses, but others supported such a vision by serving as teachers and missionaries.

Western expansion and women's civilizing imperative had an environmental and cultural impact on the peoples already living on the contested terrain of "new" lands. By the early 1800s, for example, the fur trade had transformed almost beyond recognition traditional hunting grounds and ways of living of many indigenous peoples. Their old ways of living obsolete, those Indian men who had not been relegated to reservations continued to hunt, no longer for subsistence, but for profit. In some areas, Indian women found that their traditional skills also had commercial value. Before the buffalo were hunted nearly to extinction, Indian women prepared their hides for sale and made pemmican from their meat. In the Northwest, the dependence of the Hudson's Bay Company on women to caulk and paddle the organization's canoes prompted the admiring observation that "women were as useful as men upon Journeys."[91] As the supply of animals was depleted, however, the fur trade gave way to the practice of agriculture touted by whites as one of the keys to civilization. In the process, some native women's skills lost their commercial value, although their farming and gathering abilities continued to contribute significantly to the survival of tribal peoples both off and on reservations.

In early nineteenth-century California, Indians continued to live and work on mission property, where women and girls carried out a variety of sex-appropriate tasks deemed civilized and designed to cultivate industriousness. They combined traditional ways with those imposed by the Spanish priests. In the summer months, for example, after finishing assigned labors to the benefit of the mission, such as carding and weaving cotton or wool, grinding barley, sewing, cooking, cleaning, or doing laundry, Indian girls and women received permission to gather pine nuts, chestnuts, and roots to store as a supplement to their diet in winter.[92]

In 1821 Mexico took over lands previously controlled by Spain and secularized the missions in the 1830s. Mexican colonists divided the vast former church properties and other lands for themselves, with women acquiring property through inheritance and kinship ties rather than through the grants that were deeded directly to men.[93] Bereft of what little protections the missions offered, many California Indians, like those further east, could not return to their traditional ways of life because the local environments were transformed as the result of farming and ranching, although some fled to the interior, where the changes were not yet as extreme. Even where traditional resources remained in the more transformed areas, however, access was not always guaranteed. Some of the new landowners, for example, would not allow Indian women on their property to gather the acorns or grass seed that were the former staples of their diet.[94] Indian women's provender nevertheless augmented the meager diet of the many native peoples left with no alternative but to labor on Mexican ranchos and in pueblos, but by the end of the Mexican period, the native population had dropped precipitously.[95] With the gold rush and California statehood, new waves of people poured into California, further transforming the land and making traditional food sources all the more difficult for Indian women to procure.[96] Indians in California and elsewhere were increasingly forced by the worsening conditions onto reservations, where malnutrition was common.[97]

Mexican Americans (an umbrella term that includes people of a variety of mixed Indian, Spanish, and Mexican origins) throughout the Southwest fared only somewhat better than Native Americans, with class playing a large role in their individual reactions and experiences. For example, the well-to-do Mexican wife of a customs inspector was appalled when her husband was assigned work in Goliad, Texas, a place she called a "poor, out-of-the-world, ignorant village."[98] Whatever their economic class, however, Mexicans faced much discrimination at the hands of Anglo

Americans. Moreover, Mexican women, who had been allowed under Spanish law to retain property after marriage, found that under American law, their land automatically belonged to their husbands. Some Mexican-American women took advantage of the laws of their new country to acquire rather than lose land, filing homesteading claims.[99] In Cochise, Arizona, 14 percent of Mexican land purchasers and homesteaders were women.[100]

Nevertheless, most Mexican Americans remained or became impoverished following annexation by the United States, losing whatever lands they had owned to a bewildering maze of taxes, penalties, and legal codes and costs. Even wealthy women who managed to retain their lands initially usually lost them in the end. When the good pasturelands on which her large herds had previously grazed were usurped by investors and cattle barons, Santa Fe landowner Maria Nieves Chaves was left with no alternative but to sell.[101] Floods and droughts in the late 1850s and early 1860s decimated crops and cattle herds, and stronger and better grades of stock in the Midwest further weakened southwestern cattle markets, accelerating losses.[102] Mexican-American women struggled to retain their traditional ways despite the changes to their environments made by the influx of people, crops, and herd animals. They responded to rising food prices (eggs, vegetables, butter, mutton, and flour became particularly expensive as the Anglo population grew) by raising chickens and sheep for sale or trade, but the values of even those commodities fell as Anglos switched to their preferred diet of beef and pork, products from animals that were more expensive and harder to raise.[103] By 1880, 90 percent of New Mexico's Mexican Americans had lost their lands.[104]

Priced out of ranching throughout the Southwest, some women catered to the growing Anglo population by building or working in saloons or as prostitutes. Others peddled food and beverages as street vendors.[105] By the 1880s in Los Angeles, 83 percent of the Mexican-American population lived in a central area marked by dilapidated housing and inadequate sewage and drainage systems.[106] More than 30 percent of the families in those neighborhoods were headed by women who performed wage labor for the rising Anglo population, primarily cooking, sewing, and doing laundry and other low-paying jobs that left them at the bottom of the economic hierarchy.[107] Wherever possible, women with access to even tiny plots of land used their environmental knowledge to contribute to local economies through small-scale farming and ranching, trading wool and sheepskins.[108] They continued to maintain as well the gardens vital to their families' sustenance, growing and grinding corn, a diet staple. They also carried out the tasks of their particular traditions, such as building adobe structures and weaving rugs for home use and for sale.[109]

Most migrants overlooked how their ways and numbers caused a downward spiral for local peoples. When John Downey took stock of the many changes to

California in the late 1850s, he noted only the newcomers' efforts to improve California's physical and moral environment. He wrote approvingly to a woman friend, "The face of the country is altered a good deal in the last two or three years and the manners and morals of the people are changing a good deal." He attributed much of that change to the fact that "the emigrants now to California are permanent settlers, those who can take with them wives and children." He offered assurances that, although young women had previously stayed away from California due to its "bad reputation... there [are] now [more] decent and well behaved young girls [who] are respected."[110]

The reactions of Chauncey and Catherine (Kate) Elwell to the physical surroundings of their new home in Wisconsin highlight gendered attitudes about what constituted improvement and success on the frontier. Frustrated when their Pine Tree Tavern in Rupert, Vermont, failed to thrive, the Elwells bought 160 acres in West Salem, Wisconsin, where they moved in 1852. Chauncey, as befitting a white, middle-class man, sought material gain, finding satisfaction in his achievements in the world outside the home.[111] Chauncey's letters to family back in Vermont detail the amounts of corn picked, wood chopped, water drawn, livestock sold, and, in particular, payments received. He exults over his new life in Wisconsin, telling his mother that although he is working hard, he "never felt better in his life."[112] Kate's letters, as befitting a woman whose sphere was almost wholly within the home, are more restrained in their enthusiasm for pioneer life. While she recognized that others viewed Wisconsin as a "grand project for young people to ... start for themselves ... and get a position in the place and society first and by that way can exert more influence," she frequently expressed embarrassment at her own reduced standard of living. She describes herself as so "poorly off and destitute" that she is "ashamed to describe" their home, while Chauncey claims to be "not ashamed of his honest poverty." Kate closes her detailed description of "my poorly furnished [log] home," with, "Don't laugh; it is the best we have. . . . My house is scoured and clean and all the things in order even if it is poor." Years later, however, enjoying the fruits of their slowly earned prosperity and position, Kate recognized that she suffered from what the Indians called the "too much house" of American women: "This large house is a burden, keeps one in it too much when they [could] better be in the open and viewing nature instead of cleaning, cooking, and fighting dirt all the time."[113]

Even in their early "embarrassed" financial circumstances, the Elwells were sufficiently comfortable to spare Kate much of the heavy physical labor performed by many pioneer women both in the home and in the fields. Instead, in order to be a civilizing, nurturing influence in the service of others, Kate Elwell taught the first school in West Salem, her "poor" cabin being sufficiently large to serve as the schoolhouse for twelve children during the winter of 1852–1853. She also carried out a practice very common among women on the frontier: bringing beauty and diversity to her home and garden through the exchange of seeds and cuttings with family members back east.

Beautifying the Environment

Men and women worked together to transform the environment of the West. They cleared forests, plowed the plains, irrigated arid areas, hunted animals, fished, trapped, built fences, and mined, all activities that were, to their minds, making the best possible use of the land. In short, they continued the settlement trends begun by the first European settlers on the eastern shore, resulting in profound changes to indigenous flora, fauna, and people. For many women, however, the beauty of the frontier in its natural state was its saving grace.[114]

Other women saw little in the West worthy of preservation. Julia Carpenter, who "failed in the attempt . . . to feel even a small degree of enthusiasm" about virtually everything that wasn't her eastern home commented on the "great abundance and luxuriance" of California flowers. Her admiration, however, was hardly unqualified: "The flowers of California are not as beautiful as ours, and for the reason that they grow so very large." She conceded that "the cherries are the best I ever saw, and the apricots are fine," but declared the "large and fine looking" strawberries to be "decidedly lacking in flavor," and peaches and apples inferior to those in the East, and "even the oranges" not "nearly as good as those that are raised in Florida." "Where," she asked plaintively, "are the green fields, the running brooks, the fresh sweet-scented mosses and feathery ferns that make a country drive in New England so enjoyable?"[115] While most frontier women were more appreciative of the frontier landscape, especially its wildflowers, they were also eager to improve and uplift the physical environment just as much as the moral climate.

These women saw the West as being in need of beautification. Through the exchange of seeds and cuttings with family members back east, they transformed a strange environment into one more inviting and familiar, creating a tangible tie with people and environments seemingly worlds away. By no means was this practice limited to white settlers. The few African-American women who were not relegated to domestic employment in the Far West, but settled in as homesteaders, planted flowers that linked them to the yards, fields, and families of their origins.[116]

Like Kate Elwell, Ellen Spaulding Miller, who migrated to Eau Claire from Rudolph, New York, in 1870, wrote to her family only occasionally about matters within the male sphere of politics and the local economy. She detailed instead her flowers, vegetable garden, and houseplants, which she loved nearly as much as her family.[117] She frequently requested that her family send her flower seeds, bulbs, and plants, concluding one letter, "Now write to me as soon as you get this and send me some seeds."[118] In the words of one of Miller's descendants, "What better way to stave off homesickness than to have flowers from your mother's garden growing in the garden at your new home?"[119] Ida McMechen agreed. She wrote to her mother from California of her "pots of lovely flowers and ferns . . . that will cure me of loneliness and many other discordant conditions."[120]

Plains and prairie environments posed a special challenge. Pioneer Lorena Hays didn't believe that "our hearts can love so well where there are no trees and none of the attractions of the hill and vale."[121] Even Sarah Herndon, who generally enjoyed the overland journey, complained of "the tedious, tiresome monotony of these vast extended prairies. . . . No earthly consideration would induce me to make a home on any of these immense prairie levels. How my eyes long for a sight of beautiful trees and running streams of water; how delightful to stroll in the woods once more."[122] Plains settler Mary Dodge Woodward mused, "How nice my flower garden must be looking at home," and tried to replicate her eastern garden in the Dakota Territory. However, when she attempted to cut sod in the front yard, "the soil clung to the ax like gum." She finally managed to raise a few pansies, and when she saw their

FIGURE 3.3 A mother and daughter in a flower garden. Sharing seeds created an important connection to family members a world away. (Wisconsin Historical Society 75070)

blossoms after a heavy rain they looked to her "like the faces of old, familiar friends, almost human."[123] She also found "something fascinating about gathering wildflowers. It gives one a childish delight."[124] Pioneer women sought seeds and roots to augment local plants (including berries, barks, and flowers) to be used for medicinal purposes, herbalism being another art women continued to practice within the nurturing service-centered domestic sphere.

Women on the Alaska Frontier

Western environments offered women many challenges, but an even more daunting frontier opened at the turn of the twentieth century. Alaska, purchased from Russia in 1867, remained remote and foreboding to most Americans even as the West filled, but its still intact ecosystems allowed for the development of a thriving fur trade. White trappers, traders, and early prospectors often married native women who developed a hybrid way of life combining traditional foods with European-based ways of living.[125] Like the California gold rush that preceded it, the Klondike gold rush of 1897 presented men with another "final" frontier—an opportunity to make their fortunes and prove their manhood. Gold was discovered along the Klondike River, in Yukon Territory, Canada, near the Alaska border, setting off a frenzy of migration all over the world, with most gold seekers coming up from the United States. Like the previous overland journeys, the trail to the Klondike was traveled primarily by men. Since few intended to settle in the extremely harsh conditions of the far north, most married men left their wives and families behind. Nevertheless, some wives of gold seekers did accompany their husbands on the perilous trek, braving subzero temperatures, snowstorms, blizzards, slides, and avalanches.

While her new husband sought gold, Ethel Berry spent her first two months completely alone in a tiny cabin. In their search for a major strike, she lived in a series of cabins and shacks, panning the pay dirt unearthed by other miners for a fee "by lamplight in a washtub that was also used for bathing." Despite the frigid conditions, in Ethel Berry's "Alaska uniform," heavy flannels were worn under a dress, not pants. When the Berrys struck it rich in 1897, she vowed never to return, and when asked what advice she would give to women about making the trek north, she responded, "Why, to stay away, of course." She added, "It's much better for a man, though, if he has a wife along. The men are not much at cooking up there, and that is the reason they suffer with stomach troubles. . . . After a man has worked hard all day in the diggings he don't feel much like cooking a nice meal when he goes to his cabin, cold, tired and hungry, and finds no fire in the stove and all the food frozen."[126]

Single women, as well as married women like Berry, sought their fortunes in the Yukon. Some staked or leased claims and mined themselves. Some worked as prostitutes, and many more discovered, as Ethel Berry suggested, that the domestic skills of cooking, sewing, laundering, and housekeeping were in great demand and opened boardinghouses, restaurants, and hotels.[127]

The demand for women's labor, like the gold rush itself, ended as quickly as it began. A few women who arrived early on the scene made sizeable fortunes, but the instant city of Dawson in Yukon Territory that was home to some 35,000–40,000 adventurers in 1897 was reduced to a population of 8,000 only three years later. Miners either moved on to less populated areas not already claimed or gave up and went home. In 1900, Mary McCarty, recording that it was sixty degrees below zero in Dawson, wrote to her sister back in Illinois, "Minnie, it is a good thing you didn't come in, because there are hundreds of girls in Dawson without work. Lots of them [are] going out [leaving the Yukon] over the ice. Wages have come way down."[128] In 1903, McCarty and her husband started a boardinghouse for thirty to forty men in a large tent adjacent to their cabin, but, she complained, "Everything is so high now that I don't think we'll make any money this month." Three years later, she had to borrow money from her mother.[129]

Ella Chase was one of the young women who left Dawson looking for work. She wrote to her friend Mary Zimmerman that Dawson was "just full of women out of work, and [they] are willing to work for almost anything." In more remote areas, however, Chase "had no trouble at all." She worked as a housekeeper for a wealthy mine owner and noted approvingly, "I am not treated like a servant here but as the lady of the house." Chase confided, "I don't like the Klondike very much but I am willing to stay with it for a year."[130] Two months later, although excited about staking her own claim, she advised, "Please don't have anyone to come into this country if you can prevent them for this is a hard country especially for a young single girl. . . . Oh well, I am satisfied with my lot, but if I had a little fortune I know I wouldn't stay in this country." Her brother, by contrast, "likes this country fine and says he intends to make the Klondike his future home."[131] The desire to stay or leave frontier environments frequently split along gender lines.[132]

. . .

Women's experiences in frontier environments varied dramatically, depending on factors including time period, departure point and destination, age, marital and maternal status, ethnicity, and race. Women of Indian and Mexican heritage were left scrambling for ways to survive as the lands they had settled were invaded by pioneers who eagerly remade the environment. White women enjoyed the benefits of their race, but they too struggled with a range of challenges in the West. They and their husbands responded to the environment in ways that were significantly different from each other. Girls and young single women were more likely to shed some gendered restrictions and seize the unique opportunities provided by the new environment. Most adult women, in whom the prescribed spheres were more deeply ingrained, especially mothers, clung to their established ways even in places and climates starkly different—and usually much less comfortable—than their homelands. Most white women worked to recreate in the West both the environment and the strict gender spheres of the East. The "Wild West" offered them no liberation or respite.

"The ultimate ideological convergence between nature and culture"

4 Nature's Housekeepers
PROGRESSIVE-ERA WOMEN AS MIDWIVES TO THE CONSERVATION MOVEMENT AND ENVIRONMENTAL CONSCIOUSNESS

"Feminine Dairy Wisdom"

In 1894, Adda F. Howie, Milwaukee society leader, walked away from the many comforts and conveniences of the city to return to the farm life of her childhood. She began raising what would become the largest herd of purebred Jersey cows in Wisconsin. Although the separation of prescribed gender spheres was still very much in evidence at the time of her success, Howie, rather than being vilified for making her way into the male sphere of business, money, and power, was publicly praised for her skill in bringing traditional female values into the barns and pastures of Wisconsin.

Howie took great pains to present her dairy farmer career as consistent with gender norms. She wrote a lengthy tribute to women in the history of agriculture, beginning with the Bible. She emphasized "the charms of the dairy maid [that] have been for centuries an inspiring theme for both poet and painter," praising women for their "refining instincts . . . loyally guarded as a heaven blessed gift" and their "graceful . . . and valuable characteristics," virtues that had been perpetuated through the ages. Howie criticized modern men who farmed with "reckless waste" and "hurried oversight" in their rush to make a profit. She denounced the "crudely prepared many acres" and advocated instead for the "carefully cultivated few."[1] Howie, reported the *Farm Sentinel* in 1902, put "a great deal of sentiment into her work and believes that there is more pleasure in living than in the mere sake of striving after and grasping every dollar one can hoard up. She believes one can lead an ideal life on the farm with more real enjoyment and pleasure than can be obtained by city residents."[2]

The first woman to serve on the Wisconsin State Board of Agriculture, Howie, according to the periodical *The Forecast*, had by 1925 long been "recognized universally as America's Outstanding Woman Farmer."[3] The secret to her success was her conviction that a dairy barn "should be as clean as a champion kitchen."[4] Thirty years before legislation required the sanitary production of milk, Howie's "feminine dairy wisdom" demanded that the interior of the barn and other outbuildings at her Sunny Peak Farm near Elk Grove be whitewashed, then scrubbed once a week with soapsuds and boiling water. The barn's windows were cleaned and curtained, the cattle "brushed and petted and everything done to make the barn as sanitary and attractive as possible."[5] Howie's contention that cattle "tenderly cared for" would produce higher and better yields was borne out when many of her prize cows regularly set production records. The milk from a single cow, Fernes Gypsy, produced 5,671 pounds of butter in one year. Howie's application of the domestic sphere's standards of cleanliness and comfort to the farm environment was so unusual, and ultimately so successful, that her methods were widely studied. She spoke on dairying and homemaking to agricultural experts across the country, as well as in Canada and France. The Japanese government purchased cows from Howie's herd to improve its nation's dairy stock. Adda Howie's humane and sanitary "feminine" approach to dairying, one requiring no expensive tools or equipment, markedly transformed the farm environment.

Howie's application of prescribed values to the rural environment was just one element of the wide-scale progressive reform movement that swept the nation following the end of southern Reconstruction in 1877, only to fade in the aftermath of World War I. During the Progressive Era, many middle-class female reformers, primarily but not exclusively white, claimed that male domination of business and technology had skewed America's value system.[6] Profit had replaced morality, they charged, as men focused on financial gain as the sole measure of success and progress. Men profited, for example, by selling impure food and drugs to an unsuspecting public, which included the many immigrants who made up much of the workforce of industrializing America. In the factories whose profits turned a few individuals into millionaires, working-class men, women, and children toiled long hours for low wages in unsafe conditions, only to go home to urban squalor. Nonrenewable resources were exploited with no thought to their conservation.[7] In the face of so much gross injustice, environmental and otherwise, women, long prescribed to be the civilizers of men, staged protests and organized reform efforts.

Prohibited from voting and excluded from centers of business and power, many Progressive Era women found an outlet for their energies in carrying out environmental activism.[8] In this arena, their prescribed gender role was a credential rather than a handicap. Some of these women worked individually, as scholars, nature writers, and botanists. For example, in 1903 Mary Austin published *Land of Little Rain*, which celebrated the joys and beauty of the desert, despite its lack of commercial value. Austin stressed the significance and value of a place not conducive to human habitation, urging that it be appreciated rather than disdained or, worse still, "improved."[9] Some women preferred to collaborate with others and joined women's organizations, while others operated within groups open to both sexes, such as the Audubon Society and the Sierra Club. Women like Adda Howie applied traditionally prescribed female virtues to the problems of the modern age. Their proposed solutions, including resource conservation and wilderness preservation, provide new insights into the power and limits of gender in newly industrialized society.

Challenging and Affirming Traditional Gender Spheres

The United States experienced dramatic population increases in the decades following the Civil War. Between 1870 and 1920, the population grew from fewer than 40 million to more than 105 million. Most of that growth was in the cities, as the nation changed from one with three-quarters of its people in rural areas to one with just over half in urban areas. In 1870, only fourteen cities had more than 100,000 people; fifty years later, sixty-eight cities were that large.

As the ranks of middle-class urbanites swelled, men's livelihoods changed, but old notions of gender were reinforced.[10] As men's occupations increasingly moved indoors, newly formed outdoor clubs provided reassurances of masculinity through the kinds of tests of male strength and endurance that were no longer part of everyday life. Even men who were not members of the Sierra Club (founded in 1892) or the Appalachian Mountain Club (founded in 1886) felt their masculine imaginations stirring as they read the new outdoor journals, such as *American Sportsman* (1871) and *Forest and Stream* (1873).[11] Camping became popular as a way for middle-class urban men to recapture, however briefly, elements of the pioneer experience and to demonstrate their outdoor skills.[12] Women were not necessarily excluded from these forays into nonhuman nature, but a camping vacation, unlike a stay in a hotel, spa, or resort, required that women cede control to men. Campers did not dress up for dinner, for example, nor practice other formalities of urban etiquette. Women did not abandon completely their roles as civilizers and often served to domesticate the camping experience, keeping up standards of cleanliness by doing laundry in streams and lakes for example, but they enjoyed some loosening of the restrictions of female behavior and fashion as they participated in a range of

FIGURE 4.1 Most women, like these two in 1909, favored appreciation rather than exploitation of nature. (Wisconsin Historical Society 64821)

outdoor activities.[13] Camping was celebrated in the popular press because it offered a rejuvenating alternative to the "noisy whirl of cities," providing restoration to minds "jaded" by the cares of the urban world and reaffirming "simple, elementary and healthful existence." According to *Scribner's Magazine*, ladies in particular benefited from "this perfect escape from conventionality, and restful return to nature and simplicity."[14]

Immersion into nonhuman nature sharpened the arguments for its preservation, fueling a national debate. Most Americans were so used to thinking of their nation as a land of unlimited resources that wilderness preservation and resource conservation were stoutly resisted. Many men in particular found the idea that resources should be preserved rather than exploited to be not just wasteful but unpatriotic. In 1910, philosopher William James urged in "The Moral Equivalent of War" that young American men be "a part of the army enlisted against nature." By mining coal and iron, fishing commercially, building roads and digging tunnels, and working in foundries and stokeholes, men "would have paid their blood-tax, done their own part in the immemorial human warfare against nature." As a result, "they would tread the earth more proudly, the women would value them more highly, [and] they would be better fathers and teachers of the following generation."[15] For other Americans, however, the official end of the frontier in 1890 contributed to their recognition that the nation's resources were finite, that nature must be preserved if its curative powers were to be enjoyed, and that practices like strip mining and clear cutting needed to be replaced by more sustainable methods. In addition, a spate of catastrophes originally termed "natural disasters" was increasingly recognized as the result of the mismanagement of nature. In 1889, for example, the dam hastily rebuilt in Pennsylvania in order to create a private lake resort and hunting lodge for some of the nation's wealthiest industrialists gave way. The ensuing flood killed more than two thousand people in the working-class community of Johnstown. Although the resort's elite membership called the disintegration of the dam caused by heavy rains to be an act of God, Johnstown survivors recognized that "our misery is the work of man."[16] At the dedication to a monument memorializing 777 of the victims, Pennsylvania's governor pledged, "We who have to do with the concentrated forces of nature, the powers of air, electricity, water, steam, by careful forethought must leave nothing undone for the preservation and protection of the lives of our brother men."[17] Some reformers included animals among the lives that must be preserved and protected, pointing to the specter of the buffalo, whose population plummeted from some thirty million to less than a thousand in fewer than one hundred years.[18]

Although men and women reform activists at the turn of the century sometimes operated within single-sex organizations, sweeping gender-based assertions rarely hold up under close scrutiny. The notion of a strict gender divide in which women dominated in recognizing the need for wilderness preservation and resource conservation is belied by the number of male leaders, Theodore Roosevelt, Gifford Pinchot, and John Muir among the best known, in the nascent environmental movement.[19] Conversely, even an activity as strongly linked to masculinity as hunting included women, some of whom argued for wilderness conservation out of devotion to the sport.

As more Americans bought their food in stores, by the turn of the century hunting was increasingly the domain of the elite, who hunted for sport rather than to sustain

HUNTING MOUNTAIN GOAT IN WESTERN WASHINGTON

FIGURE 4.2 Encumbered by her long skirt, a mountain goat hunter in western Washington negotiates a steep climb. (Forks Timber Museum, Forks, Washington ftm 1814)

their families.[20] Roosevelt and Pinchot were both founding members of the American Bison Society, an organization composed almost entirely of men who associated the bison with idealized qualities of frontier masculinity.[21] Similarly, other sportsmen lobbied for animal conservation laws in order to sustain a supply of game for sport hunting, an activity that for many avid hunters clearly shored up a waning sense of masculinity.[22] Sport hunting was not promoted as exclusively man's terrain, however. Women frequently appeared in the magazines and journals that popularized blood sports and were encouraged to participate to underscore hunting's long-standing status as respectable recreation.[23]

Traditional beliefs about womanhood were being challenged, however, by the realities of the new urban, industrial society. Changes in men's occupations and pastimes influenced the way middle-class Americans defined woman's proper sphere. Commerce and wage work and the moral compromises that accompanied them were considered the realm of men. The sentimental, selfless, and nurturing nature that rendered women unfit to compete in the urban business world left them ideally suited to the home, where they could remain immune to the corruptions of urban life while creating a soothing domestic atmosphere.[24] On its face, the prescribed woman's sphere limited women to activities inside the home. Taken to its logical conclusion, however, it encouraged the notion of women as uniquely qualified and obligated to lift the environmental burdens shouldered by all but borne disproportionately by marginalized populations and communities. A handbill written by social worker Susan Fitzgerald in 1915 for the Woman Suffrage Party of New York explained, "We are forever being told that the place for women is in the HOME." There they were responsible for the wholesomeness of the food and the health, safety, and morality of their children. Unregulated conditions, however, made it impossible for the woman raising her family in an apartment to carry out her role:

> She can clean her own rooms, BUT . . . if [the neighbors'] garbage accumulates and the halls and stairs are left dirty, she cannot protect her children from the sickness and infection that these conditions bring. She can cook her food well, BUT if dealers are permitted to sell poor food, unclean milk or stale eggs, she cannot make the food wholesome for her children. She can take every care to avoid fire, BUT if the house has been badly built, if the fire escapes are insufficient or not fire-proof, she cannot guard her children from the horrors of being maimed or killed by fire.[25]

According to the handbill, ineffectual city government officials—men elected by other men—were responsible for these deplorable conditions. In the words of a

FIGURE 4.3 Women pose in 1889 with Schuetzen rifles used for target shooting. Their formal dress in the requisite feminine attire lent respectability to guns and shooting. (Wisconsin Historical Society 9831)

Denver member of the General Federation of Women's Clubs, "Uncle Sam has been conducting his house on bachelor principles."[26]

The remedy was to allow women to elect city government officials who would mandate pure food and drugs, safe housing and work spaces, public health, street cleaning and sanitation, and a host of other reforms. Women wanted to do more, however, than merely elect reform-minded men. In "The College Woman and Citizenship," written for the newsletter of the Syracuse University Alumni Club, Clara Burdette described the new role for women mandated by an urban industrial society: "The woman's place is in the home. But today, would she serve the home, she must go beyond the home. No longer is the home encompassed by four walls. Many of its important activities lie now involved in the bigger family of the city and the state."[27] Journalist Rheta Childe Dorr called community "Home," deemed city dwellers "the Family," and public schools "the Nursery," then added, "And badly do the Home and the Family and the Nursery need their mother."[28] "Women are by nature and training, housekeepers," asserted handbill author Susan Fitzgerald, urging, "Let them have a hand in the city's housekeeping, even if they introduce an occasional house-cleaning."[29]

A variety of sources reinforced the message that much of women's authority as civic housekeepers came from their daily conservation of precious resources. By 1911, *Ladies' Home Journal* was reminding its readers that in a thousand ways, from reusing paper and string to using bones and meat scraps to make soup, "you have been practicing conservation all your life, doing on a small scale what the Government is beginning to do on a huge one. . . . The Government is in a way the good mother of us all."[30] Clubs were the greatest catalyst urging women to put their conservation practices and housekeeping ethic to broader use. In 1890, the General Federation of Women's Clubs (GFWC) was founded by Jane Croly, who realized the potential for power if the many women's organizations in the United States worked together to achieve common goals. While local and state clubs continued to work on relevant issues, the GFWC operated at the federal level as an "important liaison between legislators and their constituencies," challenging "government leaders to move in new directions."[31] The GFWC maintained that "conservation in its material and ethical sense is the basic principle of womanhood," adding, "Woman's supreme function as mother of the race gives her a special claim to protection not so much individually as for unborn generations."[32] In 1901, the Industrial Committee of the GFWC expressed its confidence in "the vast power of intelligent women to contribute . . . toward a peaceful and beneficent solution of the industrial problem."[33] The committee's solutions included worker protections such as the eight-hour day, workplace health and safety regulations, and the elimination of child labor.

Domestic Science

One of the central tenets of progressive reformers was that disinterested experts should be called upon to find effective solutions to the nation's political, social, and environmental problems. The work of Ellen Swallow Richards (1842–1911) helped to formalize the identity of women as experts of domestic science, conceived as a means of introducing sanitation and nutrition into curricula taught to women, yet based as much on ideology as on science. After graduating from Vassar College, Richards became the first woman to study at the Massachusetts Institute of Technology (MIT). She attributed some of her success there to the fact that she did not scorn "womanly duties": she would "tie up the sore finger" of a male professor or "mend his suspenders."[34] Richards was one of the first scientists to both perceive and publicly proclaim the human environment as a series of interconnected life systems. This insight led her to tackle many of the problems of crowded cities, including air and water pollution. Her first environmental study, a chemical analysis of local water supplies, led to the earliest state standards for water purity.

Although Richards's chemical techniques were used in municipal water and sewage treatment facilities, she maintained that the private home "was the first place to begin improving the human environment."[35] The term *oecology*, soon to be shortened to "ecology," came from the Greek *oikos*, meaning the study of the house. It was coined by German biologist Ernst Haeckel in 1873 and introduced to the United States in 1892 by Richards. Her own house was a prototype of a modern healthful home. Its innovative features included a mechanical system of Richards's own design to facilitate air circulation and ventilation. As a teacher at nominally coeducational MIT for more than thirty years, and at all-female Vassar, she urged her women students to remember "that upon the extension of scientific knowledge, and upon that application of it to daily life which trained women are especially fitted to make, depends, in decisive measure, the improvement of conditions which shall make humanity better able to meet its responsibilities to God and man."[36]

Richards viewed the home as the building block of society, and proper diet and sanitation as key to every healthy environment. The home economics movement Richards championed was "a very practical approach to educating women about the scientific basis for nutrition and sanitation" that would allow her students to build better homes and communities. Her students became the first dieticians. Richards was instrumental in the founding of the American Home Economics Association and the *Journal of Home Economics*. She was a popular speaker and published seventeen books, most of which were dedicated to educating women on the chemistry of healthful cooking and hygienic cleaning.[37] She also helped to found the Hyannis Marine Laboratory (known today as Woods Hole Oceanographic Institution), consistent with her efforts to use science to study and protect all environments. In addition to the advancements she made in sewage and water treatment, Richards's wide-ranging, internationally applied expertise included irrigation techniques and the prevention of mill fires caused by poor lubrication of machinery.[38]

Women as Municipal Housekeepers

The training provided by experts like Richards, combined with what one reformer called "their womanly instinct for caring for other people," encouraged women as the flesh-and-blood "good mothers" of America to immerse themselves in environmental protections and apply their expertise outside the home.[39] Women often led the way in urban environmental reforms, playing crucial roles in identifying and publicizing problems, proposing solutions, and demanding action.[40] In the words of celebrated urban reformer Jane Addams, "As society grows more complicated it is necessary that woman shall extend her sense of responsibility to many things outside her own home if she would continue to preserve the home in its entirety."[41] Colorado reformer Martha Conine put it more bluntly, denouncing men as "uncompromising cranks" because they were too busy studying market reports to pay attention to cleanliness, disregarding "the best interests of the city," which was "really . . . the larger home."[42] Female activists, including Addams, Richards, and their many followers, struggled with the blight of pollution, health hazards, and the physical degradation of cities.[43] Women of middle- to upper-class backgrounds were leaders in urban organizations promoting reforms that included civic cleanliness and sanitation, smoke and noise abatement, and pure food and drugs, making clear the absolute necessity of combating health hazards and pollution for the safety of all citizens.[44]

Women's educational programs to promote public health ranged from persuading citizens not to spit on city sidewalks to alerting tenement dwellers to the dangers of lead poisoning. They also addressed concerns specific to women in economically oppressed neighborhoods, revealing the hidden environmental hazards in many women's occupations. In addition, they promoted pure milk, healthful food preparation, and proper infant and child care.

In 1889, Jane Addams cofounded Chicago's Hull-House, the first settlement house in the United States. From there, she and other women served the poor, largely immigrant, local population. She was quick to point out that educating recent immigrants to the perils of carrying out traditional rural activities in modern urban settings (such as slaughtering sheep in basements and letting piles of refuse decompose in the open) was not enough to guarantee healthful living. Although Addams believed that her efforts may have "slightly modified the worst conditions," life in the inner city "still remained intolerable."[45] Rugged individualism, total independence, and self-contained living were not possible in urban environments. Women followed Addams's lead in putting pressure on city governments to do their part to ensure the health and safety of all constituents. In addition to opposing child labor, prostitution, industrial smoke, and innumerable other health hazards, Addams campaigned aggressively for municipal regulation of factories and tenements and argued for more frequent and systematic garbage collection and street cleaning.

Municipal housekeepers viewed themselves as surrogate mothers to the working class, especially its children. In 1906, Addams served as the first vice president of the National Playground Association, which sought to establish inner-city playgrounds designed to ensure wholesome, structured leisure activity for children. Proper playgrounds represented the perfect blend of nature and culture and were celebrated as the birthplace of better citizenship because creative and cooperative outdoor play taught children to appreciate nonhuman nature and to develop skills including taking turns and following rules.[46] Women advocated improving the environment of rural as well as urban schools. In her article on rural schools for the June 1910 issue of *Midland Schools: A Journal of Education*, Sarah Huftalen wrote, "Beautiful and appropriate surroundings impress themselves on the lives and the hearts of the young. Transform the schoolyard and it transforms every home in the neighborhood."[47] Jane Addams sought the advancement of municipal housekeeping at all levels and inspired thousands of women like Huftalen.

Best known as the staunch defender of women in the workplace, reformer Florence Kelley was also passionately committed to issues of public health. Five of Kelley's seven siblings died before reaching the age of six. At a time when nearly a quarter of American infants died before their first birthday, childhood deaths had many causes. Often these included infant diarrhea, a condition that occurred with greater frequency in the summer months when food and milk spoiled more quickly. Breastfeeding was in decline at the turn of the century, and milk no longer came from the family cow, but was purchased from commercial dairies. Pasteurization of milk was not standard, nor were dairy cattle routinely screened for tuberculosis. Before the toxicity of formaldehyde was understood, some dairies added it as a preservative.[48] To eliminate the huge number of deaths of urban infants and young children attributed to tainted milk, women activists began creating a model for government responsibility for promoting health care.[49] Urged on by Kelley, women waged pure milk crusades in cities across the nation, claiming authority based in science as well as their expertise as mothers.[50] Women did not limit their demands to pure milk. At its 1905 conference, the GFWC resolved to pressure Congress to pass a pure food bill and proposed to "keep hammering at the congressmen until they get what they want."[51]

Protecting Women in the Industrial Workplace

Industrial toxicologist Alice Hamilton both challenged and incorporated pre-scribed gender spheres in her efforts to promote public health. She attended med-ical school, followed by postgraduate studies in pathology and bacteriology, worked in a research lab, investigated lead factories and copper mines, and in 1919 became Harvard's first woman professor. In addition to these male settings, Ham-ilton also inhabited the mostly female world of reform, becoming a crusader for change.[52] Hamilton claimed that women were wiser and more empathetic than men. She believed that her sex enhanced her industrial work because factory owners more readily tolerated suggestions about employees' health from a woman than from a man.[53] She argued throughout much of her career that women in industry needed even more protection than men.[54] Although she claimed that "women are not physically adapted to work which requires great endurance," she did not say whether she believed this inferiority was inherent or the result of cul-tural factors, such as the routine practice of wearing corsets, which hindered wom-en's movement and diminished their lung capacity.[55] She reported that women were more susceptible to poisoning from lead and other contaminants, and suf-fered more severely.[56] She asserted unequivocally that women's reproductive ca-pacity made their exposure to industrial poisons especially dangerous, making them vulnerable to excessive menstrual flow, miscarriage, and a higher incidence of children with birth defects.[57]

Living in Hull-House, Hamilton investigated typhoid, cocaine addiction, and tuberculosis before joining the Illinois Commission on Occupational Diseases and becoming a special investigator for the U.S. Bureau of Labor. By World War I, she was the nation's foremost authority on lead poisoning, a very common occupational and domestic hazard, and a respected expert on radium. Beginning in 1917, young women were hired to brush radium powder onto the dials of watches, clocks, and various instruments to make them glow in the dark. "Radium girl" Grace Fryer worked in the Orange, New Jersey, factory of the U.S. Radium Company. To keep the paintbrush points fine, according to Fryer, "Our instructors told us to point them with our lips. I think I pointed mine with my lips about six times to every watch dial. It didn't taste funny. It didn't have any taste, and I didn't know it was harmful." Like many other dial painters, Fryer began to suffer from a series of painful ailments. Her teeth fell out, and she suffered necrosis of the jaw. The owners of U.S. Radium denied any con-nection between radium exposure and the disfiguring and painful deaths of their predominantly female workers, but Hamilton's research proved the link conclusively. She worked closely with the courts, the press, and the National Consumers League (founded in 1899 by Jane Addams and Josephine Lowell) to publicize the danger of radium and other industrial poisons.[58] She was soon internationally recognized as a pioneer in toxicology and occupational health.

Hamilton rejected the notion that industrial ills were the price to be paid for progress. She called the nation "reckless of health and life, and impatient [with] the control of the law." She termed this attitude a "relic of the youthful daring pioneer spirit," but asserted that "there is nothing admirable in allowing ignorant and helpless people to incur risks which they either do not realize, or which they are compelled to face. . . . When you really consider it," she asked, "how much can a working woman control the conditions under which she works? . . . She dare not leave her shop for fear of not finding another [so] she must take her chances" by working in an environment she knows is dangerous to her health.[59]

Labor activist Rose Schneiderman also railed against hazardous workplace environments where property was held so dear and human lives, especially the lives of "working girls," so cheap that tragedies like the Triangle Factory fire were commonplace. However, Schneiderman's immigrant and working-class origins as well as her emphasis on corporate (rather than civic) responsibility for urban suffering set her apart from most of her sister reformers. While Schneiderman defended the rights of all working women, many middle-class female reformers expressed their greatest concern for these working "girls" as the future mothers of the race. According to Elizabeth Beardsley Butler, assistant secretary of the Socialist party's Rand school in New York City, the greatest danger was not to the workers themselves, but "the threat to racial vitality by the nervous exhaustion of the girl workers." Butler accused industries of "taking young, undeveloped girls, lifting their speed to its highest pitch, and wearing them out." After a few years, the girls were destined to a fate of "unfit homes and undervitalized children."[60]

Women's Sexual and Reproductive Health

Concerns about working-class women's problems evoked different responses among middle- and upper-class urban environmental reformers. Some not only neglected or ignored problems exacerbated by class and race, but exhibited open hostility to various ethnic groups and people of color, judging them as inferiors who were as much the cause as the victims of disease and sanitation problems. These activists appeared more interested in social control than reform. They associated having large numbers of children with a lack of self-control. However, birth control advocate Margaret Sanger, herself one of eleven children, took a different view. She attributed her mother's premature death from tuberculosis to the rigors of frequent childbirth and poverty. She avoided a similar fate, attaining the skills necessary to be financially self-supporting before marrying and limiting her own family to three children. Working as a visiting nurse among immigrant families on New York City's Lower East Side, Sanger decried the toll of venereal disease, miscarriage, self-induced abortion, and frequent childbirth on poor urban women. She was appalled by impoverished women's sexual subservience to men and the resultant overcrowding and inescapable cycle of poverty. Awakened by anarchist and feminist Emma Goldman to the connection between contraception (rather than sexual abstinence) and working-class women's empowerment, Sanger believed that access to effective birth control would bring fundamental social change.[61] Sanger published a monthly newsletter asserting the right of every woman to be "absolute mistress of her own body" for which she was indicted in 1914 for violating postal obscenity laws. In 1916, Sanger opened in Brooklyn the nation's first birth control clinic, seeking to promote the socioeconomic and sexual empowerment of poor urban women by freeing them from unwanted pregnancies. Her tireless crusade to prevent unwanted pregnancies addressed one of the fundamental causes of poverty and overcrowding, making it a key component of urban environmental reform.[62]

Race, Class, and Reform

While some reformers, such as Jane Addams, came to respect the members of the working-class community with whom she worked, others did not share such egalitarianism. In late nineteenth-century San Francisco, for example, Chinese women forced into prostitution or abusive marriages sought refuge at the Presbyterian Mission Home. By Westernizing the Chinese women, the white missionaries who ran the home bettered their own situation by establishing female moral authority in the American West.[63] But the Chinese women who sought their help often did so on their own terms. While some did convert to Christianity and internalized the ideals of the Cult of True Womanhood, others used the Mission Home as a temporary refuge while they attempted to negotiate better circumstances for themselves within the traditional Chinese community.[64] As they gained more autonomy and independence, Chinese women, like women of many other races and ethnicities, would form their own self-help organizations to elevate the unique political, social, and physical environments of their communities.[65] Women of the second generation were particularly active. San Francisco's Square and Circle Club, for example, was created by seven Chinese-American teenaged girls in 1924 to raise funds for flood and famine relief in China, but soon focused on improving conditions within their own city's Chinatown, which had a reputation for vice and unhealthy conditions. The club carried out the conventional municipal housekeeping tasks, such as raising funds for community clean up and uplift.[66] It petitioned government officials for public housing and a dental and health clinic, as well as longer hours and better lighting at San Francisco's Chinese playground.[67] Some Square and Circle Club members came out of the Chinese YWCA, another women's organization dedicated to improving the local community. The theme of one Chinese YWCA oratorical competition was "Our Duty to Serve Chinatown."[68]

African-American women also strove to uplift themselves and especially their home environments. As president of the National Association of Colored Women (NACW), Mary Church Terrell celebrated "the progress of our women in every-thing which makes for the culture of the individual and the elevation of the race," praising women for their willingness "to shoulder responsibilities which they alone can successfully assume." The NACW, established in 1896, was devoted to "homes, more homes, better homes, purer homes," and shared many other goals with white women's clubs, such as sponsoring "cleanup days," and seeking to "improve sanitary conditions in the home and through better city services."[69] Because so many African Americans were relegated to the worst urban neighborhoods, the NACW sought to educate those inner-city residents on germ theory, rid their neighborhoods of flies and rats, and press local governments for sanitation services as well as trees and parks.[70] Access to nonhuman nature was a particular concern. African Americans were often banned from urban parks, lakefront areas, and other sites of outdoor rec-reation. The 1919 Chicago race riot was sparked on a hot afternoon in July by the drowning death of Eugene Williams, a fourteen-year-old African American hit in the head by a rock after he and some friends crossed into Lake Michigan waters un-officially reserved for whites only. In the wake of the riots that left thirty-eight people dead, Ida B. Wells was one of many civil rights leaders who called specifically for more parks accessible to African Americans.[71]

NACW clubs strove to end the race hatred that gave rise to such riots, and to the "hanging, shooting, and burning [of] black men, women and children" that has "become so common."[72] In addition to supporting training schools for girls, particularly in nursing, and to establishing day nurseries for wage-earning mothers, they sought the repeal of Jim Crow segregation laws. Concerns about their race, however, did not trump class-consciousness. Like their white counterparts, NACW members frequently categorized poor, urban African Americans as unfit parents, lazy, and debased, and therefore at fault for their unsanitary homes and communities.[73]

Whatever their prejudices and however imperfect their solutions, middle-class women of all races sought to improve the conditions of the working poor, becoming the forebears of modern environmental justice activists.[74] Their activism also reveals the undeniable role of gender and sexuality in toxic urban environments, as women were recognized both as uniquely affected by urban environmental dangers (at home and in the workplace) and as uniquely qualified to offer relief from some of those burdens.

The Conservation Ethic Takes Root

In 1889, a group of elite women "decided to spend some of their time in improving their minds by the study of Botany" and formed the Cambridge Plant and Garden Club (CPGC).[75] The women met every other Wednesday. One of the group's original goals was that each member would show at least one plant she had cultivated over the year, but when the serious, educational atmosphere caused the group to lose members, it added a social half hour during which the women enjoyed refreshments, awarded prizes, and exchanged seeds. The founding members spoke of the moral influence of flowers, traded recipes for fertilizers, and invited speakers from neighboring Harvard University. Professor George Lincoln Goodale introduced the group to ecology, a word so new it was not yet in the dictionary, and suggested several related courses of study.[76] Slowly, the club's members began to think more deeply about the natural world around them and their own place in it, first in rather theoretical terms, but increasingly in more personal and practical ways. For example, in 1902, members heard a lecture on "herbs for the service of man and plants as food and ornament as well as those for medicine." One member expressed her belief that "in the vegetative kingdom there are antidotes for all diseases, if only we know them." In 1914, a member noted that "trees cut down by settlers had not been replaced and recommends that it is time to begin to do so."[77]

Three years later, a guest speaker educated the CPGC about food shortages induced by the war then raging in Europe, urging members to avoid waste and to preserve food in their homes through canning and other measures. Once their own country entered the war, the club, like the nation at large, dedicated itself to food conservation. "Food will win the war," was the popular refrain of the federal government's food administration, led by future president Herbert Hoover.[78] Each CPGC member was asked to be "a committee of one in this way to win the war. . . . We must stand guard until the next crop in order to feed both the armies abroad and ourselves at home." It reveals something of club members' social class that the speaker added, "We should see to it that our maids work with us."[79]

The Cambridge Plant and Garden Club was not as political or engaged as many women's clubs during the Gilded Age and Progressive Era, but it was representative of the ways in which early twentieth-century women's interests in growing decorative plants, especially flowers, to beautify their homes gradually led them to a greater appreciation of nature, just as it had an earlier generation of pioneer women. It also imbued in women a special sense that they had a unique ability—and responsibility—to conserve and protect the natural world. Cleaning up urban environments was not enough; they saw themselves as the housekeepers of non-human nature as well.

While the Cambridge Plant and Garden Club's membership remained modest in its conservation activities, other women and women's groups were decidedly more vocal and aggressive. According to Lydia Adams-Williams, who promoted herself in 1908 as the first woman lecturer and writer on conservation, man had been too busy pursuing vast commercial enterprises to consider his environmental impact. Adams-Williams, who served as the forestry chairman of the General Federation of Women's Clubs, claimed that it fell to "woman in her power to educate public sentiment to save from rapacious waste and complete exhaustion the resources upon which depend the welfare of the home, the children, and the children's children."[80] Nature, in other words, required the nurture it had been denied.

The conservation arm of the GFWC called for fire prevention, the protection of rivers, and the preservation of wilderness, especially through the creation of regional and national parks. From New Hampshire, where women worked to preserve the White Mountain forests, to Florida, where they waged a long battle to protect the Everglades, and out to California, where they fought to save the redwoods, women played key roles in the protection of wildlife, soils, minerals, forests, and rivers. Especially during the administration of Theodore Roosevelt, they enjoyed the generous cooperation of the Forest Service.[81] However, club women found that overall "it is difficult to get anyone to work for the public with the zeal with which men work for their own pockets."[82]

Nonetheless, regional and state member clubs pursued a variety of conservation goals. Gertrude Hollister, president of the Colorado Federation of Women's Clubs, demanded that the nation "enter upon a great conservation policy" because it had "wantonly wasted everything." Remarkably prescient in her views, Hollister predicted the coming of "solar ray power that cannot be exhausted" and declared, "The important work is to grow a moral responsibility in America that shall make it disgrace and dishonor to wrest from our soil everything and return to it nothing." Like many women, she presented nature as a feminine force, promoting "protection of our mother earth" who is "as dear as life to us."[83] The preservationist ideology of the Colorado Federation of Women's Clubs led to the creation of the Committee for the Restoration and Preservation of the Cliff and Pueblo Ruins of Colorado. Its members were instrumental in the creation of Mesa Verde National Park, signed into law by Theodore Roosevelt in 1906.

Women dominated the teaching profession by the late nineteenth century, confident that they were the natural educators of children and that nature study would benefit both their pupils and the nation's dwindling natural resources. Club women worked with teachers to promote resource conservation in the classroom, urging children to clean up local streets, plant trees, and not litter.[84] Teachers saw firsthand that because of the "concrete and clutter of urban life," city schoolchildren were "alienated from their human birthright of trees, flowers, and fields." When asked by a New York City schoolteacher to name a sign of spring, one child replied, "Yes Ma'am, I know when spring is here because the saloons put on their swinging doors."[85] The Fresh Air Movement sponsored racially segregated rural retreats and visits to farms for poor, urban children. African-American women's clubs raised funds so that black children might experience a wholesome and pure environment to build them up "and make a fresh start for another year's battle."[86] Marion Crocker's passion for conservation work was ignited by a story told to her by a teacher who had witnessed slum children's first exposure to the countryside. A little boy looked around in amazement at the various wonders of nature, saying, "Apples on trees, by God!"[87]

Fannie Parsons, a pioneer of school gardens, reasoned that lack of contact with nature made children hard and unfeeling, and thereby especially vulnerable to urban vices and crimes. Nature study and gardening would create "responsible and active citizens."[88] A leading light in the bird preservationist movement, Mabel Osgood Wright, argued that "a broad campaign of public education must serve as the foundation for conservation efforts."[89] Bird Day, established in 1894 by Charles C. Babcock, a Pennsylvania superintendent of schools, quickly caught on in schools and communities nationwide.[90] To facilitate Bird Day and other aspects of conservation education, Wright produced a steady stream of nature books for children aimed at "the average child in the public school system," who, she claimed, might be one of the many uncultured immigrants who posed a special threat to American wildlife.[91]

Although Wright celebrated specifically "the moral comprehension of the animal-loving Anglo Saxon," African-American educator Booker T. Washington shared Wright's desire that outdoor life be cherished. Washington argued that education was the training needed to "get real happiness out of the common things of life— everything: out of grass, out of trees, out of animals." He especially urged the African-American woman, whose "planning and thinking is done for her" by the supervisor at her industrial job, to pursue outdoor occupations (like raising poultry) in order to develop "self-reliance, independence, and initiative."[92] Washington also urged his female students to make "the farm home the most attractive on earth," in order to stem the tide of young men abandoning rural life for the cities.[93]

FIGURE 4.4 During a field trip dedicated to plant study, three African-American women supervise students of the Whittier School of Hampton, Virginia. (Frances Benjamin Johnston, Library of Congress, Prints and Photographs Division, LC-USZ62-38147).

Clara Bradley Burdette of Pasadena, the first president of the women's California Club, spoke plainly of the gendered divide that existed across the nation on issues of natural resource conservation: "While the women of New Jersey are saving the Palisades of the Hudson from utter destruction by men to whose greedy souls Mount Sinai is only a stone quarry . . . the word comes to women of California that men whose souls are gang-saws are mediating the turning of our world-famous Sequoias into planks and fencing worth so many dollars."[94]

By 1910 there were hundreds of women's conservation clubs with a combined national membership, according to activist Lydia Adams-Williams, of one million.[95] Most male conservationists were happy to exploit to the fullest the prescribed notion that women were, unlike greedy materialist men, motivated purely by good, specifically by the desire to uplift and improve society. In particular, women were presented as the guardians of natural resources that needed to be protected for the enjoyment of all rather than sacrificed for the enrichment of the powerful few. Congressman Joseph Ransdell, chair of the National Rivers and Harbors Committee, identified himself in 1910 as "a representative of the men who need and wish the help of women," declaring, "We know that nothing great or good in this world ever existed without the women."[96] However, the very class of women who were joining reform groups and clubs appeared in some cases to be contributing more to resource depletion than preservation.

The Campaign to Save Birds

The Progressive Era was marked by elaborate designs in women's clothing and accessories. Women's hats featured lavish displays of feathers. "The extremely softening effect," proclaimed one fashion magazine, "is ever desirable, especially for ladies no longer young."[97] Sometimes the entire bird was reconstructed, making it appear as if the wearer had a living bird, often roosting in a nest, perched on her head. Other times only select feathers were plucked, sparing the life of the bird, but only temporarily. Frequently these harvests rendered the birds flightless, guaranteeing a quick kill for the nearest predator.

Women's conservation efforts extended to the protection of both domesticated and wild animals, particularly birds. In 1886, nature writer Sarah Orne Jewett published the short story, "A White Heron," an early, impassioned plea for plume-bird conservation.[98] By 1910, the activities of the Massachusetts Audubon Society (established in 1896 by Boston socialite Harriet Lawrence Hemenway in response to the slaughter of the heron exposed by Jewett) were augmented by those of the 250 women's clubs that were active nationwide, dedicated specifically to the protection of birds and plants. Marion Crocker, who strove to alert club women to the dangers of soil erosion, took up as well the campaign to dissuade women from wearing feathers in their hats. The significance of the campaign to save the birds is revealed when placed in historical and ecological context: It has been conservatively estimated that in the late nineteenth century, five million birds a year were killed throughout the world for their feathers.[99] In addition, some bird species, such as the passenger pigeon, were hunted for sport and for pig feed. There had been an estimated nine billion passenger pigeons in the United States prior to European colonization, but the last passenger pigeon died in captivity within weeks of the death of the last Carolina parakeet, which was hunted into extinction for its striking plumage in 1914.[100] Resistance to further extermination of birds, Crocker insisted, was vital to the preservation of the human race. Before the widespread use of insecticides following World War II, birds provided virtually the only check on the insect population that threatened crops prior to harvest.[101] Warned Crocker, "If we do not follow the most scientific approved methods, the most modern discoveries of how to conserve and propagate and renew wherever possible those resources which Nature in her providence has given to man for his use but not abuse, the time will come when the world

THE CRUELTIES OF FASHION. "FINE FEATHERS MAKE FINE BIRDS."
SEE PAGE 162.

FIGURE 4.5 Women had long been chided for their dedication to feather decorations for their hats. This 1883 drawing depicts the shooting of a bird and the nest of orphans left behind. A taxidermist stuffs the bird, which then appears on the hat of a stylishly dressed young woman. The caption emphasizes "The Cruelties of Fashion." (*Frank Leslie's Illustrated Newspaper*, November 10, 1883, page 184.)

will not be able to support life and then we shall have no need of conservation of health, strength, or vital force, because we must have the things to support life or else everything else is useless."[102]

Speeches on the floor of the U.S. Senate fueled Crocker's and Lydia Adams-Williams's assertions that men could not be trusted to carry out the crucial task of saving the birds and, ultimately, humanity. Missouri's James A. Reed responded to a 1913 bill introduced to protect migratory birds by asking: "Why should there be any sympathy or sentiment about a long-legged, long-beaked, long-necked bird that lives in swamps and eats tadpoles?" He urged, "Let humanity utilize this bird for the only purpose that the Lord made it for . . . so we could get aigrettes for the bonnet[s] of our beautiful ladies."[103] To the horror of those who saw clearly the crucial role that pest control provided by wild birds played in national and international economies and ecosystems, Reed dismissed the protection of birds as trivial, born out of "an overstrained, not to say maudlin sympathy for birds born and reared thousands of miles from our coast."[104]

The notion that women were especially suited to carry out the campaign to save the birds was reinforced by the photographs and stories of birds by popular novelist and nature author Gene Stratton Porter.[105] Most of Porter's nine nature books featured her signature close-up photographs of birds in their natural habitat. Porter claimed that the female domestic sphere was responsible for the patience, sympathy, and attentiveness necessary to capture those photos. In particular, Porter credited her photographs to a "mute contract between woman and bird. . . . Your nest and young will be touched as I would wish some giant, surpassing my size and strength as I surpass yours, to touch my cradle and baby."[106] All Porter's books suggested women's special affinity with nature, reinforcing that it was their obligation to be at the vanguard of bird preservation.

Crocker and her fellow reformers sought legislation protecting the birds but took more immediate action as well. When the powerful millinery industry deflected their criticisms by proclaiming that it was merely acceding to the demands of women, they countered by educating the female hat-buying public. In describing the practice of using birds' bodies to decorate women's hats, the chair of the conservation committee of the Colorado Federation of Women's Clubs observed, "Each beautiful head, wing, or breast mutely declar[es] the cruel death of the bird—the mercenary spirit of men and the vanity of women."[107] Some bird protectionists deemed feather wearing to be the antithesis of true womanhood, for it "demoralized and degraded womankind and made a travesty of the better instincts of motherhood."[108] Most admonitions appealed to maternalism. Aigrettes, for example, were "harvested" during the breeding season, when the feathers were at the height of their beauty, leaving the parents dead and the young to die of starvation. "Remember, ladies," urged a California Federation of Women's Clubs newsletter, "that every aigrette in your hat costs the life of a tender mother."[109] Crocker's approach was unusual for a female Progressive

Era reformer, because although she appealed specifically to women, she chose not to play the maternal card. She stressed the necessity of birds in interrelated plant and animal kingdoms, reminding her listeners of the crucial roles birds played in agriculture and pest control. "This is not sentiment," she stated flatly. "It is pure economics."[110]

Crocker's pleas for women specifically to take action, combined with the campaigns of various other women conservationists, ultimately led to a variety of legislative successes, indicated by the plea from a Colorado legislator to the president of the General Federation of Women's Clubs: "Call off your women. I'll vote for your bill."[111] In October 1913, a new Tariff Act outlawed the import of wild bird feathers into the United States, and in 1918 the Migratory Bird Treaty Act established protections for birds that migrated between the United States and Canada.[112] Women continued to wear hats, but milliners throughout the United States and Europe bowed to the legal and societal pressures to dramatically reduce the use of feathers as primary decoration, although they did continue to use naturally shed feathers, particularly ostrich and peacock. Thus, prior to achieving suffrage, women were able to successfully wield legislative influence and, by preserving millions of birds, protect complex and vital environmental relationships from ruin by a powerful American industry. Using similar rhetoric and techniques, women worked to protect forests, lakes, rivers, and a host of other natural resources.

Women's Conservation Efforts Undermined

Pioneering educator and psychologist G. Stanley Hall claimed that many nature study textbooks, like those by Mabel Osgood Wright, suffered from "effeminization" and charged that "caring for nature was female sentiment, not sound science."[113] Some men risked being scorned for holding "unmanly" views in their promotion of resource preservation. Wilderness preservation icon John Muir was a prime target, rendered both impotent and feminine in a contemporary cartoon that depicts him, elaborately clothed in a dress, apron, and flowered bonnet, fussily (and fruitlessly) attempting to sweep back the waters flooding Hetch Hetchy Valley.[114] By 1910, Muir's crusade to preserve the glories of the valley in its natural state rather than inundate it to accommodate San Francisco's need for water was supported by 150 women's clubs nationwide but opposed by Gifford Pinchot. Although Pinchot, who was serving as the first chief of the Forestry Service, was a self-avowed conservationist, he escaped the charges of effeminacy heaped on Muir by making it clear that he measured nature's value by its service to humanity. "Wilderness is waste," he famously proclaimed. "Trees are a crop, just like corn." He dedicated his agency to "the art of producing from the forest whatever it can yield for the service of man."[115]

Sweeping Back the Flood

FIGURE 4.6 John Muir ridiculed as effeminate for his preservationist efforts. (*San Francisco Call*, December 13, 1909, Library of Congress, Chronicling America)

When the U.S. House Committee on Public Lands debated the flooding of Hetch Hetchy, Pinchot's testimony, that use should take precedence over beauty, was echoed by former San Francisco mayor James Phelan who targeted Muir in arguing, "To provide for the little children, men and women of the 800,000 population who swarm the shores of San Francisco Bay is matter of much greater importance than encouraging a few who, in solitary loneliness, will sit on the peak of the Sierras loafing around the throne of the God of nature and singing His praise."[116] San Francisco's city engineer, Marsden Manson, spoke for many when he described Hetch Hetchy's defenders in thinly veiled homophobic terms as "short haired women and long-haired men."[117]

Theodore Roosevelt framed his support for conservation in terms of benefiting human rather than nonhuman nature. In 1907 he addressed both houses of Congress to gain support for his administration's effort to "get our people to look ahead and to substitute a planned and orderly development of our resources in place of a haphazard striving for immediate profit."[118] It is a testament to Roosevelt's hypermasculine persona that he could so successfully sow the seeds of conservationism within a male population deeply suspicious of any argument even remotely linked to sentimentality.[119] Writer George L. Knapp, for example, termed any call for conservation "unadulterated humbug" and the dire prophecies "baseless vaporings." He preferred to celebrate the fruits of men's unregulated resource consumption: "The pine woods of Michigan have vanished to make the homes of Kansas; the coal and iron which we have failed—thank Heaven!—to 'conserve' have carried meat and wheat to the hungry hives of men and gladdened life with an abundance which no previous age could know." According to Knapp, men should be praised, not chastened, for turning "forests into villages, mines into ships and skyscrapers, scenery into work."[120] Such beliefs were reinforced in the press. The Houston *Post*, for example, declared, "Smoke stacks are a splendid sign of a city's prosperity," and the *Chicago Record Herald* reported that the Creator who made coal "knew that smoke would be a good thing for the world."[121] Pittsburgh city leaders equated smoke with manly virtue and derided the "sentimentality and frivolity" of those who sought to limit industry out of baseless fear of the by-products it released into the air.[122]

Despite such assertions, the reputation of women as sentimental preservationists is belied by the facts. Although many women had been persuaded that the beauty of Hetch Hetchy Valley was more valuable than the water its damming would redistribute, they were by no means opposed to all irrigation projects. One of Harriet Strong's patented designs, a system of flood-control dams and reservoirs, won two medals at the 1893 World's Columbian Exposition. Strong had no engineering or business school training and had been an invalid. Following her husband's suicide in 1883, she saved the family ranch by pioneering new methods of irrigation and introducing new crops. She ultimately devised a plan to conserve and distribute water for irrigation and flood control in the Los Angeles basin. Although more than fifty years earlier Thomas Starr Jordan had argued that the transformation of the arid West would be achieved through the educated, organized, and moral labor of men, Strong attributed her successes to the application of her domestic skills to the marketplace. This was something, she argued, that all women could do because they were "excellent, experienced managers" and "naturally more honest than men."[123] She held public office in her irrigation district and testified before Congress in 1918 on her plan to fill the Grand Canyon with water, an idea that presaged the Bureau of Reclamation's Colorado River Project, a vast water system incorporating five states. "Preserve the water power in the hands of the government for the benefit of the people," she demanded in testimony she labeled as based in "the ethical" rather than "the law."[124] The vice president of the GFWC agreed. In urging club women to sign a petition for a national irrigation bill, she advised, "You need to study forestry that you may appreciate what you have and struggle to keep it and not permit men for the almighty dollar to denude your lands and turn them back to a desert. Forestry and irrigation go hand in hand and make possible for portions of this United States the homes and school houses and churches for your sisters and their children."[125] Despite these and other efforts, women's deepening understanding of the need to harness various resources for the good of people and their environmental programs and goals were increasingly dismissed by influential men as sentimental rather than scientific.

Women reformers who wished to continue their crusades to protect the environment faced considerable challenges. Although men and women worked together between 1880 and 1905 to form Audubon societies and to pass legislation to preserve avifauna, this bridging of the gendered divide proved temporary.[126] Once conservation and forestry came of age as technical professions, women were not welcome.[127] Men, whose environmental concerns focused on economics, jobs, and efficiency, gained more support and funding than women who focused on environmental justice and public health.[128] Although male environmentalists were gratified by the moral authority women's activism brought to conservation and preservation concerns, they ultimately rejected female incursion into the world of masculine authority.[129] Men feared being tarred with the brush of sentimentality or by the homophobic rhetoric that denounced as effeminate concerns based on anything but pure science.[130] As smear tactics ultimately weakened many progressive reforms, including environmental causes, men eased or forced women out of positions of authority. The founders of the first national smoke-prevention organization, for example, required that members have a professional interest in the issue, and after 1912, the American Forestry Association ceased to print articles or news items on the work of women in forestry.[131] Moving beyond the Progressive Era, women were not only pressured into resigning from leadership positions in environmental organizations, but into quitting various outdoor activities as well.[132]

Women received a reprieve from their exile into indoor domesticity during World War I. Upon entry into the war, Americans were urged "in the name of humanity" to conserve wherever possible.[133] The production of foodstuffs was so important that women's potential contributions could not be discounted. The U.S. government not only urged conservation measures by women but it also created the Women's Land Army, employing more than fifteen thousand urban women between 1917 and 1921 to carry out the agricultural work left by men who had gone to war.[134] This government-sanctioned return to the outdoors did not, however, extend into the postwar period.

In some ways, U.S. entry into World War I was the culmination of progressive ideals. Woodrow Wilson's pledge to "make the world safe for democracy" and his determination to make this the "war to end all wars" by establishing "peace without victory" are consistent with the progressive belief that dispassionate experts could create lasting solutions to complex problems. Although the progressive light that had burned so brightly was not doused entirely in the war's aftermath, the belief in reform had weakened considerably. Warren G. Harding's winning 1920 presidential campaign slogan promised not a recommitment to progressive reform, but a "Return to Normalcy." This call to abandon the mania for reform resonated strongly for a variety of reasons. A generation that had devoted great passion and energy to a host of reforms only to find that many problems (including war, poverty, and pollution) stubbornly remained found itself with diminished optimism and drive. Some Americans simply grew tired or were war-weary. For others, the commitment to progressive reform was a victim of its own success: many progressive reformers took justifiable pride in their accomplishments and felt less of an imperative to continue. And the younger generation that came of age enjoying the benefits garnered by progressive reformers saw fewer reasons to commit their full energies to reform agendas.

. . .

During the era of reform at the turn of the twentieth century, the same gender prescriptions that had originally relegated women to the domestic sphere propelled them into public life and activism. Near the end of the progressive period, Clara Burdette took stock of women's environmental victories that were the result of that activism: "A lot of money-making wood-sawyers would have split up our giant Sequoias, pride of the world, into toothpicks and lead pencils if we women hadn't caught them at it and raised such a shriek that the men couldn't help hearing. . . . We are reforesting the denuded mountains, and we are pumping up our rivers into irrigating ditches and canals." She observed sardonically, "Ever so many things like these we are doing, and the best of it all is, we have got the men interested by letting them believe they thought of it first and are doing it all, and we are getting ever so much help and good work out of them just by letting them hold the lines and shout at the

FIGURE 4.7 In this propaganda poster commissioned by the National War Garden commission, Lady Liberty sows the seeds of victory, encouraging Americans to consider "every garden a munition plant," and to support the war effort by growing and preserving their own food. (James Montgomery Flagg, U.S. Food Administration, National Archives 512498 4-P-59)

team, while we drive." Burdette claimed not to worry that women might have been too successful at this technique that also often resulted in the judgment that women were superfluous to Progressive Era reforms. "Seriously speaking," she declared, "we believe that . . . organized womanhood . . . has so impressed itself on the commonwealth that good men court our helpfulness and active strength in all reforms that make for higher moral standards and the protection of the home; and that bad men fear the power of an organized force that demands through public opinion pure homes, clean life, clean politics, and pure religion, the only sure basis for continued prosperity."[135] Although many of the movements seeking the preservation and conservation of natural resources were dominated by white middle- to upper-class women such as Burdette, women of racial and ethnic minorities also used gender to uplift urban environments—and, like their more privileged sisters, elevated themselves in the process.

The perceptions of gender that initially empowered many women in the Progressive Era ultimately proved to be a double-edged sword, as using those perceptions as the basis for their environmental expertise made women vulnerable when men came to distance themselves from female sentimentality.[136] The dilution of the progressive reform spirit, brought on by a mixture of success, fatigue, and the grim realities of World War I and its aftermath, exacerbated women's decreasing power and visibility in many mainstream environmental protection movements and organizations.

5   Reasserting Female Authority

WOMEN AND THE ENVIRONMENT FROM THE 1920S
THROUGH WORLD WAR II

The Campaign for Children's Environmental Education

Mary K. Sherman, chair of the GFWC's Department of Applied Education, wrote a pamphlet in 1922 to promote extensive natural science and nature study in the elementary grades. Conforming to gender norms, she couched her argument in a maternal emphasis on future generations: "It is said that the thinking of one generation is the practice of the next. A first-hand, intimate and scientific knowledge of nature's methods and the earth's products gained now will serve the nation well a generation hence when the question of the intelligent handling of our natural resources will be second in importance to none."[1] Her pamphlet came to the attention of John James Tigert, U.S. commissioner of education, who pledged the support of the Bureau of Education. He added, "I appreciate the work of the General Federation of Women's Clubs in its efforts to awaken a general understanding among parents of the value of science teaching of the child, and urge that the work be continued."[2] As long as women limited their activism to gender-appropriate topics, such as educating children, planting trees, and cultivating gardens, they merited the approval of influential men such as Tigert.

In 1924, the GFWC celebrated Garden Week to promote understanding of natural science: "Work in the garden, where the gardener comes in direct contact with Mother Earth is a splendid antidote for our complex civilization which seems to be neither wholly sane or wholly safe. . . . And the home garden with all its lessons in nature and science will go a long way toward solving the leisure time problem of people and the playtime of children—a vital problem in the life of every community." At least 600 member clubs in Ohio planted trees—greater participation than in any other state—and more than 400,000 Detroit schoolchildren competed in a GFWC contest for the best garden plan. Club members repeatedly emphasized the importance of their activities to not only promote the constructive use of leisure time, but also to develop good citizens: "Work in a garden is good for everyone but for children it cannot be overestimated. For in addition to learning how to produce food from the soil and gaining definite ideas about thrift and industry, garden work trains the child in observation, concentration, and responsibility. And while the child is getting actual food from the ground, he will be learning things, the value of which cannot be measured in dollars and cents and bushels and pounds."[3]

The GFWC made grand claims indeed for the value of garden labor: "The garden is the place of places where the child should learn his first life lessons. Here he will learn how nature sets in operation for man all the elemental forces of the universe—the orderliness of nature, the regularity of the seasons, and relation of seed time to harvest. And under sympathetic guidance he will come to feel that nature is a spiritual as well as a material force—with a soul as well as a body—and that if he speaks to her, she will understand." In addition to being touted as the key to understanding the entire universe, garden work inculcated more plebeian virtues as well, such as a proper work ethic. The child gardener learns "that he must not be a parasite—whether it is in the home, the city, or the state he feeds upon. He will also learn that honest productive labor with the hands is as honorable as the labor with the brain."[4] Gardening also had great political value, serving as "a bulwark against socialist tendencies, [for a] man or a woman with a garden does not have time to be a discontented reactionary."[5]

President Calvin Coolidge praised Mrs. Thomas O. Winter, president of the Washington, D.C., branch of the GFWC, for her chapter's role in making Garden Week a truly national affair. Secretary of Commerce Herbert Hoover echoed Coolidge's praise, but like Coolidge, he ignored the federation's assertions that their work was important to the economic and spiritual future of the nation. Instead, Hoover reinforced the "woman's place is in the home" adage by noting only, "The home cannot be considered apart from its setting in nature."[6]

The Home as Industrial Workshop

Women gained the vote at the end of the Progressive Era, but gendered spheres stubbornly persisted. Pushed out of leadership roles in mainstream clubs and organizations dedicated to the protection of natural resources during the waning years of the

Progressive Era, women influenced the environment by voting for political candidates who shared their values, and through the few avenues, such as children's education and leisure activities, left open to them by their prescribed gender roles.

In 1926, nine-tenths of the General Federation's three million club members were homemakers. They frequently used their status as wives, mothers, and housekeepers to argue for their political goals, but they refused to accept that their influence and power were entirely circumscribed by the boundaries of home and hearth. Mary Sherman, who had advanced from chair of the Applied Education Department to become president of the GFWC, tried a different tack to aid women and strengthen the effort to regain women's authority in mainstream society. She charged that "the general contention that 'women's place is in the home'" was but an "erroneous assumption." According to Sherman, "There has been an implication that the home as the woman's realm is detached from the world of industry." She wondered "how we ever acquired the habit of thinking of our homes as being equipped or operated without the active and constant cooperation of manufacturers, engineers, public utility officials, and the commercial concerns indispensable to distribution of facilities." Sherman's complaint was that too many of the new labor-saving devices were being incorporated into the industrial world while "the primary essentials of efficient homemaking—running water, and the fundamentals of household sanitation—proper sewage and drainage disposal—lag so far behind the march of progress."

In an address delivered before the all-male Central Supply Association, Sherman noted, "The General Federation has for more than thirty years advocated a policy of conservation of the nation's material resources as well as of its human energies. Our forests and our coal mines are among our most valuable national assets. The fight of science and industry to eliminate waste in the use of natural resources has been going on for years. In this fight the 26,000,000 American homes must join because the fight cannot be won unless they do. The human energy and material resources spent daily in these family workshops outrank consumption by any other industry or group of industries." Sherman was specific in her demands: "You, the men of each community, must work with us to secure the best available [fuel supplies], under existing regional and local conditions." Moreover, "as a measure against the weakening of home influence we want better equipment—and fundamental to such equipment is running water and sanitary sewage disposal . . . [so] in communities where no waterworks exist, we want you to join us in the campaign, based on the slogan 'running water for every American home.'" Finally, Sherman demanded that the U.S. government recognize homemaking as central rather than peripheral to the efficient and profitable running of the nation by mandating that it be listed as an occupation to be counted in the census.[7]

The category of homemaker was added to the U.S. census in 1930. Despite this recognition of homemaking as a legitimate occupation contributing to the economy, the drive to modernize businesses to increase profits continued to outpace efforts to bring water and sewer lines to every American home. Even GFWC efforts that honored traditional domestic boundaries and female values were rebuffed wherever they came into conflict with the potential for industrial profits. For example, the federation's National Committee for Restriction of Outdoor Advertising, dedicated to preserving scenic and civil beauty, met with furious opposition, which charged club women with "injuring a big industry." Mrs. W. L. Lawton, chair of the national committee, fired back, "No industry has the right to appropriate the beauty of our countryside or of our towns for the benefit of its own private pocket, and then claim that we cannot retake our own simply because it will interfere with its usurpation."[8] Seattle poet and naturalist Ella Higginson echoed Lawton's concerns, charging that men who valued profit over beauty wanted to "tear down our forests, rip open our mountain sides, blow out our stumps with giant powder, dam up our water ways, and see in our wide, sea-weeded tidelands only 'magnificent places for slaughter houses, b'God,' and in our blue, shining sea only so much water that at ebb tide would 'carry off the filth, y'know.'"[9] Protests by women like Lawton and Higginson were in keeping with women's aggressive, more encompassing reform efforts of the Progressive Era, but they met with greater resistance in the 1920s. Only more benign efforts, especially ones that in no way threatened male power and profits, made progress.

Outdoor Good Manners

One example of such an effort was the tremendously successful Outdoor Good Manners (OGM) campaign, conceived in 1925 by Cora Whitley, who succeeded Mary Sherman as chair of the GFWC Conservation of Natural Resources Committee. The need for such a program developed soon after Woodrow Wilson signed the National Park Service into being in 1916. As travel to Europe was restricted during World War I, the railroad slogan "See America First!" first coined in 1906, gained new popularity. Increasing automobile ownership contributed to greater park use, culminating in littered campgrounds and polluted rivers and streams throughout the park system. According to Whitley, "Motoring, from the standpoint of the conservationist, has been like the ancient goddess Siva, both the destroyer and life-giver; the automobile carries the crowds to the distant solitudes where formerly wild life, birds, flowers, blossoming shrubs were comparatively safe." Rather than limit visitor access, Whitley promoted user education. The OGM campaign applied domestic standards of hospitality and housekeeping to the outdoors, resulting in what scholar Rebecca Conard calls "conservation with a decidedly feminine twist."[10] According to Whitley, children must be taught "that to abuse the hospitality of parks and forests is just as really a breach of good manners as would be such conduct in a home where one has been treated kindly."[11] The OGM's campaign urging Americans to "leave the woods and parks as beautiful as you found them" and to "take your indoor manners with you when you go outdoors" was widely disseminated with the support of national organizations, including the American Forestry Association, the American Nature Association, and the Parent Teacher Association. OGM also spread the message of the U.S. Forest Service to "always leave a clean camp and a dead fire."[12] Whitley's altruistic campaign, with its emphasis on housekeeping, gained far more support within the mainstream press than GFWC efforts that used women's authority to threaten masculine autonomy over business and profits. Pleased to find an organization that valued their efforts, by 1929 female membership in the National Parks Association outnumbered that of men.[13]

Scouting Activities for Girls

Camping and visiting national parks brought relief to urbanites exhausted by the artificial environments and hectic pace of city life. Getting back to nature allowed people who were subordinate to others and felt stifled by routines to practice the wilderness skills associated with rugged individualism. For most Americans, however, these rural retreats could at best be enjoyed only a week or two a year, and many parents worried about the fate of children increasingly removed from the frontier past. Urbanized America allowed children little access to unfettered nonhuman nature. Some adults tried to expose boys to conditions that would arm them with manly skills and values and prepare them for strenuous and successful careers.[14] The Boy Scouts of America began in 1910, augmenting the ranch camps in the Far West that promised to inculcate urban boys with traditional masculine frontier values and skills. When their daughters were excluded from such experiences, some women, themselves barred from membership in many outdoor associations, began creating outdoor activities and organizations for girls.[15]

Charlotte Farnsworth, preceptor of New York City's prestigious Horace Mann School, helped to establish the Camp Fire Girls in 1911. Like many other youth group organizers, Farnsworth embraced outdoor activities because she believed they would reinforce, rather than challenge, the separate gender spheres. Farnsworth stated unequivocally that girls "are fundamentally different from boys in their instincts, interests, and ambitions." Luther Gulick, another key cofounder of the Camp Fire Girls, spent twenty years investigating anatomy, physiology, psychology, ethics, and religions in his effort to understand what it meant to be manly and womanly and saw the Camp Fire Girls as a "clearer vision of this question." He believed that to copy the Boy Scouts "would be utterly and fundamentally evil. . . . We hate manly women and womanly men, but we all love to have a woman who is thoroughly womanly, and then adds to that a splendid ability of service to the state."[16] Gulick put a new spin on the traditional perception of womanhood, claiming that "women had acquired the undesirable trait of independence because they had been sequestered in their homes" and therefore needed to be taught collective obedience to best prepare for gainful employment, efficient homemaking, and public service.[17] Until 1943, although women served on the board of directors of the Camp Fire Girls, all of the executive directors were men who shared some version of Gulick's basic belief that the organization's primary goal was to discourage female independence. They sought instead to imbue girls with the values and skills that would allow them to excel in women's traditional sphere. The original Camp Fire Girls handbook demanded service but cautioned that "service is meaningless unless it grows out of or grows into love of others. A mother who takes care of her child because it is her duty is missing what is most vital: to realize the wonder of a little child, and to know that the greatest service one can render God and man is to watch the unfolding of a soul and with kind understanding help it to grasp the right meaning of life."[18] Members were routinely reminded in one signature song that "love is the joy of service so deep that self is forgotten."[19]

Camp Fire Girls were to cultivate a variety of domestic as well as outdoor skills. They should be able to "mend a pair of stockings, a knitted undergarment and hem a dish towel," "make ten standard soups," as well as "make an ascent of two thousand feet and return to the starting level" and "identify and describe twenty wild birds."[20] As future mothers, they should also be able to "name the chief causes of infant mortality in summer" and "make a set of baby clothes."[21] Camp Fire Girls were trained to be dedicated wives and mothers and to serve others in their communities as well. Members could earn special honors by making dolls and toys for hospitals and settlement houses.[22]

A year after the founding of the Camp Fire Girls, Juliette Gordon Low established the Girl Scouts of America. Low was convinced of the value of sports for women, the advantages of outdoor exercise, and the wisdom of preserving the environment. Boy Scouts and Girl Scouts were taught the value of routine, patriotism, and skills necessary to outdoor living, but they were expected to take very different things from their activities and study of nature. According to the original handbook for the Girl Scouts, *How Girls Can Help Their Country*, "The Scout movement, so popular among boys, is unfitted for the needs of girls." A different system was needed to give "a more womanly training for both mind and body."[23] A Girl Scout's first duty was to "Be Womanly," for "none of us like women who ape men." Scouting adhered to a strict gender divide: "For the boys it teaches *manliness*, but for the girls it all tends to *womanliness*. . . . If character training and learning citizenship are necessary for boys, how much more important it is that these principles should be instilled into the minds of girls who are destined to be the mothers and guides of the next generation."[24]

To Boy Scouts the outdoors was a stage on which to rehearse manhood, while "one of the most important principles to be instilled" in Girl Scouts was "strict and prompt obedience to laws and orders."[25] Where Boy Scouts were taught to be aggressive in order to become providers and fighters, Girl Scouts were told to go about their business "quietly and gently," to "never draw attention to themselves unnecessarily," to display "moral courage," and "to never marry a man unless he is in a position to support you and a family."[26] The first wish of the Girl Scout was "to make others happy."[27] The task of the Girl Scout was to become a proper mother to the next generation of workers, fighters, and scouts: "When you are grown up and have children of your own to bring up you will have to know what food to give them, how to look after their health, how to make them strong, and how to teach them to be good, hardworking, honorable citizens in our big growing country."[28] Even something as seemingly gender-neutral as the campfire, the heart of communal rituals for both sexes, held entirely different meanings for children in scouting organizations. Boys were told that fire stood for the camaraderie of the battlefield, factory, and office.[29] Girls learned that fire represented hearth and home: Without the fire of domesticity brought by woman's "magic touch," a house is "dark, and bare and cold."[30]

Girls Scouts were told that they were the natural leaders in conservation: "Women and girls have it infinitely more in their power than men have to prevent waste. . . . The real test of a good cook is how little food she wastes."[31] Trained to think about future generations, Girl Scouts were urged to apply to natural resources the principles of conservation practiced at home, recognizing that "in this United States of ours we have cut down too many trees and our forests are fast following the buffalo."[32] After an initial emphasis on dominating the environment by routinely mandating activities like chopping down trees, Boy Scout leaders acknowledged that women were correct that nature needed to be conserved, but they stereotyped women as too sentimental and selfish toward nature to conserve it properly.[33] In 1916, National Boy Scout commissioner Dan Beard celebrated the Boy Scouts' dedication to bird protection, contrasting it to the actions of women, who sought only the "upholstered skins of these poor birds as ornaments for their hats," ignoring

FIGURE 5.1 Fire was crucial to the mission and philosophy of the Camp Fire Girls as well as the Girl Scouts. This formal studio portrait of the Camp Fire Girls of Menomonee Falls, Wisconsin, shows the troop and its leader (top row, center) all dressed in the organization's traditional "Indian maiden" costume and posed in front of a mock fire. (Menomonee Falls Historical Society)

the leadership women had provided in the Save the Birds campaign of the previous decade.[34]

While the Boy Scouts were taught that conservation was the rightful domain of the male sex, girls in outdoor organizations were learning about females' special powers, abilities, and rights. Girls' alleged innate qualities left them uniquely qualified— and obligated—to conserve, protect, and defend parks and forests. Camp Fire Girls earned special honors when they contributed to their community via street cleaning, beautifying yards, conserving streams, birds, trees or forests, or improving parks and playgrounds.[35] Girls in scouting organizations eagerly joined other groups dedicated to nature appreciation, like hiking clubs and urban improvement societies promoting sanitation and health education.[36] The natural nurturers, women and girls who contributed to the uplift of society and the protection of the natural world would themselves be nurtured.[37] This message found ready adherents, and in its first ten years membership in the Camp Fire Girls jumped from 60,000 to 160,000.[38] The Girl Scouts of America experienced a similar explosion, with membership rising from 70,000 at the end of the 1910s to 200,000 one decade later.[39] Both organizations gave girls outdoor experiences and fostered their environmental awareness.

Support for girls' outdoor organizations came from a variety of influential sources. In 1898 Lou Henry became the first woman to earn a degree in geology at Stanford University, and the following year she married future president Herbert Hoover. Lou Henry Hoover loved the outdoors and was very influential in her support of the Girl Scouts, serving as the organization's president from 1922 to 1925. She believed that scouting made girls better homemakers, citizens, and friends and that it encouraged keener minds and stronger characters. Like many former scouts, however, she reserved her greatest praise for the sheer joy provided by life experienced out of doors: "To me the outing part of scouting has always been the most important. The happiest part of my own very happy childhood and girlhood was without doubt the hours and days, the sometimes entire months, which I spent in pseudo-pioneering or scouting in our wonderful western mountains with my father in our vacation times. So I cannot but want every girl to have the same widening, simplifying, joy-getting influences in her own life."[40] Lou Hoover was particularly successful as a fundraiser for the organization, helping to dramatically increase participation, including that of African-American and other nonwhite girls, as well as girls from impoverished families. Hoover and others helped to establish the foundation for an organization that would grow over the decades and ultimately expose millions of American girls to the joys of outdoor activities while honing their environmental consciousness and authority.

The Rise of Summer Camps

Lou Hoover designed Rapidan Camp, the rural getaway in Virginia that served as the presidential retreat during her husband's administration, which they subsequently donated to the National Park Service. In accordance with the Hoovers' wishes, it was leased for many years to the Boy Scouts as a summer camp. Organized summer camps like Camp Rapidan gained popularity in the 1920s because they provided children with several weeks of sustained exposure to outdoor living. They followed in the footsteps of the Fresh Air Fund, which had been giving camping and other outdoor opportunities to New York City's low-income boys and girls since 1877, the year of a particularly virulent tuberculosis epidemic. Fresh Air camps, funded first by donors, then by the *New York Tribune*, and finally the *New York Times*, offered a respite from urban living to those least able to afford it. However, in general, the private summer camps that proliferated in the 1920s catered to a decidedly upper-class clientele. These camps were designed to remove elite boys from the feminizing, softening influence of home and teach them the virtues of hardiness and resourcefulness that emerge from confrontations with nature.[41] Camps for well-to-do girls were created as well, including Camp Chonokis, which opened on the shore of Lake Tahoe in 1927. The camp's brochure was typical in that it promised to provide its campers with "all the joys and advantages of outdoor life" and to promote interest in health, cooperation, sports, and nature.[42] Its founders promoted ladylike conduct ("let's keep our manners high") but also implanted an appreciation for nature and inspired women to support conservation ideals.[43]

Camp Moy Mo Da Yo, on Pequaket Lake, near Cornish, Maine, drew "only girls of the highest ideals" from Boston, New York, Philadelphia, and Chicago. Catering to girls aged eight to twenty, the two-month-long camp cost $400 in 1928 (roughly $5,000 in 2012 dollars). Its founder, F. Helen Mayo, presented her camp as "the missing link in education—a safety valve with which to offset the tremendous strain of the winter spent in the city." Mayo marketed the camp experience as one that should be repeated through several years, for "the environment, associations, and training provided at camp cannot possibly be supplied at any home, or at any summer hotel." Girls "need to live in the woods, and do the 'woodsy thing,'" she advised in the camp's brochure, "and thus avoid the 'nervous breakdowns' so prevalent among girls in their teens" who cracked under the myriad pressures of urban life. Moreover, Camp Moy Mo Da Yo offered "the all important training in the more intangible things that make or mar the success of the individual: reliability, co-operation, punctuality, courtesy, consideration of others, initiative, attitude toward work and play, and execution of work." In addition to the "art of living together," the camp provided "intimate and sympathetic appreciation of growing things."[44]

Mayo's promotional literature was clearly designed to appeal more to parents than to the campers themselves. What the campers gained from their experiences at the various elite girls' camps is more difficult to discern. The author of an unsigned three-page memoir, "Camp Moy Mo Da Yo to Medical School," attended the camp for ten consecutive summers, beginning in 1928. Her typescript describes the ladylike camp uniform as restrictive. The canoeing safety test required two fully clothed campers to deliberately tip over their canoe. Their uniforms were so heavy and cumbersome that campers had to remove their clothes in order to be able to upright and bail the canoe and then re-enter. The author makes no mention of the camp's efforts to inculcate her with proper female values. Her camp memories revolve around swimming, horseback riding, and canoeing, outdoor activities she enjoyed thoroughly. Moreover, it was a female doctor's visit to the camp to teach the girls how to do blood counts using a microscope that inspired the memoirist to become a doctor.[45]

Alumna of Walden Camp, founded in Maine in 1916, also remembered their camp experiences with great fondness. The twentieth anniversary edition of the camp publication *Splash* included "The Husbands' Lament," a humorous poem in which

FIGURE 5.2 Relieved of the pressures of urban life, girls at Camp Moy Mo Da Yo learn teamwork through canoeing. (Camp Moy Mo Da Yo records, Maine Women Writers Collection, University of New England, Portland, Maine)

husbands of former Walden campers mourn the depths of their wives' attachment to the summer camp:

> We try to make them happy,
> But they tell us with a sigh
> That tho' they love us dearly,
> It's still Walden til they die.[46]

The camaraderie with other girls, the guidance of much-admired young counselors, the outdoor activities, and even the food all earned nostalgic praise. Regardless of the camps' stated purpose to inculcate prescribed female values, judging from the enthusiasm of women who attended a variety of camps in the 1920s, the girls' camp experience offered more freedom, independence, and a welcome exposure to outdoor activities than did their regular urban lives.

Camp life offered subsequent generations of urban girls exposure to the outdoors and the cultivation of empowering skills. In the summer of 1946, sixteen-year-old Virginia Remy and a friend were so fascinated by everything associated with the West that they persuaded their parents to let them take the train across the country

FIGURE 5.3 Camp life continued to offer urban girls exposure to the outdoors and the cultivation of empowering skills. In this 1946 photograph, sixteen-year-old Chicagoan Virginia Remy rides a horse at the all-girls Woods Ranch near Bisbee, Arizona. (Author's collection)

in order to experience the "frontier" firsthand. The two teenagers spent six weeks at a ranch camp near Bisbee, Arizona, an experience they repeated the following summer. Exhilarated by a climate and lifestyle so different from her year-round life in Chicago and delighted by the natural world to which she was exposed, Remy later attended the University of Arizona, earning a degree in geology. She ultimately settled in California, taking up a lifetime of wildlife study. She credits the ranch camp experience with instilling a love of the natural world that has lasted nearly seventy years.[47] Although such camp experiences remained beyond the economic reach of most girls, scouting organizations continued to spread in American cities and suburbs.

Rural Girls

Outdoor organizations and camps for urban children were designed to give boys and girls some of the contact with nature absent from city life. In rural areas, however, the social, economic, and physical environment posed different challenges. The Southern Women's Education Alliance (SWEA), a national organization created in 1914, recognized that a large proportion of youth faced "heavy handicaps in their rural environment; handicaps of health, education, training, recreation, job finding, and other aspects of human welfare," with rural girls coming up against even more obstacles than rural boys.[48] The field notes of a SWEA social worker reveal some of the barriers to attaining a formal education in Virginia in the 1920s. The leading cause was "lack of appreciation at home of the value of an education," with education for girls valued even less than for boys.[49] A principal in Chesterfield County, Virginia, reported that "more girls are leaving school than boys" and gave as the top two reasons lack of supervision during the day and financial necessity.[50] A North Carolina principal observed, "The influence of a static home environment is often responsible for the [female] child's lack of educational ambition. She cannot break away from the family inertia . . . the price of struggling with the conservative country environment is often more than the child will pay." Another "great problem," he added, "lies in getting teachers . . . to be willing to put up with rural life."[51]

Additional disadvantages within the rural environment arose as centralized and sophisticated schools designed to improve the quality of education replaced one-room schoolhouses, forcing farm children to commute long distances, by foot or by horse, or to board in town. Notes of "typical North Carolina girls" targeted for SWEA assistance describe the reasons provided by the family of a sixteen-year-old girl who left one of the more centralized schools despite good performance and regular attendance: they needed her help in picking cotton and stripping tobacco, and she "has enough education now to be a good mother and stay on the farm. Girls don't need to know nothing." The teacher acknowledged that the girl, one of nine children, was needed at home, but added that the girl was "afraid she can't keep up in 'town' school." The social worker concluded that "shyness and lack of understanding at home of educational values are evidently deterring forces."[52] The mother of another North Carolina schoolgirl was typical in her resistance to having her daughter "board in town" some eight miles from their home, adding that there was "not much use" for her daughter's education, as she would get married "anyway," a measure of the limited opportunities for most rural women and the lack of familial support for their intellectual advancement.[53]

A variety of government programs arose in response. When the Department of Agriculture created the Cooperative Extension Service in 1914, it included clubs for rural boys and girls that quickly became known as 4-H Clubs (for Head, Heart, Hands, and Health). The 1920s saw not only the growth of SWEA to aid rural girls, but also the development of women's branches of the federally initiated agricultural demonstrations designed to aid farmers. As men were instructed in new methods to increase the quality and quantity of farm production, agricultural and home economics demonstrations were designed to help farm women improve their households and family care, including beautifying their homes and gardens. The Negro Cooperative Extension Service had its own Home Demonstration Service, which instructed women on improving their homes as well as preparing eggs, butter, and preserved fruits for sale. In 1923 alone, one hundred salaried African-American women worked in one hundred counties in eleven states, seeking to uplift lower-class African Americans by modeling home improvement, particularly for the exteriors of their houses.[54] The resultant flowerbeds were rare spots of beauty in hardscrabble lives, lives that became even harder as the decade drew to a close.[55]

During the Great Depression of the 1930s SWEA recognized that a rural girl suffered a grimmer economic outlook than a rural boy, because her labor was in even less demand. Nevertheless, the organization expanded in the 1930s to include boys as well as girls, becoming the Alliance for Guidance of Rural Youth (AGRY), dedicated to serving rural youth because "the seeds of national disaster lie in frustrated rural youth no less than in the urban."[56] AGRY encouraged rural youth to lay the foundations for their future by developing the skills necessary for nonfarming occupations. Work in local mills was aggressively promoted. Although young women who worked in mills in the mid-1800s felt increasingly estranged from the natural world, AGRY declared such sentiments relics of the past.[57] A 1937 script for the AGRY radio program *American School of the Air*, entitled "Young Workers Who Make Our Clothes No. 1," featured the testimony of a girl detailing the improvements that made mill work superior to labor in the natural world: "Mechanical ventilators and humidifiers keep the air fresh and clean. The temperature is mechanically controlled. In hot summer months, it is actually more comfortable in the mill than it is outdoors." Despite her complaints about the noise and speed of the machinery, the girl added, "All things considered . . . I'd take the work of the mill every time in preference to farm work. The mill is sociable. You have a chance to see and talk with your friends every day, even when you're working. There's always something going on." Most important to AGRY, she concluded, "Mill work opens up paths into more fields than does farm work."[58] Such opportunities were not available to thousands of Depression-era youths, however, leaving girls even more than boys with few options but to conserve and appreciate as best they could what little came their way.

The Civilian Conservation Corps

In the United States, as in much of the industrialized world, technology and urbanization were perceived as alienating men from nonhuman nature. That change, along with the expansion of women's political authority, especially once they gained the vote, posed a threat to traditional notions of power as based in a masculine domination of the natural world. The Great Depression only confirmed widespread anxieties about cities, decadence, and the dissolution of manhood.[59] As the unemployment rate rose to 25 percent for white workers and even higher for nonwhites, President Franklin D. Roosevelt formulated the New Deal and put into place a variety of programs intended to generate relief, recovery, and reform. Unmarried young men could enroll in the Civilian Conservation Corps (CCC), where they provided the manual labor for a variety of conservation and natural resource development projects across the nation. The CCC was typical of the New Deal in that it was based on the philosophy that jobs programs should privilege men over women. Separate branches of the CCC were established for African-American and Native American men.

At the insistence of First Lady Eleanor Roosevelt, residence camps were also created for unemployed women, but female campers were prohibited from reforestation and environmental projects and focused instead on learning and practicing housekeeping skills. Women could only stay for two to three months in the camps, whereas men were recruited to the CCC for a year. The CCC ultimately employed some three million men, while only 8,500 women experienced life in residence camps. In addition to their room and board, CCC men were paid thirty dollars a month as compensation for their labor, of which twenty-five dollars was automatically sent to their families. Women, seen more as charity cases despite their labor, were "given" an "allowance" of fifty cents a week.[60] Eleanor Roosevelt regretted that more extensive opportunities were not offered for women, but few in government shared her view that women should participate equally in voluntary service and education programs. Even within the Roosevelt administration, the women's camps were referred to derisively as "She She She," a parody of "CCC."[61]

CCC workers planted trees, fought forest fires, built roads, controlled floods, and created park facilities and campgrounds. By building up male bodies, the CCC was believed by New Deal administrators to also be strengthening the state. In their view, educated men had been building their minds rather than their bodies. Worse still, the degeneracy of the cities had produced a vast urban poor: unwashed, uneducated, and physically unfit. The solution was to place unemployed urban young men outdoors where they could commune with nature and develop virility and toughness. "The greatest achievement of the CCC," according to one administrator, "has not been the preservation of material things such as forests, timber-lands, etc., but the preservation of American Manhood." The type of male the CCC boasted of producing bears a striking resemblance to the ideal masculine images celebrated in Nazi Germany, the Stalinist Soviet Union, and other Western nations in the 1930s. The male body, its muscles honed by strenuous activity in the outdoors, represented strength and health, a welcome reassurance in a time of tension and uncertainty, and an affirmation of traditional gender spheres.[62]

CCC work included the development of public campgrounds. Camping provided a vacation alternative for families no longer able to afford hotels or resorts, and magazines like the *Ladies' Home Journal* encouraged women to "pack up your family and go!" First Lady Eleanor Roosevelt joined the chorus, writing in *Women's Home Companion* in 1934 of the relaxation and enjoyment found in the outdoors. She acknowledged that women might find the "first day or two a little difficult," for the gender spheres were generally upheld even in this very different environment. Women continued to carry out the brunt of the domestic work of cooking and laundry, tasks made more difficult by the primitive conditions.[63]

Women's Conservation Skills in the Great Depression

In the 1930s most Americans were preoccupied with trying to eke out a living during what was for many a decade of desperation. Urban dwellers unable to keep up mortgage payments or make rent found themselves homeless. Hoovervilles, settlements of tents and of shacks made of corrugated metal or other scavenged materials, sprang up in vacant lots in or near virtually every major American city. Widespread privations brought renewed admiration for the self-reliant pioneers who were able to live off the land rather than depend on the vicissitudes of banks and the stock market. The first five of Laura Ingalls Wilder's *Little House* books celebrating pioneer life and values were released in the 1930s and became immediately popular. Farmers, previously held in disdain by many urbanites as "hicks from the sticks" for their crude lives without electricity, flush toilets, and other conveniences of modern living, were spoken of with envy, as they at least had the ability to grow their own food.

On the Great Plains, however, farmers had replaced the naturally growing drought-resistant grasses with plowed fields, giving no thought to soil conservation. Land under cultivation increased in the 1920s in response to high wheat prices during World War I. Without the native root system to keep the topsoil in place, several years of drought in the early 1930s combined with high winds resulted in dust storms so severe that vast clouds of dense dust from the Midwest "rained" upon East Coast cities and fell on ships far out in the Atlantic Ocean.[64] Approximately one million acres were denuded, leaving thousands of families destitute. Outside the boundaries of the Dust Bowl, economic conditions were frequently no better as farmers were rendered homeless by foreclosures brought on by stubbornly low crop prices.

With all but the lowest paying jobs considered the rightful province of men, many women eked out incomes vital to their families' survival by working as farm laborers, domestic workers, and waitresses—jobs that did not meet requirements for the benefits provided by the newly formed Social Security Board. Homemakers relied heavily on conservation abilities touted for decades as being part of the natural female sphere. A 1936 California Federation of Women's Clubs' Bulletin on Conservation reminded members, "Every club should be active in this section. Conservation of our natural resources vitally affects our daily life."[65] Women at all economic levels practiced conservation within their homes. When her family was forced to go on welfare, Violet Cottrell, a Puyallup, Washington, homemaker, was proud that her thrift contributed to their survival: "Gardens became *very* prevalent and very important in our life to get nutrition. . . . I canned and preserved up to a thousand quarts of food every year, so . . . that my children . . . my first thing, had good nutritious food. I've always felt good about that because I did do these things. Underwear was made from flour sacks."[66] Women preserved whatever foodstuffs they were able to grow in their gardens; they darned, mended, and remade clothes rather than buy new ones for themselves and their families.

As items like meat, coffee, and sugar became unaffordable, women made meals out of inexpensive ingredients and saw to it that not even the crumbs were wasted. Dorothy Tolley recalled that on their farm in Springdale, Arkansas, her mother "just used everything and saved everything. There was not a scrap of material that wasn't used; . . . food was stretched to the very limit. She could do more with . . . [a] chicken than anybody I've ever seen." Even so, the Tolley family's diet during certain periods of the Depression consisted almost exclusively of pancakes and navy beans.[67] Mary Raymond of Wyoming, a child during the Depression, remembers craving candy made with sugar, a luxury her family could not afford. The family kept both dairy cows and bees, however, so her mother made candy out of honey and sour cream to sate her daughter's sweet tooth, a feat of resourcefulness Raymond recalled with admiration and gratitude some fifty years later.[68]

Environmental Activism

The Great Depression offered women a mixed bag: difficult privations along with a newfound respect for their role as conservationists. Women continued to participate in conservation and preservation movements throughout the 1930s, frequently through garden design and nature writing, making a variety of botanical and literary contributions. Even women authors not traditionally associated with "nature writing" incorporated alternative conceptions of nature into their works.[69] Less literary women found acceptance for their outdoor activities by forming female leagues and societies promoting wilderness preservation, especially the protection of wildflowers.[70] Despite their activism, women's participation in mountaineering, organized hiking, and similar mainstream outdoor group activities continued to decline in the 1930s, as did their leadership positions in the Sierra Club and other alpine organizations.

As the Depression challenged women's thrift and skill at conserving household resources, club women strove to reassert women's authority and power as protectors of the environment. The Conservation Division of the Wisconsin Federation of Women's Clubs (WFWC), for example, called for women to not only save and recycle within their own homes, but also to campaign for forest preservation, protection against water pollution, and more humane animal traps to replace the "steel trap atrocities."[71] Noting approvingly that "a growing appreciation of state and national forests is very apparent," WFWC members saw themselves as part of the national conservation effort and sent representatives to visit CCC camps and other government-sponsored environmental protection programs.[72] When storms of dust and sand ravaged grain seedlings in many Wisconsin counties in 1934, women of this largely agricultural state advocated soil erosion prevention practices, such as halting the cutting of trees, shrubs, and vines along streams and fences so that their roots might keep soil in place. WFWC conservation chairwoman Genevieve Branstad recommended that each member of the state's club study the newly published *Rich Land, Poor Land* by Stuart Chase "in order to create enthusiasm and interest in the Conservation of Natural Resources."[73] "Nothing is more important than conservation," Branstad urged. She noted that the conservation division's "most important [work] of all" was its outreach to schools, fostering an environmental ethic and respect for nature in the next generation, most notably through its "Trees for Tomorrow" program. To that end, Branstad also worked with the Eastman Kodak Corporation on a campaign that urged parents to buy their sons cameras and encourage them to shoot pictures instead of guns.[74]

Leaders in the conservation division included Wilhelmine La Budde, conservation chairwoman serving Milwaukee County in the 1930s and liaison officer between the WFWC and the U.S. Forest Service. La Budde frequently cited gender differences throughout her twenty-five-year crusade to protect her state's natural resources and beauty. She believed that women were "far-sighted and able to look more clearly at how actions today will impact the environment in the future."[75] She noted in the *Milwaukee Journal* in 1932 that women had only begun to interest themselves in conservation, "yet in that brief time they have better grasped its real significance than all the [male] rod and gun clubs." Men continued to think "dully" of conservation in terms of preserving enough wildlife to facilitate fishing and hunting, she charged, while "women think of it in terms of land use, scenic beauty, agriculture, and the national environment." "When men get over quibbling about shooting and join with the women in fostering major projects," La Budde scolded, "we'll get more real conservation."[76] She criticized the Cecil and Nelson Timber Company for its plans to clear-cut the old growth in the Flambeau forest for immediate profit. La Budde suggested to the company the wisdom of a longer-term alternative: "You know, there are other things, aesthetic values in life, which *cannot* be figured in dollars and cents." La Budde consistently pushed for safeguarding the future, noting that conservation "must enter into the very fiber of our daily life and its essence must permeate every civic activity." Only by conserving resources at every turn "can we hope to project nature's wonderful heritage into the distant future of the human race."[77]

La Budde successfully pushed for conservation education in all Wisconsin public schools and held office in many conservation organizations.[78] Under her leadership the WFWC submitted a formal resolution to the governor, the Wisconsin Conservation Commission, and the legislature endorsing the creation and maintenance of shelter belts (plantings placed along main highways) to check the drying and eroding winds created primarily by the large-scale removal of timber. By 1953, Wisconsin had 6,000 miles of such plantings.[79] Women like La Budde used their status as conservationists to successfully pursue their goals through the political process.

Even isolated rural women found new avenues for environmental activism during the Great Depression. During the first third of the twentieth century a variety of technological advancements revolutionized work and home life on the farm. Nearly 90 percent of all urban dwellers had electricity by the 1930s, but for a variety of political, economic, legal, and technological reasons, rural areas lagged behind. The electricity brought by the New Deal's National Rural Electrification Program powered many labor-saving devices on farms and brought electric lights into barns, farmhouses, and schools. Delores Levy remembered when the "miracle" came to her family's dairy farm in Superior, Wisconsin: "One minute they were in the Stone Age, the next, heaven itself." The family wandered from room to room in the house "to see the glory of it all. They were awestruck at such nighttime splendor." In the barn, even the animals shared "in the delight of daylight brought by the hand of man."[80] Another New Deal agency, the Tennessee Valley Authority (TVA), promoted economic development, including rural electrification, in ten states. Tressa Waters, of Lebanon, Tennessee, recalled the thrill when electrification came to her family's home, noting, "We were so proud of TVA."[81]

On newly electrified American farms a radio was the most popular appliance purchase. For most farming families, radio quickly became the prime source of news, information, and entertainment. Especially for women with no nearby neighbors, radio brought much-needed adult companionship as they carried out chores that consumed an average of sixty-five hours or more each week.[82] Radio programs in rural areas provided local news and farm and weather reports, while other broadcasts were national, enjoyed by rural and urban audiences across the country. Such broadcasts opened the world to previously isolated Americans, spurring rural imaginations and offering unprecedented opportunities. As one farm woman, Sibylle Mitchell, pointed out in 1937, "A few years back a lot of us had never heard an opera—probably thought we never could hear one. Now we have opera brought right into our kitchens." Mitchell and her friends agreed that "the farm woman—the woman in the home, appreciates all the valuable information she gets over the radio."[83] As radio became a staple in farm homes, it quickly became a tool for rural women to cultivate and spread their authority as conservation experts.

In Wisconsin, the Dane County agent's office began a radio program in 1937 called *The Dane County Farm*. The broadcast provided advice and information on matters including agriculture, education, and home economics. The agent's office soon received requests for a rural "ladies" program as well. Assistant county agent Bill Clark explained, "When we started planning a women's program all of a sudden it came to me—We men can't advise women! Being a married man myself, I knew women didn't think the same way as men do." Rather than hold auditions, the county agent's staff invited "five outstanding rural women to try their hand."[84] Although the show was heavily scripted, every effort was made to present the women selected as old friends enjoying a spontaneous conversation on issues of importance to rural girls and women and fostering community by sharing common interests and helping to solve common problems.

May 25, 1937, marked the premiere of the *We Say What We Think Club* (WSW-WTC), a program produced exclusively for women on radio station WIBA out of Madison, Wisconsin. This virtual club was hardly exclusive; any woman within the listening radius of the station's 1,000 watts of power (expanded to 5,000 watts in 1942) was encouraged by the women's easygoing banter and conversational style to feel that she, too, was a part of the discussion and that these women were her friends.[85] Listeners, as unofficial members, were invited to mail in their comments and suggestions so that they could say what they thought. Although a run of only six months was originally anticipated, the cast members broadcast their thoughts for fifteen minutes once a month for a total of twenty years.[86] The scripts of their monthly "meetings" provide an unprecedented window into the thoughts of rural Wisconsin women. Unlike many other talk radio programs targeting "ladies," the WSWWTC did not feature professionals offering women lessons in refinement or advice on relationships or homemaking tips on "making do." It featured Wisconsin homemakers who, as the program's title suggests, were determined to speak their minds, revealing a range of female interests and emphasizing female authority. Current events topics ranged from world affairs to rural education, and included the need for recreation, the dangers of marijuana, the importance of religion, and the value of women's clubs, gardening, and rural mail delivery services.

Noting in their very first broadcast that "radio has become the most powerful educational weapon for good or for evil in this country," the club's cast members took themselves seriously as a positive moral influence and strove to educate their female listeners. They emphasized the important role of mothers in monitoring

FIGURE 5.4 In 1950, the leaders of the We Say What We Think Club (from left, Selma Sorenson, Sibylle Mitchell, Isabel Baumann, Grace Langer, and Ruth King) celebrate their thirteenth anniversary on the air. (Wisconsin Historical Society 40679)

their children's listening habits, stressing that, even more than newspapers and movies, radio had become such an intimate part of rural lives "that every woman should be alert to the kind of things her children and family listen to."[87]

The need for a "women only" approach reflected the increasingly separate gender spheres that emerged on farms as material conditions improved. As in pioneer days, men and women on farms still worked together at various seasonal tasks when all hands were needed. However, even as they continued to share the burdens of farming, the two sexes increasingly worked apart. Women were occupied primarily within the house, where they cleaned, cooked, canned, and sewed, and also in the immediate yard, where they raised flowers, fruits, vegetables, and chickens. Some tasks formerly relegated to women, such as milking cows, were becoming part of the male sphere of crops and animals in fields and barns. Although childrearing tasks continued to be shared by both parents, the day-to-day supervision of babies, toddlers, young children, and older daughters was almost exclusively the mothers' domain. Fathers spent more time with sons old enough to take on significant tasks in the world of male chores and responsibilities. The rural leaders of the WSWWT Club viewed these separate spheres as both sensible and natural, noting on air, "In the division of labor (and this has never been written into the law, but it is true just the same), between man and woman it has been woman's function to feed, to prepare food, and to do the other things necessary for the maintenance of the family unit."[88]

The cast of the *We Say What We Think Club*, like women all across the country, grew up in a culture still steeped in the belief that the masculine drive caused men to focus more on the short term—to support their families by exploiting nature in order to acquire profits.[89] The perception of women as naturally driven by altruism and the desire to protect future generations also persisted. The club's program reveals the cast's fundamental belief that the environmental forces of rural living shaped their own lives and the lives of their listeners and that it was woman's obligation to protect that environment.

In "Do I Want My Daughter to Marry a Farmer?" the club's third broadcast, the five cast members read aloud from letters elicited from country mothers and daughters. Throughout the show's long tenure, its cast seldom discussed rural life on its own merits but presented it in contrast to urban living. In this early broadcast, the club leaders and their listeners, clearly stung by assumptions that farm life was inferior, bristle with righteous indignation. Rejecting the stereotype of farm women as worn-out drudges who envied urban women's lives of convenience and sophistication, they firmly established the superior features of their rural lives, claiming their rightful place as actors in the national scene, not wishful onlookers. Club members, including the listeners at home as well as the women in the studio, agreed that "the girl who marries a farmer . . . will know and love nature . . . will know the joy of labor and tasks well done," and "will live a full, well rounded life."[90] While they acknowledged that no lifestyle guaranteed happiness, they were confident that rural women had many advantages over their urban sisters of the same socioeconomic class, and that this superiority was the result of environmental differences.

Wisconsin farmers suffered from the persistently grim agricultural prices that plagued all farmers during the Depression but did not endure the conditions of the Dust Bowl. Almost exclusively white, they also suffered none of the discrimination experienced by farmers of African, Asian, and Mexican descent. With running water, electricity, and other conveniences increasingly available to farming families by the late 1930s, the WSWWTC asserted that the Wisconsin farm wife "has more opportunities to enjoy life in her work than the city wife who has to coop herself up in some small, hot stuffy apartment in a large city."[91]

City life came under particularly harsh criticism when the subject turned to children. "The farm," the club members agreed, "surely is a far healthier place to rear children," with "wide open spaces . . . in preference to the dusty, busy city streets." Heavy street traffic was a menace that "almost drives mothers insane," but "one can't expect to keep a child cooped up in a wee little apartment most of the day." Even mothers fortunate enough to find urban housing with a yard or garden had to worry about potential damage to plants, shrubs, and lawns that would alienate the rare landlord willing to rent to families with small children in the first place. By contrast, "there is no better place to raise children than in the country," they proclaimed repeatedly. They scoffed at the notion that farm children lacked educational opportunities. The one-room schoolhouse of yore had been replaced by a consolidated system guaranteeing all Wisconsin students the same high standards of education: "School advantages are so much improved now that no one has to fear disadvantages in the country."[92]

Marriages too, they announced confidently, fared better in the country. Although club members acknowledged that rural women spent most of their time in the home, away from the outdoor world of men, they emphasized the shared time that remained. As a result of a lifestyle that still mandated the occasional sharing of chores and afforded the opportunity to share meals and interactions at various times throughout the day, country couples enjoyed "a spirit of comradeship that never seemed to exist in town." Club members perceived urban gender spheres as nearly totally separate, with the middle-class women exclusively in the home all day while the men worked their desk jobs in the city. "On the farm," they bragged, "the woman works hand in hand—side by side with her husband." They agreed that such physical comradeship became spiritual in nature, for "nothing binds people so close together as common interests and common work."[93]

Club members emphasized that "farm people have a better outlook on life" because they were in tune with the rhythms of nature rather than consumed by the artificiality of urban living. "We love to watch things grow, little calves, pigs, lambs, baby chicks—to say nothing of enjoying fragrant spring flowers, the garden, the sun, the song of the birds, the hum of the mower and binder. We would feel so cramped up in the city." A club member quoted approvingly from a letter by a listener who had lived in town for eighteen years before moving to the country: "I feel now as if I never actually lived before I moved on the farm." Rural living, club leaders agreed, imbued a spirituality unique to farming families because "they are closer to nature, their religion seems to mean more to them, and there is a genuine spirit of friendliness and helpfulness between most neighbors in farm communities."[94] In support of the contention that "the farm woman has an opportunity to put more into life than the average city woman," club leader Isabel Baumann noted the balm for the soul supplied by the act of gardening, concluding, "A garden—and I say this in utter reverence—is an ambassador of God, and can do miracles."[95]

Women embraced their role as natural conservers during the Great Depression. Despite male efforts to minimize female participation and influence in formal organizations, even full-time homemakers continued to see environmental issues as part of their rightful sphere and to include environmental activism in their various individual, club, and volunteer activities.[96] The Depression-era broadcasts of the WSW-WTC reflect women's mindfulness of the role of physical environment in shaping their own lives.

Women as Workers and Conservers in World War II

The coming of World War II lifted the nation out of the mire of the Great Depression. The shortage of male labor created by wars traditionally allowed women access to jobs for which they had previously been declared unfit. As the nation faced a critical labor shortage after men left farms by the millions to either serve in the military or take on jobs in the lucrative war industries, the Women's Land Army was reestablished. More than three million women, primarily urban homemakers, office workers, and students, contributed on the agricultural front, doing the heavy farm work previously carried out almost exclusively by men.[97] Wartime labor shortages made work for women widely available in previously all-male venues, including forest management, sawmills, and logging camps, bringing women new environmental insights as well as authority.[98] Women who moved into the previously male-dominated world of forest fire lookouts, for example, termed themselves "housekeepers of the forest."[99]

FIGURE 5.5   After the United States entered World War II, women were once again called upon to serve in the Women's Land Army. (Hubert Morley, 1944, Government Printing Office, National Archives 515177 44-PA-1511)

Many of the more than six million women who entered the paid workforce for the first time took up industrial jobs previously held exclusively by men. They worked in munitions factories, built planes, ships, and tanks, and carried out other types of construction and manufacturing jobs closed to them in peacetime. Women's labor was vital to many wartime industries and disproved the various claims that they were physically or mentally unfit to carry out industrial labor.[100] Women like Emma Harris were grateful when the call came for female shipyard workers: "There was so much dedication and patriotism, and I too wanted to do my part . . . Well, I had a mortgage to pay, plus two small children to support, so I was eager to go for such a well paying job." Harris worked as a welder, first on ships and then on amphibious tanks. More than fifty years later she recalled, "How noble we felt to help out!" and asked, "Anyone need a welder?"[101] Women were proud to be publicly recognized as competent, and to be performing industrial work critical to the war effort. Sonya Jason painted glider wings before becoming a "gunpowder girl." The war work changed her, and millions of women like her, profoundly: "Few of us who'd had the satisfaction of doing a demanding job well and earning a 'man's wage' could ever be so constrained again, even by the men we loved."[102]

Women who did not enter the paid workforce also earned praise for their contributions to the war, especially their continuation of the conservation measures they had adopted during the Great Depression. Evie Foster of Anchorage, Alaska, learned from government-sponsored home extension agents to "not waste anything" during wartime: "We would remake our adult clothing into children's clothing and use the scraps for quilts, or braided or crocheted rugs . . . we learned quite a few savings methods."[103] The *We Say What We Think Club* reflected the national belief in women as conservers. The club leaders suggested how listeners could benefit themselves and their country in wartime: by wasting nothing, purchasing only essential goods and raising everything else, and spending any surplus money on war bonds.

The club leaders recommended substitutions for some rationed or scarce items, and "extender" ingredients to make rationed ingredients stretch farther.[104] Even as they acknowledged the difficulties wartime rationing posed to the proper feeding of families, leaders of the WSWWTC could not resist emphasizing some of the decidedly practical advantages of rural life during wartime. City women could buy only what their local stores were able to stock and were further restricted by the rationing system. Farm women, on the other hand, could plan at planting time for the year ahead as long as "the weather man is at all cooperative." Because of the ability of farm women to raise and preserve their own food, the WSWWTC reminded listeners in 1943, "You won't need to spend sixteen of those precious [ration] points for a No. 2 can of tomatoes."[105] Like lots of other wartime media aimed at women (including other radio programs and magazines), the WSWWTC praised women for the contributions they were making to their country through their inherent penchant for conservation, even as it offered advice on how to improve and enhance this "natural" tendency.

The National Federation of Women's Clubs urged its members to broaden their commitment to conservation during wartime: "Every clubwoman should know the conservation status and needs of her own community," because "conservation of our Natural Resources is the HOPE of the FUTURE."[106] The WSWWTC's 1943 "What Women Should Know about Soil Conservation" broadcasts bolstered this message, educating women about conservation issues and encouraging them to see environmental concerns as uniquely illuminated by the female perspective. "After all," reasoned club leader Sibylle Mitchell wryly, "we women do know something about dirt." The women's claim to authority on the subject of soil conservation was born of their own experiences, but also based in sophisticated analyses of the results of past and current farm practices. The club's broadcasts reflected a growing appreciation of the importance of resource conservation fostered by their perception of rural women as especially attuned to nature, morality, and spirituality.

Cast members of WSWWTC were careful to make the point that the dirt they and their listeners cleaned out of their houses and turned over while tending their flower and vegetable gardens was not the same as the dirt in the fields cultivated by their husbands. Women's gardens usually were heavily fertilized and consisted of comparatively small plots of level land. As such they were less subject to erosion and soil exhaustion. "Perhaps," worried Isabel Baumann, "the very fact that we [women] are working in gardens blinds us to some of the things which are taking place in the fields which are devoted to crops."[107]

Baumann, who had just completed teaching a unit at her local rural school on the county's soils, led the club's first discussion on soil erosion. She noted that there were thirty-six different kinds of soil in their county alone, but emphasized that the character of the soil, whether sandy, loamy, or otherwise, was not the farmer's only important consideration. Astonished that in less than a century of farming, "between 25 percent and 75 percent of [the original topsoil] the Good Lord gave us is gone," Baumann detailed how traditional farming methods had led to such devastating erosion. "We Americans," co-leader Selma Sorenson put it succinctly, "haven't used soil wisely."[108]

The problem, chimed in Ruth King, was that twenty-five years ago "most farmers thought all they had to do to get good crops was to keep a herd of dairy cows and put the manure back on the land." As a result, explained Baumann, in order to get good crops "today we have to use lime and . . . commercial fertilizer and . . . plow under crops for green manure and . . . a lot of things that farmers didn't have to do 25 years ago." The biggest problem, according to Baumann, was soil erosion, which she took care to explain in terms the club's listeners could easily understand: "Our forefathers who came to Wisconsin started farming in a climate which was unfamiliar to them. Ours is a violent climate. The rains which fall in this climate are sometimes hard torrential rains. When these rains fall, the soil can't soak them up so they run off the land and carry dirt with them. This process is known as erosion."[109]

"Surely we can't stop the rains," exclaimed co-leader Grace Langer. "No, we can't," agreed Baumann, "but as farmers and farmers wives, we can do something about the erosion."[110] In their next broadcast, Selma Sorenson presented in greater detail why erosion was a relatively new problem, emphasizing that it was a result of modern farming techniques, not the state's natural flora. "When our fore fathers took this land away from the Indians," she explained, "trees and grass provided the land with a protective coating. Leaf mold in particular readily absorbed ground water, keeping the soils moist and secure. Clearing the trees and leaf mold and plowing the grass removed that protective cover."[111]

Like their husbands, the club women were concerned about the short-term repercussions of soil loss: the high labor and material costs of extensive fertilization. But the women, in keeping with their role as the civilizers of society and as mothers in tune with generational concerns, also emphasized more long-term and ethical considerations. They called for a "return [of] some of our land to its natural condition . . . which Nature originally provided." They warned that if such calls were not heeded because of men's relentless quest for immediate profit, lesser crop yields would lower the standard of living for future farming families. Haunted by the specter of the Dust Bowl, club leaders took the long view, asserting that every farm woman was obligated to know about the dangers of soil erosion, as well as the strategies to combat it, for the sake of "our children and our children's children on the farms."[112]

Farm women, the club broadcast urged, should make every effort to learn more about soil conservation. Ruth King stressed the shared power and authority in farm marriages: "Where changes in farming practices are to be made, it is just as necessary to get the cooperation and approval of farm wives as it is the husbands . . . [because] we farm women . . . generally have as much to say about our husband's business as he does."[113] Accordingly, the club leaders offered their listeners remedies for soil erosion. They described and advocated for each remedy in considerable detail: the cover principle, entailing keeping steep land in grass and trees to avoid runoff; the contour principle, planting crops in rows along the contours of intensively farmed sloping land; and the principle of diversions, constructing terrace ridges to divert surplus water in fields during hard rains. "MAN MUST WORK WITH NATURE," they concluded, taking on the voice of authority, "NOT AGAINST HER."[114]

Racism and Environment

Many Americans remained too preoccupied with the struggle for daily survival to commit as much thought as the WSWWTC to solving long-term environmental problems. Racial minorities suffered disproportionately socially and economically in both rural and urban environments during the Great Depression and did not benefit to the same degree from the recovery brought about by American entry in World War II. In addition to significantly higher rates of unemployment, African Americans in cities and on farms lived under the specter of lynching and segregation. They were relegated to the poorest-paying menial jobs. Throughout the South, educational opportunities, medical care, and other social services were inadequate at best, and in some areas entirely absent. African Americans, especially men, were subject to arrest and hard labor for crimes such as vagrancy, or leaving without permission the farms on which they were contracted to work.[115] Urban African Americans continued to have limited or no access to public parks, pools, and playgrounds, and even those who could afford camping or resort vacations were usually only welcome at racially segregated destinations like Michigan's Idlewild.[116] Other races suffered as well. In 1937, for example, the California club women who campaigned tirelessly to save the redwoods and who preached on the unifying virtues of womanhood amended the constitution of the City and Country Federation of Women's Clubs to bar non-Caucasians from membership. In the pages of the *Chinese Digest*, activist P'ing Yu bitterly denounced their elitism and racism: "It's just this high and mighty 'holier than thou' attitude of 'working for' rather than 'working with' people that makes this world so divided in spirit."[117]

Japanese Americans suffered uniquely when they were forced from their homes and relocated into the harsh, isolated environments of inland concentration camps. Those environments played a role in both the subjugation and resistance of the inmates.[118] Almost half of the Nisei (second generation) had made their livings in agriculture. They had leased or purchased some of the poorest lands along the West Coast and made them profitable. But the desert lands on which they were interned, whipped by strong winds and subject to freezing in the winter and temperatures over 100 degrees in the summer, proved a trial to even these skilled, experienced farmers. Internees with no agricultural experience faced challenges greater still. A physician noted as the ground was being broken at the Manzanar camp in southern California's Owens Valley, "We've had a few cuts and sprains and lots of blisters—after all, you expect that when you try to make sagebrush cutters out of good lawyers and accountants."[119] Despite the land's bleak appearance, internees who arrived at Manzanar in the spring of 1942 found the rich alluvial soil to be enormously fertile after they designed and constructed an irrigation system to divert water from a local creek to augment the water they received from the Los Angeles aqueduct. Working in shifts because they initially had only one plow, they immediately planted guayule, a rubber plant crucial to the war, and had raised sufficient food by September to feed their own camp and ship excess crops to other camps with poorer soils.[120] By 1943 Manzanar internees were also raising cattle, pigs, and poultry. Despite the challenges caused by local climates and conditions, all ten internment camps ultimately produced impressive crop surpluses, even as the Caucasians running the farms formerly belonging to the Nisei could not match the original owners' productivity levels and had to be federally subsidized.[121]

Men carried out most of the heavy agricultural labor, but all able-bodied adult internees were expected to work for low wages. Crude living quarters and communal dining meant that Japanese-American women who had previously spent much of their day cleaning and cooking now had time for both paid labor and leisure and artistic pursuits.[122] Women created rock gardens, victory gardens, and cactus and flower gardens for their own pleasure and as a way of practicing traditional arts.[123] To make their physical desolation and isolation more bearable and spend time outdoors they also participated in uniquely American activities, including forming softball teams and leagues.[124] The wide range of paying jobs available offered women unprecedented opportunities. They could sample work as cooks, clerks, bookkeepers, teachers, secretaries, and the many other positions necessary to keep their self-contained community running smoothly. Women at the Colorado River Relocation Center in Poston, Arizona, made adobe bricks as well as sewed clothing and uniforms and served as librarians.[125] Such opportunities in no way compensated for the privations and indignities suffered by Americans imprisoned exclusively on the basis of race. A woman interned at Heart Mountain in Wyoming allowed to travel four hours by train and bus to go shopping realized the toll the camp environment took on the human spirit: "It was quite a mental relief to breathe the air on the outside.... You just can't imagine how full we are of pent-up emotions. ... A trip like that will keep us from becoming mentally narrow. And without much privacy, you can imagine how much people will become dull."[126] After her release from the Topaz camp in Utah, Marii Kyogoku found herself "still thrilled every time I see a street lined with trees."[127] Despite its many privations, the internment experience did provide job training and experience, allowing some Japanese-American women to achieve a sense of independence and self-confidence. The style of living mandated by the camps also broke down certain traditional customs like arranged marriages and reshaped the lives of many Japanese Americans when they reentered mainstream society.[128]

. . .

As the backlash against women's leadership pushed them out of many reform movements in the 1920s, reducing their authority in mainstream environmental groups and organizations, the only campaigns and organizations granting any female authority that were widely welcomed were based on traditional female values. The privations of the Depression and the shortages and rationing of the war years so altered American society that the role assigned to women once again shifted. The experience of Japanese Americans in internment camps reveals how being placed in a harsh or unfamiliar social and physical environment made carrying out aspects of traditional gender roles impossible. For most Americans, however, the social upheaval of the Depression and war years brought new power to the traditional gender stereotypes and prescribed spheres that had reemerged in response to women's gains in power during the Progressive Era. Because the traditional women's sphere emphasized frugality and conservation, these qualities were central to women's environmental roles and relationships that were cultivated by the GFWC and celebrated and encouraged during the 1930s and 1940s. Like the Progressive Era women before them, women in groups like the GFWC and the WSWWTC used arguments based within the women's traditional sphere to empower themselves. They argued that environmental problems and resource shortages concerned not just the men in the fields and factories, but also their wives and daughters in the home—women who, by virtue of their gender, had a unique and important role to play in resource conservation and preservation. As World War II created a different kind of national crisis, women were encouraged to see themselves as independent and valuable individuals taking their rightful place as workers and conservation activists.

"'Papa' does not always know best"

6 Middle-Class White Women in the Cold War

"We Beat the Monster"

On December 1, 1973, the members of the Wisconsin-based League Against Nuclear Dangers (LAND) staged a highly publicized release of red balloons tagged with postcards describing the various radioactive substances they represented. The balloons' finders returned the postcards to LAND from as far away as West Virginia, vividly demonstrating the traveling range of airborne contaminants. The league, made up of Wisconsin homemakers without any experience as activists, had formed earlier in the year in opposition to a proposed nuclear power plant in Rudolph, Wisconsin. Although the plant's proponents touted nuclear power as a safe, cheap, and virtually unlimited source of energy, LAND members remained unconvinced. They expressed publicly their concerns about the potential for accidents and leaks that could result in an uncontrolled spread of toxic radiation, leading to a range of problems—from the mutation of human genes to the death of all forms of life.

The organization's members were white middle-class women in their thirties or forties; most were raising children and were not employed outside the home.[1] They were, claims one scholar, "naturals" for activist work because their prescribed gender role as the primary caregivers to their children had previously involved them in

```
                    UNITED NUCLEAR OPPONENTS
Are opposed to nuclear power plants anywhere in Wisconsin.
Balloons like this one have been released from four
potential nuclear plant sites in this state.

We would like to know how far this balloon went before
you found it.  PLEASE FILL IN THE CARD AND MAIL IT BACK.
              TAHNK YOU VERY MUCH.

WHERE FOUND    Newcomerstown  Tuscarawas   Ohio
               CITY            COUNTY        STATE
WHEN FOUND     Dec. 22, 1973
                    DATE
MAIL TO:  LAND, BOX 240,  RUDOLPH , WISCONSIN  54475
  released  Dec 21, 73  from
                          Rudolph  WER
```

FIGURE 6.1 A postcard attached to one of the balloons released by LAND to represent nuclear dangers was found in Newcomerstown, Ohio, revealing the potential range of airborne toxins. (Wisconsin Historical Society 89868)

broad humanistic and nurturing issues, their interactions with other activists were minimally contentious, and their dearth of conventional power left them with little to lose.[2] Ridiculed for their lack of scientific credentials (many, like co-chair Naomi Jacobson, had no college education) and dismissed by utility officials as "illogical, emotional housewives," LAND members educated themselves about nuclear hazards. Most significantly, they worked to educate and gain the support of the entire community, not just appeal to those perceived to be in power.[3] Accordingly, they did not limit their activities to the tools of traditional male-dominated efforts: petitions, reports, graphs, and charts. They also employed innovative methods of consciousness-raising that required no specialized knowledge to appreciate, using familiar songs and popular advertising techniques to make their position accessible to anyone. Of the written materials they did circulate, many were based on information provided by prizewinning scientist and biostatistician Rosalie Bertell, a Roman Catholic nun who preferred not to tackle government and industry but "to work directly with people and support them with scientific information" written in clear, layperson's terms.[4]

One of LAND's unconventional methods of attracting attention was inspired by the wildly successful Burma-Shave advertising campaign, in which some 7,000 sets of verses (for a total of approximately 42,000 signs) were posted along highways in forty-five states.[5] A classic Burma-Shave series of signs, with each verse on a separate sign and the signs spaced over the course of several miles, read:

In School Zones
Take it Slow
Let the Little
Shavers Grow

with the final sign always reading "Burma-Shave."[6]

In one of the many different LAND versions, the signs read:

Nuclear Leaks
Can Cause
Human Freaks

LAND also produced antinuclear lyrics to familiar songs, such as, "We'll All Go Down Together When It Comes," sung to the tune of "She'll Be Coming 'round the Mountain When She Comes."

In 1980, the Wisconsin Public Service Commission bowed to widespread opposition, much of it generated by LAND, and canceled plans that had grown to include eight proposed nuclear power plants. "We won, we beat the monster," noted one LAND member, adding that the "sweet victory" was "gratifying for all the small Davids to confront Goliath and come out on top."[7] By the time the group formally disbanded in 1988, the world had witnessed events that proved LAND's concerns were well-founded: the partial core meltdown at Pennsylvania's Three Mile Island nuclear power station in 1979 and the Chernobyl accident in the Ukraine in 1986. Many former LAND members became active in state, national, and international groups concerned with nuclear issues.

Old and New Gender Prescriptions during the Cold War

LAND's balloon release was a tactic modeled by the larger counterculture that emerged in the turbulent 1960s in response to the conservatism of the immediate postwar period. Through speeches, strikes, boycotts, marches, teach-ins, sit-ins, and a variety of other actions, protesters sought an end to the war in Vietnam and demanded new freedoms, including rights for students and racial minorities. The balloon release was one of many events staged by white middle-class women across the nation as environmental concerns increasingly entered American consciousness in the early 1970s. The emergence of women's groups like LAND, however, represented less a sharp departure from some of the prescribed gender values that resurfaced in the immediate postwar period than a culmination of those values, especially women's dedication to home and family, including future generations.

Yet the nascent feminist movement also fueled women's activism. The women's groups and activities that developed throughout the Cold War era reveal how race, class, age, and marital and maternal status, as well as gender, all shaped responses as Americans recognized both old and new sources of environmental degradation imperiling their nation.

Not all environments, however, were deemed equally worthy of preservation and protection. During the materialism and suburbanization of the immediate postwar period, the contributions and rugged individualism of pioneer women so admired during the lean years of the Great Depression and World War II once again faded from memory. Even as Girl Scouts, Camp Fire Girls, and girls' summer camps glorified the values and skills of women in the rural past, contemporary rural women faced increasing scorn and derision in popular culture. In 1945, Betty MacDonald published *The Egg and I*, which she called "a sort of rebuttal to all those recent successful 'I-love-life' books by female good sports whose husbands forced them to live in the country." Unlike the books by Laura Ingalls Wilder, which appealed to Americans struggling a decade earlier by romanticizing pioneer hardships and environments, MacDonald's was "a bad sport's account of life in the wilderness without light, water, or friends."[8] When MacDonald's husband wanted to take up chicken farming in the Chimacum Valley of Washington's Olympic Peninsula in the late 1920s, she readily agreed, for "Mother had taught me that a husband must be happy in his work."[9] Two children later, MacDonald discovered that she hated even baby chickens, was lonely, and "seemed to have married the wrong man."[10] She left her husband in 1931, returning to Seattle where she eventually remarried and wrote her account of her four years in the wilds of the Pacific Northwest.

First serialized in the *Atlantic Monthly*, *The Egg and I* sold more than three million copies in hardback and was eventually translated into thirty-two languages. MacDonald's humor was edgy rather than warm, and she acknowledged openly that she was ill suited for country life because she thought it was "stretching this wifely duty business a bit too taut to be . . . carrying water, baking, canning, gardening, scrubbing, and taking care of the chickens and pigs [while living in] the day of the magnetic eye, automatic hot water heaters, and television."[11] Although like Wilder she wrote in detail about the natural beauties of the local environment, she also wrote derisively of the privations she endured as a woman in a house without electricity or running water. In addition to the usual domestic chores, made all the more difficult without the labor-saving technologies available in cities, she was shocked that rural life required her to share in more "masculine" labors like hauling water, wood, and chicken feed and pulling stumps. To MacDonald, the phrase "Woman's work is never done" "signified the dinner dishes which I washed and dried while

Bob smoked his pipe and took his ease."[12] She highlighted the gender differences between herself and her husband, noting that he "never seemed to be lonely, he enjoyed the work, he didn't make stupid blunders . . . [but] then, of course, he wasn't pregnant."[13]

MacDonald expressed great disdain for the local Indians and wrote in biting wit of neighboring farmers and ranchers, including the "shiftless, ignorant, and non-progressive" Kettle family, made up of Ma, Pa, and their fifteen children.[14] MacDonald's former neighbors were not amused. Several claimed that MacDonald's book subjected them to ridicule, and they sued her for libel. MacDonald countered that the unflattering characters were composites of rural life rather than individual portraits. On the basis of this sweepingly negative stereotype, MacDonald was exonerated.[15] Most Americans shared her dim view of the woman who "didn't have lights, water, radio, toilet, bathtub, movies, neighbors, or money, and she just LOVED it."[16] When *The Egg and I* appeared as a feature film in 1947, the hapless characters of Ma and Pa Kettle were a huge hit. Marjorie Main was nominated for an Academy Award for her portrayal of Ma Kettle. An additional nine "Ma and Pa Kettle" films were made over the next ten years, cementing into popular culture the notion of people in contemporary rural environments as unsophisticated, lazy, and simple-minded yet conniving, holding nothing in common with the noble pioneer past.

People in contemporary rural environments were not the only ones who suffered a setback in the public's esteem. The social upheaval of World War II had loosened the grip of traditional gender stereotypes and prescribed spheres, but with the Cold War came new, stricter, and more rigid prescriptions, many prohibiting women from claiming any environmental authority. In 1950, when American Elizabeth Cowles became the first Western woman to penetrate deeply the southern approach to Mount Everest, *American Magazine* assured its readers that Cowles was a mother and "so completely feminine that one would never suspect her stamina and spirit of adventure," noting approvingly that she carried a lipstick in her knapsack.[17] Women mountaineers whose achievements had been celebrated in previous decades were described derisively by noted writer, explorer, and preservationist Paul Gayet-Trancrède as sexually perverse "lonely crows who, aping men, haunt the huts and great mountain faces and ply the harsh tools of the mountains, baring their faces to the winds in ecstasy and straining to their bosoms the unfeeling rock with the ardor of lovers." "True women," he continued, "are too tender for the rigours of the mountain, and men will not accept that they should penetrate their domain."[18] In the Sierra Club and other major conservation organizations, even women who had previously held office were routinely relegated to voluntary rather than paid positions.

Women were not totally without environmental influence. In 1961 Sylvia McLaughlin, whose husband was a University of California, Berkeley, regent, became alarmed by a plan to fill in some of San Francisco Bay's shoreline zone for development. With Esther Gulick, married to an economics professor, and Kay Kerr, wife of the university's president, McLaughlin called together representatives of the Sierra Club, the Audubon Society, and the Save the Redwoods League. The male leaders listened sympathetically, but upon hearing that the women's "chief concern was the beauty of the bay," recalled McLaughlin, they "filed out and wished us luck." The three women launched the Save San Francisco Bay Association, later shortened to Save the Bay. Fifty years later, the organization's director noted that in 1961, "This was a man's world. Men were running the conservation organizations.... That's why most people gave them [the three women] little chance of succeeding." Despite its lack of male leadership, the association's founders used ties created by their husbands to other men in power, including members of the state legislature, to beat back the developer's plans, a victory that led to the creation of the Bay Conservation and Development Commission that same year.[19] Save the Bay showed that women could still play a role in environmental protections, but a role circumscribed by renewed gender-based constraints.

The perception of communism as a powerful threat to American freedoms and ways of life produced a pervasive fear used to justify a return to traditional definitions of true womanhood and, in concert with race, class, and sexual orientation, shaped the gendered responses to environmental issues. Patriarchy, Christianity, and especially the heterosexual nuclear family were viewed as not only socially desirable, but also as politically necessary if the nation was to survive—and to triumph over—the communist menace.[20] The nascent lesbian and gay organizations that emerged out of World War II found their hopes for growth and acceptance dashed in postwar America. Having taken a few tentative steps toward promoting self-acceptance, homosexuals were forced into retreat as they were widely declared to be rare but inherently subversive and therefore especially dangerous to the American social fabric. The ideal American family, glorified as the greatest bulwark against communism, featured a husband and father who produced the family's single income, and in a revival of republican motherhood ideals, a wife and mother whose sole occupation and concern was caring for her family, especially catering to her husband's needs and raising a large brood of obedient and patriotic Americans.

Economics as well as politics shaped the way women viewed themselves in the Cold War period and redefined their relationships with the environment. During World War II some six million women joined the workforce, a rise of 50 percent, increasing the overall percentage of women in the labor force from 25 to 36 percent. The same full-time homemakers who had to be coaxed into the paid workforce by wartime propaganda were told at the war's end that the best way to serve their country was to leave those jobs, which rightfully belonged to male breadwinners (especially veterans), and to return home.[21] Many women left their fulfilling and relatively well-paying defense-industry jobs reluctantly. However, the majority of women in paid positions, making up 32 percent of all American women, remained in the workforce in the immediate postwar period. Working-class women forced out of war-production jobs could not afford to stay at home full time, turning instead to the only employment available to them, frequently "pink-collar" jobs (in beauty and hair salons, and food and beverage service), at substantially reduced pay. Some middle-class women eagerly threw themselves into the prescribed domestic sphere. Most, despite wartime propaganda, had never left it. Others who had taken on paid work missed the sense of self-worth engendered by their wartime contributions, but rejoiced at no longer having to carry out paid labor while simultaneously shouldering all domestic burdens.

Woman's proper and natural place was, once again, decidedly within the home rather than outdoors, where, according to one wag, they "are apt to blither—at things that slither."[22] Women were nevertheless encouraged to participate in family camping trips (where even in the wilderness they could enjoy the homey comforts provided by newly available products, including air mattresses, improved sleeping bags, kerosene camp stoves, freeze-dried foods, and nylon tents).[23] Women's role on the campground, as in the home, was to see to the health, happiness, and safety of others, ensuring that their husbands and children were well provisioned and cared for throughout the trip. From war's end through the 1950s, Americans were assured that the family camping trip, rock climb, or any similar expedition reinforced the natural social hierarchies by placing women in supportive roles.[24]

Growthmania

Although women did not disappear from wilderness entirely, the rightful place for those who were white and middle class was determined to be the suburbs, which offered family living in a tamed and domesticated wilderness: groomed, individual park-like settings, featuring fenced lawns and carefully delineated flowerbeds. The rise of suburbia was an outgrowth of the unprecedented material wealth available to many middle-class American families, even those on a single income, in the immediate postwar period. The new veterans' benefits made available through the G.I. Bill combined with another innovation, the thirty-year mortgage, making home ownership a reality for many for the first time. The exodus from congested urban areas into the comparatively wide-open spaces of the new suburbs mushrooming across the nation was termed "white flight" by social critics. Among people of color, African

Americans in particular were excluded from suburbia through racial discrimination in hiring as well as housing. Such exclusionary tactics were by no means reserved to the American South, but were practiced routinely throughout the nation. Those prohibited from leaving urban neighborhoods by the color of their skin and limited financial resources were soon faced with crumbling infrastructures and a variety of environmental hazards.

Suburbs transformed the physical American environment. Agricultural lands were rezoned for commercial and residential use. Undeveloped lands were deforested, leveled, and turned into tract housing. Fueled by low gas prices, the previous emphasis on public transportation, including urban bus, subway, streetcar, and trolley systems, was rapidly transferred to individually owned cars that offered personal freedom. The exhaust produced by the new onslaught of cars, mixed with the emissions from local industries and backyard incinerators, dramatically increased air pollution. As suburban areas expanded, their residents, who enjoyed the exposure to trees and lawns that their homes afforded, grew alarmed by the potential threats to these constructed environments.[25] In the Los Angeles area, a rapidly growing urban and suburban population was settling in a geographic basin that frequently suffered from temperature inversion, leaving smog trapped close to the ground. In October 1954, dense smog shut down schools and industry. Local homemakers called for women's groups to take direct action. They criticized elected officials and businessmen for their reluctance to implement costly smoke conversion measures. One woman publicly derided men's inability to look beyond short-term expenses, arguing that conversion costs would be "cheaper than doctor bills."[26] In one of the first of many well-publicized, all-women protests against pollution, several dozen homemakers from Pasadena, eleven miles northeast of Los Angeles, staged a march.[27] Calling themselves the Smog-a-Tears, the women wore gas masks and carried banners criticizing the government's inaction. Theirs was a gendered appeal, as they highlighted their traditional "nature's housekeepers" role. One woman, for example, carried a broom in addition to a sign reading, "We Want a Clean Sweep of Smog." They emphasized their maternal concerns as well; the youngest marcher, three-year-old Agatha Acker, wore a gas mask, as did the doll she carried.[28] Winds finally dispersed the heavy smog, but local women, working individually and in groups, including the Pasadena Council of Women's Clubs and the Los Angeles Council of Women for Legislative Action, continued to marshal their forces to combat air pollution to ensure the welfare of the people of greater Los Angeles rather than sacrifice it on the altar of corporate profit.[29] Their efforts were outstripped, however, by the rapid increase of cars and roads. In California and across the nation, the network of freeways created by the 1956 Interstate Highway Act was augmented by countless new roads, highways, and expressways.

That the huge increase in the numbers of cars traveling the new roads linking suburban living spaces to urban work places dramatically increased air pollution is not surprising. But some of the greatest sources of suburban America's contributions to environmental devastation are far less routinely recognized. The suburban lawn, for example, led to sharp increases in water consumption and the use of chemical fertilizers and pesticides, with runoff contaminating groundwater.[30] By the end of the Cold War, yard waste was the second most prevalent item in urban landfills. In California alone, gas-powered lawnmowers were producing as much pollution per year as 3,500,000 cars each driving 16,000 miles.[31]

More than the lawns of the new suburban developments took their toll on the postwar environment. Middle-class Americans, forced to make do with precious little during the Great Depression and faced with rationing and shortages during the war years, proudly acquired a vast array of consumer goods. Homemakers, repeatedly urged to conserve during the war years, were encouraged to consume. Suburban homes featured not just a car in the garage and an electric refrigerator and range in the kitchen, but new mass-produced products including furnishings, clothes, barbecues, books (newly affordable due to the widespread availability of the paperback), and children's toys. By 1955 half of all American households owned a television, one of the most powerful tools promoting what was later termed derisively "growthmania," an obsession predicated on the assumption that "more [more goods, more living space, more people, more profits] is better." The gendered consumption patterns of the many newly available and heavily advertised consumer goods aligned suburban women in particular with environmentally harmful practices. Homemakers freed from traditional burdens, including hand-washing laundry and preparing foods from scratch, found their hours spent on household chores actually increased by time spent shopping in vast new stores, distant from home and reached only by automobile, and keeping in good repair the large number of products considered essential for suburban living. New standards of cleanliness and appearance necessitated carrying out this work through the use of a wide range of chemical compounds inside each suburban home, garage, and toolshed.

Advertisers enticed women to not only perfume themselves, but also to buy pleasingly scented household products, neglecting to warn that the artificial fragrances added to many cosmetics, cleaners, and polishes were made from petroleum, which does not degrade in the environment and can have toxic effects on both fish and mammals, including humans. Women's daily use of ozone-depleting aerosol hairsprays and deodorants became a new norm. Women were also pressured to consider the cleanliness of their families' clothes to be an important measure of their own success or failure as homemakers. Yet the phosphates routinely found in detergents, as well as disinfectants and deodorants, unbalanced ecosystems by fostering dangerously explosive marine plant growth. Quaternium 15, an alkyl ammonium chloride, was used in detergent as a surfactant (bringing foreign matter—in this case dirt—to the surface), releasing as a by-product formaldehyde, a potent toxin. Women encouraged to beautify the inside of their homes and keep them in pristine condition routinely used solvent-based paints, primers, and varnishes that emitted volatile organic compounds (VOCs), contributing to the destruction of the stratospheric ozone layer and playing a significant role in the creation of the greenhouse effect. The pesticides and weed killers touted as essential to women's beautification of their lawns and gardens, especially through the cultivation of colorful flowerbeds, made their way into the groundwater. The result was serious health problems in humans, including disruptions in the endocrine system, cancer, infertility, and mutagenic effects.[32]

In 1947, Mrs. L. Eugene Emerson of the Cambridge Plant and Garden Club returned from a meeting of the Garden Club Federation Conservation Day with news of hope for elm trees that had been dying of a Dutch fungus carried by beetles: the pesticide known as DDT (dichlorodiphenyltrichloroethane). The following year Paul Hermann Müller, the inventor of DDT, was awarded the Nobel Prize in Physiology or Medicine for this pesticide that had been so effective in clearing South Pacific islands of malaria-carrying insects, saving untold numbers of American troops during World War II. It was also effective as a de-lousing powder. Following the war, DDT was celebrated by gardeners and farmers as an especially efficient way of eliminating crop-destroying pests. By 1955, members of the Cambridge Garden Club, like gardeners across the nation, routinely dusted young leaves on houseplants and in gardens with DDT to control pests. Chemical agents like DDT were not feared for their potential toxicity, but embraced as miraculous, time-saving solutions to age-old problems. Prior to 1962, most Americans welcomed advances in chemistry with open arms, especially as benefits included the increased yield, appearance, and quality of produce, meat, poultry, and dairy products at their local supermarket.

Women's Antinuclear Activism

Women lacked awareness of the toxins within their homes and gardens but were increasingly cognizant of some of the poisons multiplying in the atmosphere. The atomic bombings of Hiroshima and Nagasaki at the end of World War II generated powerful fears of nuclear weapons with their potential to destroy life upon impact and through lingering radiation.[33] Those fears intensified as other nations gained nuclear capabilities. The testing of new weapons proved that the results could not always be controlled, as when the force of the explosion in a 1954 American test was more than twice that predicted. Many scientists dismissed claims that radioactive nuclear fallout, particularly strontium-90, produced by these tests was appearing in both breast milk and commercially produced cow's milk and thereby finding its way in significant amounts into the mainstream population.[34] Others, led by Nobel Prize–winner Linus Pauling, predicted that the effects of current fallout would be damaging for generations to come. In January 1958 Pauling presented to the United Nations a petition approved by more than 9,000 scientists in 48 nations demanding an immediate halt to all nuclear testing.

Women contributed significantly to the nascent antinuclear movement. In St. Louis, a group of women, Eves against Atoms, provided the foundation for the Citizens Committee for Nuclear Information (CNI). The same year that Pauling submitted his petition to the United Nations, CNI's Dr. Alfred Schwartz proposed that children's baby teeth formed from 1948 to 1953, the first years of the fallout era, be collected and tested to form a baseline for levels of radioactivity. The project, directed by internist Louise Reiss, sought 50,000 baby teeth and attracted instantaneous worldwide attention. Its emphasis on babies and milk raised the concerns and involvement of mothers everywhere, placing them at the center of this high-stakes scientific controversy and the movement to stop nuclear testing.[35]

Dire predictions concerning babies, children, and future generations made nuclear testing increasingly a women's issue. The rapidly expanding military-industrial complex intensified their fears, as huge government contracts were awarded to businesses to develop new defense technologies and weapons, both chemical and conventional. On October 30, 1961, as part of the escalating weapons race, the Soviet Union tested the largest nuclear weapon ever. Two days later, some fifty thousand primarily white, middle-class American women abandoned their homemaker duties to march and demonstrate in major cities and suburban communities. In this Women Strike for Peace (WSP), founded by congresswoman Bella Abzug and activist Dagmar Wilson, protesters demanded that the world "End the Arms Race—Not the Human Race."[36] According to Wilson, "In the face of male 'logic,' which seemed to us utterly illogical, it was time for women to speak out."[37] WSP activist Blanche Posner put the strike in starkly maternal terms: when mothers were putting their children's breakfasts on the table, they saw "not only Wheaties and milk, but they also saw strontium 90 and iodine 131 [released by the nuclear blast]. . . . They feared for the health and life of their children."[38] An organization of the same name grew out of the original strike.

The Cuban Missile Crisis in October 1962 made the potential for nuclear disaster impossible to dismiss as female hysteria. Schoolchildren practiced nuclear attack drills and panicked suburbanites dug bomb shelters in their backyards. WSP held demonstrations throughout the crisis, wielding signs promoting peace and urging

President John F. Kennedy to "Be Careful" as they emphasized that their maternal concerns about children and the future trumped any and all political issues. Kennedy claimed that the women's generational arguments influenced his thinking, culminating in the U.S.-Soviet treaty to ban all but underground testing of nuclear devices: "I have said that control of arms is a mission that we undertake for our children and our grandchildren, and that they have no lobby in Washington. No one is better qualified to represent their interests than the mothers and grandmothers of America."[39] Although WSP members were denounced as "annoying lapel-pullers" by some members of Congress, their persistent outspokenness also generated votes for the treaty as politicians attempted to avoid "fall out from mothers."[40]

Silent Spring, The Feminine Mystique, and Women's Environmental Activism

In the early 1960s two books on entirely different topics transformed the consciousness of untold numbers of American women. These works, written in nontechnical terms for general audiences, were wildly popular. In many ways they grew out of

FIGURE 6.2 Women Strike for Peace: Congresswoman Bella Abzug (wearing one of her many distinctive hats) marches in front of the White House with other women calling for an end to the nuclear arms race. (Dorothy Marder, Women Strike for Peace, Anti-Nuclear, Swarthmore College Peace Collection)

earlier concerns about the personal and environmental impact of women's postwar patterns of consumption, and out of women's antinuclear activism. They set into motion a series of diverse groups and movements that changed the way many women viewed themselves and their relationships with the physical world.

In 1962, pioneer ecofeminist Rachel Carson dramatically challenged conventional notions of progress and celebrations of prosperity with *Silent Spring*, first serialized in the *New Yorker* and then published in book form. Carson, who held a master's degree in zoology from Johns Hopkins University, had spent many years working in the U.S. Fish and Wildlife Service. A best-selling author, she had the platform to challenge authority—both the scientific authority that assured the public that chemical pesticides were safe and, implicitly, the patriarchal authority that relegated women to second-tier positions in professional science. In 1954 Carson proclaimed women's "greater intuitive understanding" of the value of nature as she denounced a society "blinded by the dollar sign" that was allowing rampant "selfish materialism to destroy these things." Even as she stressed the role of natural beauty in human spiritual growth, she defended the presence of emotion in science and nature writing.[41]

Silent Spring, a *New York Times* best seller and Book of the Month Club selection, reached a wide audience, much of it female. In Wisconsin, for example, it was read aloud in October 1962 in ten half-hour installments on the 1:00 p.m. "Chapter a Day" program of the Wisconsin State Broadcasting Service.[42] The many home-makers listening heard Carson use some of the established beliefs about gender to give credence to her message. Chastening "man" for his "arrogant" talk of the "con-quest of nature," Carson warned that the power to achieve that boast had not been tempered by wisdom. She noted that Americans were stunningly oblivious to the dangers of their embrace of all things chemical. Specifically, she questioned the wisdom of the governmental "fathers" concerning industrial waste and the vast reli-ance on pesticides, especially DDT. In arguments featuring the traditional female emphasis on beauty, spirituality, and future generations, assertions highly reminis-cent of those put forward by Lydia Adams-Williams and other women conservation pioneers, Carson pointed out long-term consequences of chemical use that far out-weighed short-term private profits and consumer benefits. She also appealed to Americans to consider their individual roles as users of chemicals, and to recognize themselves as contributors to the problem. Her chapter "Rivers of Death," for example, makes plain the detrimental impact of toxic chemicals on the environment, whether generated by homemakers or by corporations.

In *Silent Spring* Carson evaluated not only the intended but also the unintended consequences of using toxic agents. She noted that DDT killed the targeted insects along with most life forms with which it came into contact. DDT, she cautioned, was effective to a fault, remaining toxic for weeks and months, even in diluted form. Following its initial surface application, it seeped into the soil and water and ultimately into the food chain, ingested by birds and animals, including humans, in whose tissues it caused cancer and genetic damage. The book's title came from the opening chapter, "A Fable for Tomorrow," predicting an American town where all life, from songbirds to children, had been silenced by this killer unwittingly unleashed by the scientific community. "Future historians may well be amazed by our distorted sense of proportion," Carson warned. "How could intelligent beings seek to control a few unwanted species by a method that contaminated the entire environment and brought the threat of disease and death even to their own kind?"[43]

Silent Spring, which attacked the government's misplaced and ineffectual paternalism, appeared just one year before Betty Friedan's assault on patriarchy, *The Feminine Mystique*. Friedan contested the popular image of the postwar, American middle-class homemaker as a woman who found total fulfillment in serving the needs of her husband and children and in volunteering within her local place of worship and community. While acknowledging middle-class women's material comforts, Friedan emphasized the mysterious sense of longing and dissatisfaction felt so intently by many full-time homemakers that they self-medicated, finding solace in alcohol and "mothers' little helpers" (tranquilizers). This "mystique," insisted Friedan, was not a symptom of selfishness, nor was it an indication that women did not love their husbands and children. Rather, such feelings were the natural response of human beings exclusively dedicated to fulfilling the needs of others, cut off from opportunities to explore their own desires, gifts, and talents. Friedan's middle-class female readers across the nation breathed a collective sigh of relief that they were not alone in their vague but pervasive, often crippling, sense of dissatisfaction. Friedan's urging that women throw off patriarchy contributed to Carson's message that they no longer assume "that someone was looking after things—that the spraying must be all right or it wouldn't be done."[44]

Both Friedan and Carson were roundly criticized. Many middle-class husbands, exhausted from shouldering alone the financial responsibilities for their families, expressed outrage over Friedan's claims that their wives were somehow oppressed. *The Feminine Mystique* nevertheless helped to put into motion the modern feminist movement: a rejection of prevailing gender spheres in favor of the political, economic, and social equality of the sexes. Exposure to feminism in turn made women all the more receptive to *Silent Spring*'s message of individual as well as corporate and governmental responsibility for the environment.

Gender and sexuality influenced the critical response to *Silent Spring*.[45] Carson was dismissed as "hysterically overemphatic" in a scathing review in *Time* magazine. The many denunciations within the popular press of Carson's work as overly emotional played to the stereotype of women as unscientific and inherently hysterical.[46] Also, most members of the scientific community, especially those in the chemical industry, initially dismissed her. Ezra Taft Benson, an elder in the Mormon church who had been secretary of agriculture in the Eisenhower administration, suggested to Eisenhower that Carson's unnatural status disqualified her from making inquiries in the first place. Although Carson had helped raise her two nieces and formally adopted one of their sons when he was five years old, Benson persisted in viewing her as childless and wondered "why a spinster with no children was so worried about genetics." He settled on the only possible conclusion. Most likely unaware of Carson's long-term intimate relationship with Dorothy Freeman, Benson accused her not of lesbianism, but of "probably [being] a Communist."[47]

Carson and Freeman strove to protect their privacy, labeling certain letters in their voluminous correspondence as destined for "the strong box," their code for eventual destruction.[48] Although few people knew of their relationship, the popular press focused on Carson's marital status. She was variously described as "unmarried," "never married," and "a shy female bachelor." Her obituary in *Time* quotes a friend who called her a "nun of nature." Even without charges of lesbianism, such references served to "desex Carson and brand her as not-quite-woman" in an age in which marriage and motherhood were upheld as woman's highest calling and defined femininity.[49] At the same time that Carson's "unnatural status" was used by some of her critics to undermine her credibility, others criticized her for being all too stereotypically female. After a California reader complained to the editor of the *New Yorker* that "Miss Rachel Carson's reference to the selfishness of insecticide manufacturers probably reflects her communist sympathies," its author added, "As for insects, isn't it just like a woman to be scared to death of a few little bugs!"[50]

Carson's defenders, however, openly defied disparaging arguments that hinged on widely held perceptions of gender and sex. One woman's praise for Carson denounced the highly touted postwar notion that "Father Knows Best" (the title of one of the era's many popular TV shows in which a happy, nuclear, middle-class family is shepherded through life's little hazards by a wise and benevolent patriarch): "'Papa' does not always know best. In this instance it seems that 'papa' is taking an arbitrary stand, and we, the people are just supposed to take it, and count the dead animals and birds."[51] When Carson was compared to Carrie Nation, the hatchet-wielding temperance advocate, the *New York Times* later noted, "This comparison was rejected quietly by Miss Carson, who in her very mild but firm manner refused to accept the identification of an emotional crusader."[52] Although Carson continued to use stereotypically feminine "sentimentalist" arguments to make her case, her calm demeanor throughout the controversy deflected charges of female hysteria or overwrought overemotionalism.

Given her public respectability, Carson's critics also found it hard to persuade others to reject this considerably renowned and gifted scientist who lyrically translated scientific complexities into terms accessible to a nonscientific audience.[53] When several reputable male scientists rose to defend *Silent Spring*, President Kennedy ordered his Science Advisory Committee to investigate. Carson's painstaking research was impossible to refute, and her emphasis on the interconnectedness of all life could not be dismissed as feminine romanticism.[54] No longer would the traditionally male emphasis on immediate, market-driven needs to exploit natural resources for profit automatically trump women's "sentimental" emphasis on future generations. The understanding that any disturbance to the web of life has consequences throughout was accepted by most as a scientific reality. Through her refusal to adhere to prevailing gender stereotypes of female subservience to male wisdom, Carson made the public aware that attempts by the scientific-industrial complex to manipulate and control nature were to the ultimate detriment of all.[55]

Just as *The Feminine Mystique* did not single-handedly launch the modern feminist movement, *Silent Spring* cannot alone be credited for inspiring the modern environmental movement. Both books, however, did build on women's environmental activism of the early postwar period and played major roles in making the vital need for reform clear to large sectors of the general public.[56] Many of the women awakened by Friedan's work to take themselves seriously took their first steps into finding a larger place in the world by responding to Carson's call, written in terms they could understand about a cause with which they could identify, to question authority. *Silent Spring* was cited in educational pamphlets written by women and in their letters to editors and petitions to politicians. Despite frequent condescension from men in positions of power, women contributed significantly to a variety of existing environmental organizations whose membership remained open to both sexes, such as the Sierra Club.

Carson's message also inspired the formation of untold numbers of local grassroots groups and movements that continued to multiply in cities, in suburbs, and on college campuses throughout the 1960s. Lorrie Otto, for example, a mother of two, had long shared Carson's concerns about DDT. In the 1950s Otto was disturbed by the number of robins she saw suffering from convulsions and dying after the trees in her Milwaukee suburb had been sprayed with DDT.[57] Although not yet forty, she was originally dismissed by local officials on the basis of both age and sex, called "just an old lady in tennis shoes, not to be taken seriously." Otto persisted in her efforts to stop the routine spraying of DDT, and with the publication of *Silent Spring*, public opinion shifted in her favor.[58] In 1968, as part of a group of other concerned citizens, she took advantage of an obscure state statute that allowed residents to petition for public hearings if they believed their water quality standards were being violated.

The group sought and organized scientists, attorneys, and witnesses from the United States, Canada, and Sweden to present evidence at lengthy and well-publicized hearings in Wisconsin that DDT was a water pollutant. The hearings led directly to Wisconsin banning DDT in 1970, making it the first state to do so. Two years later, the rest of the nation followed Wisconsin's lead.

Groups like Otto's found an ally in Wisconsin senator Gaylord Nelson, who on March 25, 1963, told his fellow senators, "We need a comprehensive and nation-wide program to save the national resources of America. We cannot be blind to the growing crisis of our environment. Our soil, our water, and our air are becoming more polluted every day. Our most priceless natural resources—trees, lakes, rivers, wildlife habitats, scenic landscapes—are being destroyed."[59] This call was soon taken up in the White House. During Lyndon Johnson's election campaign in 1964, his wife, known as Lady Bird, recalled that during their travels throughout the country, "Our eyes frequently met the majesty of America's splendor, but, sadly, all too often the evidence of neglect and abuse."[60] She agreed with her husband that "once our natural splendor is destroyed, it can never be recaptured," and that "once man can no longer walk with beauty or wonder at nature, his spirit will wither, and his sustenance be wasted." The two were committed to promoting environmental awareness and protection.[61] With Secretary of the Interior Stewart Udall, Lady Bird Johnson formed the Committee for a More Beautiful Capital in 1965 as she sought to make Washington, D.C., a model for urban beautification and revitalization efforts nationwide. Perhaps best known for her advocacy of the Highway Beautification Act, which favored the planting of flowering bushes rather than billboards along the nation's highways, the First Lady campaigned aggressively for environmental reform. She emphasized that "beautification" did not mean merely cosmetic appearance, but involved "much more: clean water, clean air, clean roadsides, safe waste-disposal, and preservation of valued landmarks as well as great parks and wilderness areas." Like many a female advocate of environmental protection before her, she couched her arguments not in terms of politics or economics, but emphasized instead the impact on subsequent generations: "Beautification means our total concern for the physical and human quality we pass on to our children and the future."[62]

As environmental issues won increasing recognition and support at the federal level, they were decreasingly dismissed as the exaggerated concerns of a fringe element of overwrought alarmists. On April 22, 1970, the first Earth Day ushered in a new national sensibility. The brainchild of Gaylord Nelson, this "teach-in" honoring the earth and calling for its protection was celebrated by some 20 million environmental activists, protesters, and concerned citizens across the country. Students responded with particular enthusiasm, rallying at more that 2,000 colleges and universities against pollution and population growth.[63] Concerns about the environment were shared by both the counterculture and the conventional. Environmental organizations and cooperative movements continued to develop on campuses and in communities nationwide. The environmental movement quickly became mainstream. The activities of the newly created U.S. Environmental Protection Agency and the international organization Greenpeace quickly came to dominate much of the media's attention on environmental issues. Meanwhile, smaller exclusively women's groups continued to be active, many of them made up of mothers of the student activists, women who believed that their traditional role as housekeeper, nurturer, and caregiver made them uniquely qualified to contribute.

Individually and in groups, women campaigned to ban the bomb, clean up rivers, save forests, and stop pollution. Women's organizations that were particularly active in promoting environmental awareness and protection included the League of Women Voters, the American Association of University Women, the Federation of Women's Clubs, and the Garden Club of America.[64] Traditional gender stereotypes were used to undermine the authority of these and other women's organizations. For example, a project engineer dismissed a member of a group of California homemakers opposing the construction of a highway with a sneering, "Get back in your kitchen, lady, and let me build my road!" Clean air activist Michelle Madoff anticipated just this kind of demeaning rejection when she stated, "I didn't want to go and testify and be branded another idiot housewife in tennis shoes, as we're referred to—you know, uninformed, emotional."[65] Instead, women responded to the challenge of raising the nation's environmental consciousness in a variety of ways. Some downplayed traditional gender stereotypes and joined new, ostensibly gender-neutral mainstream organizations. Others embraced their traditional role as women, but redefined it in a variety of creative ways to shore up their authority as environmentalists, frequently borrowing tactics such as political theater and other innovative protest methods used effectively by the student and antiwar movements.[66] The balloon release staged by the League Against Nuclear Danger was only one of many protests inspired by the larger counterculture but also shaped by perceptions of gender.

A growing chorus of American voices expressed concerns that not all scientific advances conformed to the DuPont chemical company's slogan "Better Living . . . through Chemistry."[67] As the feminist movement spread, critics began to recognize that rampant consumerism was rapidly depleting natural resources and poisoning the environment, with women uniquely responsible and at risk. According to activist Donna Warnock in *What Growthmania Does to Women and the Environment*, women were taking some of the basic tenets of the new feminism and applying them to issues of consumption and environment. Feminist values, according to

Warnock, were reflected in ecology and energy conservation: "The road to women's liberation lies not only in ousting the patriarchy, but also in rejecting its inequitable and environmentally and socially disastrous production system which is based on man's dominion over women and the earth and the illusion of infinite resources."[68] Warnock quoted Elizabeth Dodson-Gray, who warned that more chemicals were found in the average modern home than in the chemical labs of the past and that "many homemakers know little about these chemicals and even less about their toxic and polluting effects." In addition, Dodson-Gray cautioned, the relentless pressure to acquire more goods carried with it hidden emotional costs to women: "I am told by this consumer culture that my identity is defined by what I have—house, car, fur coat, kitchen appliances, cosmetics and clothes. I am supposed to feel good about myself if I have lots of expensive things. Meanwhile the TV ads make me feel that I must as a woman be lovable at all costs (men achieve, but women must be loved), and I only will be if I use the latest in floor wax, cosmetics, and vaginal deodorants."[69]

Women for a Peaceful Christmas: Attacking War and Waste

Many middle-class women rejected the notion that their identity could be found only within what Betty Friedan called the "comfortable concentration camp" of their suburban homes. They sought empowerment through contact with nature. In the 1980s, travel agencies such as Seattle's Womantrek and Minneapolis's Woodswomen encouraged women to challenge themselves physically and nurture themselves emotionally by camping, hiking, and rafting with other women. Yet middle-class women did not have to tramp through the woods to play a role in forging a new environmental consciousness. In 1937 the Conservation Division of the Wisconsin Federation of Women's Clubs included in its goals for the year the passage of a law making the disposal of Christmas trees less wasteful, lamenting,

> A sight that almost makes me weep
> Is a Christmas tree on the rubbish heap.[70]

Nearly thirty-five years later, another group of Wisconsin women took up with a vengeance the attack on wasteful practices associated with Christmas. In the autumn of 1971 about a dozen Madison homemakers, including Nan Cheney and Sharon Stein, began a unique effort to remake American culture: Women for a Peaceful Christmas (WPC).[71] Its founders had previously worked in various political campaigns but found that altruistic letter-writing campaigns by women within the domestic sphere who were perceived as "just housewives" produced no results.

During the 1960s and 1970s, women frequently combined a call to save the earth with their demands to stop the war in Vietnam. In 1967 a group of mothers seeking to encourage all women, not just students, to take an active role in eliminating war founded the nonpartisan, nonprofit group Another Mother for Peace. Los Angeles artist Lorraine Schneider designed the group's logo: a sunflower and the slogan, "War is not healthy for children and other living things." The simple, childlike drawing and words became an iconic image, appearing on bumper stickers, posters, key chains, t-shirts, and medallions and was a fixture in peace demonstrations and marches.

The founders of WPC were inspired by the nationwide Women's Boycott for Peace held in June 1971, organized by women in Ann Arbor, Michigan.[72] Citing Gallup poll figures that 78 percent of American women wanted the United States to withdraw its forces from Vietnam by the end of the year, these women decided to "speak in a language all men can understand: refuse to support a wartime economy."[73] "Money talks," noted Judy Olson, one of the WPC founders. "This is our non-violent form of pressure." Added co-founder Sharon Winderl, "We do not want to support the economy which is killing our sons."[74]

Members of WPC wanted more than "just" peace. They sought a "re-ordering of national and personal priorities," beginning with a turning away from waste and conspicuous celebrations.[75] "What we're really aiming for," explained Nan Cheney, "is a change in attitudes. We're trying to raise people's consciousness about the wartime economy and what they can do to control their own consumption of resources."[76] According to Cheney, "If our economy is based on dishwashers that must be replaced every five years and automobiles that we can't find parking places for, then something is seriously the matter with our values."[77]

WPC lamented the fact that Christmas in particular had become for the middle class "a time of tremendous waste of resources, with mountains of wrapping and packaging materials thrown away and tremendous pressure to buy badly-made toys."[78] Commercialism had distorted the message of peace, love, and joy, persuading consumers that "peace is the product of exploitation, that love is measured by material possessions, and that joy abounds in compulsive consumption."[79] Their goal was not a holiday boycott, but rather they offered alternatives designed to make celebrations "more meaningful, less commercial, less wasteful, and more peaceful." Their suggestions ranged from gift ideas (including handcrafts, environmentally friendly canvas shopping bags, and organic cleaning products) to substitutions for energy-consuming outdoor Christmas lights. "If you don't want your Christmas celebrations to be controlled by the monoliths that corrupt governments and pollute

FIGURE 6.3 Lorraine Schneider's logo design for Another Mother for Peace. (Sunflower design and words by Lorraine Art Schneider, image and text®© 1968, 2003 by Another Mother for Peace, Inc. (AMP) www.anothermother.org. Use permitted by AMP)

environments," WPC urged women, the sex that did the vast bulk of holiday shopping, "take matters into your own hands. Don't buy the pre-packaged, disposable Christmas! Make your own."[80]

Under the slogan "No More Shopping Days 'til Peace," WPC organized ostensibly powerless homemakers into a quiet revolt against what it called "an economy which thrives on war and the destruction of our earth's resources." According to the press, its membership entertained "no illusions of making much of a dent in an economy that encourages over consumption," and yet in five months' time, the movement had spread to almost every state, with members ranging in age from teenagers to grandmothers. The Wisconsin chapter was inundated by more than 15,000 queries and requests for its informational "starter" packet, buttons, and bumper stickers. Their message spread rapidly, aided by press coverage varying from church bulletins to national publications, including *Women's Day, Christian Science Monitor,* and *Newsday,* as well as support from the National Association for the Advancement of Colored People (NAACP). WPC members in Madison held an annual Peace Fair promoting environmentally friendly ideas and gifts and celebrating women's ability to "however infinitesimally, slow down the breakneck speed of American consumerism" and preserve precious natural resources.[81]

WPC denounced traditionally commercial Christmas celebrations as "wasteful of the earth's energy and resources, and encourag[ing of] a thing centered, rather than a people-centered way of life."[82] When asked during the group's fourth year of operation if its goal was to undermine the "American Way of Life," founder Jan Cheney

FIGURE 6.4 A button carrying the message of Women for a Peaceful Christmas. (Wisconsin Historical Society 66617)

responded, "I hope so. We have to rethink the way we live. I can't believe we're so dependent on [frivolous, manufactured] 'things' that we can't learn to make useful things, instead of what Madison Avenue tells us we want."[83] Simplified, environmentally friendly alternatives allowed individuals "to decide what's really important in life and what just gets in the way."[84] One headline summed up the group's emphasis on the long-term goal of controlling waste: "Christmas Can Be Saved for Future Generations."[85] As the war in Vietnam came to a close, the focus of Women for a Peaceful Christmas shifted increasingly to environmental issues. Mindful of worldwide food and energy shortages, pollution, and economic uncertainty, its members campaigned especially against waste.[86]

The Cambridge Plant and Garden Club Goes Global

Middle-class women across the nation during the 1970s were refusing to allow lingering gender-based stereotypes of their political passivity to prevent them from achieving environmental reforms, especially at the local level. Women dominated the leadership and ranks of a variety of community efforts designed to protect the environment. The threat of nuclear destruction was so pervasive that even members of the traditionally apolitical Cambridge Plant and Garden Club felt compelled to action. Following World War II, club members at first continued to be occupied primarily with plant life in their own homes and gardens and with planting trees and flowers in their local community. Even as larger issues of resource conservation increasingly crept into the club's discussion and lecture topics, as late as 1972 it was noted in the official minutes that while the club's advancements in "knowledge of horticulture and civil responsibilities" were of "greatest importance," "we do enjoy all those exquisite and delicious teas."[87] The club's increasing concern with environmental damage, however, was evidenced by its emphasis on rejecting highly toxic household cleaners and pesticides in favor of returning to old-fashioned herbal preparations and other environmentally benign alternatives. Instead of using commercial toxic sprays to kill ants, for example, members were urged to use cinnamon, cream of tartar, red chili pepper, sage, or "that old gift perfume you don't like," while an oven could be cleaned with salt, baking soda, water, and "elbow grease."[88]

Global issues and concerns ultimately disrupted the focus on local gardening that had dominated the club for nearly a century. Noted one member, "I came to look upon dear old Earth (and Cambridge in particular) as besieged, crying for rescue. Enjoying the beauty of plants and gardens was no longer enough for me."[89] According to the club's executive committee in 1983, it was time for their "distinguished and honored institution . . . to take a public stance against the greatest environmental threat the world has ever faced—the possibility of nuclear war." From 1981 through 1986 the club produced *Conservation News and Action*, sending this monthly newsletter to garden clubs throughout New England and the Midwest. Openly political, the newsletter described problems including acid rain, pollution, population growth, the growing energy crisis, and most of all the nuclear threat—not only the possibility of atomic war but also the problem of nuclear waste already generated. *Conservation News and Action* also provided detailed advice on the content of letters that members were encouraged to send to Congress to influence governmental actions and policies.

Not all club members were pleased by this new political, global focus and activist stance. Republican members bristled at the newsletter's negative depiction of the Reagan administration, especially its build-up of nuclear weapons, and urged that the club refrain from political discussions or activities that were not directly related to its primary focus on plants and gardens. More activist members insisted that the

"In regard to our 'favorite-bird' poll, some of you will be thrilled to learn that the chickadee is leading by seventy-five votes."

FIGURE 6.5 Helen Hokinson drew hundreds of cartoons for the *New Yorker* and other magazines depicting middle- to upper-class women, especially club members, as earnest yet silly and self-aggrandizing as they squandered their considerable energies and resources on superficialities frequently related to birds and flowers. (*New Yorker*, Media ID #111014)

nuclear focus and political activities were wholly suitable for the club: "The environmental consequences of nuclear weapons, nuclear power, and nuclear waste are all bad for the garden."[90]

In 1983 seven club members traveled to the World after Nuclear War conference in Washington, D.C., where they heard leading activists and scientists, including Paul Erlich, Carl Sagan, and George Woodwell, discuss from a global perspective the environmental consequences of nuclear weapons. Upon their return, believing that "individuals can make a difference," members of the group crafted a talk, which they presented to various gardening organizations. The talk's title, "Non-Trivial Pursuits: Working to Prevent Nuclear War," made it clear that "exquisite and delicious teas" were no longer preoccupying the membership. Following the Chernobyl disaster in 1986, the talk was revised and given a title whose play on words indicated the club's new emphasis on global, political concerns rather than personal and local ones: "Waste Watchers International, Working Out Ways to Protect the Environment."[91] This new club image, noted one member, contrasted with the frivolous Garden Club Woman, for decades the subject of Helen Hokinson's cartoons in the *New Yorker*, who spent her time "poking and meddling, 'beautifying,' tricking and decking things out . . . when we have emerged to do so *many, important* things."[92]

. . .

The patriarchal resurgence that marked the immediate postwar period relegated white middle-class women to support-staff status in their environmentally unfriendly suburban homes. A sense of celebratory materialism culminated in "growthmania" and a flood of enthusiastically embraced chemical products. In the early 1960s, however, *Silent Spring* and *The Feminine Mystique* revealed the costs of the degradation of both the environment and women, leading to a variety of movements in which issues of women's rights and empowerment overlapped with those of environmental protection. Across the nation, countless homemakers who identified primarily as wives and mothers dedicated to the service of others joined students in seeking alternatives to common and untenable practices affecting the environment. Issues and activities emerging from perceptions of gender, race, and class combined to transform American ideas about women's proper roles—and influenced attitudes and actions that would ultimately help shape the future of the planet.

"A city built from the ground up by feminist values"

7 Women's Alternative Environments
FOSTERING GENDER IDENTITY BY STRIVING
TO REMAKE THE WORLD

"The Planet Should Be Like This"

Since 1976, the Michigan Womyn's Music Festival (MWMF) has been held every August, welcoming women of all nationalities, ages, races, sexualities, and physical abilities.[1] In 1982 it moved to a private rural setting of more than 650 acres, where it consistently attracts thousands of women each year and has been "celebrated for decades as a must-see destination for activists in lesbian cultural production."[2] Diversity is strongly valued. In addition to the Womyn of Color Tent, features include networking spaces for teens, Over-40s, Jewish Womyn, the Deaf, and womyn from other countries, plus dances, musical performances, a film festival, a crafts bazaar, and a wide array of workshops. Tickets are priced on a sliding scale to encourage attendance by women of all economic levels. The festival's emphasis is on community. There are no onlookers, only participants: "Each womon [staying the entire week] does two shifts in a community area during her stay (one for the week-enders), adding her own splash of color to the fabric of the Festival. Every womon's personal involvement forms the foundation of the Festival spirit, built on the energy, ethic, good fun and challenge of living and working together."[3]

Working together to create a truly alternative environment involves respecting the earth and leaving the lightest possible footprint. Central to the MWMF experience are the "forest, meadow, and sky [that] stretch out in all directions."[4] Participants are required to be "land stewards" and honor nature as a partner rather than a backdrop. This involves living simply for the duration of the festival, thereby consuming fewer resources and creating as little waste as possible. This creation of an "ecology consciousness," one participant reflected in 1983, offers a "real hands-on experience in 'what are we doing here? How are we living here?'" and a lesson in "how fragile the ecology is . . . thru more than a textbook." She spoke for many participants when she emphasized her "vested interest in more and more women feeling connected to the land."[5]

For many participants an environmental consciousness is further fostered by woman-centered spiritual practices emphasizing women's "oneness" with the earth, the moon, and natural cycles. From its earliest beginnings, part of the MWMF's radical mission, according to feminist scholar and festival regular Bonnie J. Morris, "was its safe space for woman-identified and woman-centered spiritual practice. Events and Goddess rituals . . . allowed women who had been hurt by their exclusion from (male-only) religious office or women recovering from male-dominated fundamentalism to find themselves in feminine images of the divine."[6] Groups at the festival, including the Salsa Soul Sisters, made up of African-American lesbians, strove to position their tents correctly so that they would align properly with the rising moon for their nighttime rituals.[7]

Despite the fleeting nature of the festival itself, the sense of community, sisterhood, and environmental awareness it instills is permanent. "We go," asserts Morris, "because festivals offer the possibility of what our lives *could* be like year-round if we lived each day in a matriarchy actively striving to eliminate racism and homophobia . . . [while] living tribally."[8] One 1990 participant said simply, "The planet should be like this."[9]

Many of the leading feminist organizations of the 1960s and 1970s, including the National Organization for Women, were not particularly focused on environmental concerns. And yet directly and indirectly, the feminist movements of those decades contributed significantly to the environmental justice, ecofeminist, and alternative community movements of subsequent decades. Certain feminist-inspired women's groups rejected sexism and incorporated environmental consciousness and greater sustainability into their efforts to model new and more egalitarian ways of living. Their actions demonstrate the role that environments can play in fostering gender identity. Like straight feminists, many lesbians were more focused on gaining rights for themselves than on protecting the environment. However, certain lesbian groups reveal how nonhuman nature as well as built environments played extremely powerful roles in fostering lesbian consciousness and community, particularly as some lesbians traded the comforts of gay bars and vacation settings to participate in far more environmentally focused short-term and long-term experiments in alternative living. They also reveal the role that sexual identity played in the creation of some of the new forms of environmental thinking and living that served as early models of how egalitarianism could extend to include nonhuman nature. Although still seen as extremists even in the increasingly environmentally conscious 1960s, 1970s, and 1980s, some of these groups' innovative practices and experiments in sustainability— recycling, use of alternative energy sources, emphasis on organic production, and the healing powers of nature—would become mainstream in the early twenty-first century.

Early Lesbian Environments

Place has played an important role in the creation of lesbian identity and community. Although modern urban environments, with their softball fields and lesbian bars and bookstores, are conventionally perceived as conducive to lesbian life, pockets of safe spaces for women who loved women existed earlier, even in conservative areas, such as the rural South. Prohibited from frequenting white establishments by virtue of their race and economic status, rural African Americans danced, drank, and socialized to blues music in ramshackle jook joints, also called barrelhouses, frequently located in wooded, remote areas away from disapproving eyes and ears. These informal nightclubs offered great sexual freedom. Many female African-American blues performers were openly lesbian, and their songs celebrated sexual love between women.[10]

Most lesbians, however, associated sexual freedom with urban rather than rural life. To Mabel Hampton, a young African-American lesbian who moved from Winston-Salem, North Carolina, to New York City's Harlem in 1920, the idea that nonurban, outdoor settings might prove valuable in creating and fostering a positive lesbian identity would have been an anathema. For Hampton, there could be no more nurturing and empowering environment for working-class lesbians of color than the open atmosphere of Harlem. "I never went in with straight people," she recalled decades later. "I do more bother [have more contact] with straight people now than I ever did in my life." She summed up her memories of the clubs and nightlife available to openly lesbian women with a wistful, "[you had] a beautiful time up there—oh, girl, you had some time up there."[11]

In Hampton's heyday, it was indeed urban environments, with their potent combination of proximity and privacy, that promised the greatest liberation for most homosexuals. The very notion of homosexuality as a lifestyle grew out of the urban centers of newly industrialized nations. Many cities included a more "bohemian" area in which people who were considered to be outside mainstream society found a home. There they enjoyed the chance to experience nightlife in clubs featuring lesbian entertainers, some of whom got their start in the jook joints of the rural South.

Private parties were far more common than nights on the town, however, because they were cheaper and provided both safety and privacy. During non-work hours, "I didn't have to go to bars," Hampton recalled, "because I would go to the women's houses."[12] During periods when she was not working at the Lafayette Theater, Hampton and her friends "used to go to parties every other night. . . . The girls all had the parties."[13] As Hampton recalled, lesbians "lived together and worked together. When someone got sick the friend [lover] would come and help them—bring food, bring money and help them out. . . . I never felt lonely."[14]

Early twentieth-century urban environments, with their occasional lesbian bar and clusters of same-sex living spaces, including the YWCA and other women-only boarding and rooming houses, offered the greatest potential for freedom and opportunity for lesbians. As Hampton noted, "In a small town you wouldn't have a chance to get around and meet [gay] people. Now in New York, you met them all over the place, from the theater to the hospital to anything," concluding, "Yes, New York is a good place to be a lesbian."[15] Urban lesbians in bohemian environments created informal communities, providing places to connect with each other as well as generate emotional and financial support and solidarity.[16]

An Early Alternative Environment

Urban life offered only fleeting and furtive opportunities for white middle- and upper-class lesbians to find each other and to carry out relationships while remaining at least nominally in the mainstream. Some of these women began to seek out environments more conducive to living fully as they desired, such as Cherry Grove on Fire Island, a narrow sand spit about thirty miles long between the Atlantic Ocean and the southern coast of Long Island.[17] It was a relatively easy commute from the New York metropolitan area and served as the perfect antidote to the many sections of the city that seemed huge, dirty, crowded, overwhelming, and, in general, overwhelmingly homophobic. Few cars were allowed on the island, and it was "so wooded, and so beautiful, [with] a canopy of trees wherever you'd walk."[18] One-time resident Natalia Murray recalled coming there in 1936: "[In] this place, so close to New York, you can breathe the fresh air; when we found it it seemed so secret, [so] wonderful." Its lack of electricity and running water dictated a simpler lifestyle. Island life allowed people to "breathe freer."[19] In addition to its refreshing physical characteristics, Cherry Grove was already home (at least in the summer and on weekends) to the same kind of arts and theater crowd that had helped to cement Harlem's bohemian reputation. The energetic white women who flocked to Cherry Grove "enjoyed independent incomes, professional occupations, or both. . . . Most were connected to or identified with the theater world," making them, in the words of Murray, "Interesting, talented people . . . who had so much fun!"[20]

Being near the beach contributed to a more relaxed dress code. For women, time at Cherry Grove meant discarding the constraints of mainstream society, sometimes literally: "We could throw off our girdles, dresses, heels," elements of the uniform virtually required of middle-class women. Lesbians gloried in being able to "wear slacks and to be with and talk to others like [themselves]," providing "a simply extraordinary feeling of freedom and elation."[21] Free from straight men's unwanted sexual attention, urban lesbians could temporarily step out of the closet into a glorious, natural setting.[22] In Cherry Grove, women could walk alone, even at night, without fear of violence, harassment, or arrest.

The results of all this freedom were more often personal than political. Unlike the lesbians who would seek alternative environments in the 1970s and 1980s, the "fun gay ladies" of Cherry Grove were not consciously political or inspired to activism. They sought only a safe place where they could openly express their sexual identities and be their authentic selves.[23] Despite their appreciation of the natural beauty around them, they were not especially concerned with environmental protection. The negative environmental impact of the lack of indoor plumbing, for example, was never mentioned. Their goal was not to improve, let alone remake, the greater society, but simply enjoy a respite from its incessant expectations that all women be heterosexual and conform to the demands of patriarchy. They didn't come to transform Cherry Grove physically, to "civilize" the land, to tame or develop it. They sought privacy and were content to live in relatively simple dwellings that blended with the natural setting rather than dominating it. Compared to the elaborate housing developments that were to come, their environmental impact was relatively small.

The early lesbian residents of Cherry Grove frequently spoke of it as another world, including Peter Worth, a lesbian who gloried in being, for once, in the majority: "This was my world and the other world was not real."[24] Although the lesbians at Cherry Grove were able to shrug off the homophobia of that other world, its racism and classism remained; they did not reach out to their working-class sisters or to lesbians of color. As one resident recalled, "In those days, the Grove was like a very private gay country club."[25]

Beginning in the 1950s, the tenor of Cherry Grove changed. Early in the decade a younger generation, still middle-class but more committed to butch and femme identities, took up residence. After electricity and running water were installed in

1961, construction of new homes doubled, then tripled. An overwhelming percentage of the buyers of the newly constructed homes were gay men, who, by virtue of their sex, had more purchasing power than most women. "The [lesbian] old-timers looked on aghast as the 'unspoiled' natural setting of their 'gay country club' was 'raped,'" as Natalia Murray put it.[26] The passage of the National Seashore Act in 1964 froze the limits of Cherry Grove, prohibiting further sprawl, but by then its transformation into what resident and film historian Vito Russo called "a Coney Island of [male] sex" was already complete.[27]

The lesbian "country club" became a gay man's "sexual social club."[28] As Cherry Grove became a playground almost exclusively for gay men during the 1960s, virtually all of the original "gay ladies" of Cherry Grove moved on, many to become part of the "Bermuda shorts triangle," so named to indicate the imaginary line between their apartments in Manhattan and their summer cottages in the Hamptons or near Westport, Connecticut. Significant numbers of lesbians of all classes began returning to the Grove only in the 1980s when women gained increasing purchasing power.[29]

Despite the near total absence of communal activism, the history of Cherry Grove between the 1930s and the 1960s offers a glimpse into a pioneering experience that highlights the way that living a simple, more sustainable lifestyle in a natural setting can contribute to an exhilarating rejection of society's condemnation of lesbianism. Cherry Grove was valued as an apolitical alternative environment, offering its lesbian residents a sense of belonging in a more natural world and a respite from the artificial urban jungle of patriarchy, misogyny, and homophobia. It became the work of a later generation of lesbians to tackle the myriad problems of that other world head on, and to create alternative environments—not simply as respites from the real world that offered unique opportunities to build lesbian communities—but as viable models of just how that other world might be recreated socially, politically, and environmentally.

The New Environmental Consciousness

Following the first Earth Day in 1970, concerns about the environment were increasingly incorporated into politics and popular culture. That same year, in response to the growing public demand for clean water, air, and land, the federal government established the Environmental Protection Agency to assess and prevent damage to the American environment.[30] Television commercials decried littering and waste, reminding members of the public of the importance of individual actions to the fate of the planet. The nation's newfound environmental consciousness led many Americans to modify, but not transform, their existing lifestyles. Membership in established organizations like the Audubon Society grew dramatically. Within the counterculture, new environmental groups and organizations proliferated and environmental objectives were added to the agendas of already-existing groups and organizations promoting peace, civil rights, and economic justice.[31] Some of these groups, often strongly influenced by gender and/or gender-related concerns or identities, modeled intriguing possibilities of new ways of living on and with the earth.

Ecofeminism

Some women and a few men became ecofeminists, uniting their interests in fostering feminism with protecting the earth. Grounded in the movements launched to no small degree by the writings of Rachel Carson and Betty Friedan, ecofeminism unites environmentalism and feminism and holds that there is a deep bond between the oppression of women and the degradation of nature.[32] Some ecofeminists argue that, because of that connection, women are the best qualified to understand and therefore to right environmental wrongs. In most parts of the world, women are the ones who are "closest to the earth," the ones who gather the food and prepare it, who haul the water and search for the fuel with which to heat it. Everywhere they are the ones who bear the children and, in areas made toxic by pollution, suffer increased rates of miscarriage, stillbirth, and damaged children. Brazilian ecofeminist Gizelda Castro echoed the sentiments expressed by Lydia Adams-Williams nearly a century earlier: by dedicating themselves to the pursuit of immediate profit, "Men have separated themselves from the ecosystem." Castro concluded that it therefore falls to women to fight for environmental justice and to save the earth.[33] Within the United States, a variety of mutually exclusive forms of ecofeminism rival for dominance. One branch emphasizes the power of goddess mythology. Practitioners of Goddess Spirituality seek to reclaim ancient traditions in which, they assert, a Mother Goddess (rather than a Holy Father) was revered as the great giver of life. Some argue that, despite the efforts of the patriarchal Judeo-Christian tradition to eradicate this belief, all women, especially mothers, are the natural guardians of "Mother Earth."

Their horrified feminist rivals counter that these kinds of claims perpetuate old gendered stereotypes and are a violation of the egalitarianism of true feminism. In the words of Bella Abzug, congresswoman and cofounder in 1990 of the Women's Environment and Development Organization, "It's OK to show your emotion and come in as a mother . . . to say that this is going to hurt my children, but it's not good enough."[34] This latter school of ecofeminists insists that nature should not be anthropomorphized into a mother to be protected but instead respected as a nonhuman, nongendered partner in the web of life. Its adherents argue that women and nature are mutually associated and devalued in Western culture and that it is strictly because of this tradition of oppression that women are uniquely qualified to understand and empathize with the earth's plight and to distribute more fairly its resources. These ecofeminists see anthropocentrism as severely damaging to the earth and analyze it as one of many strands in a web of unjust "isms," including ageism, sexism, and racism, that must be destroyed in order to achieve a truly just world. In the words of 1980s activist Donna Warnock, "The eco-system, the production system, the political/economic apparatus and the moral and psychological health of a people are all interconnected. Exploitation in any of these areas affects the whole package. Our only hope for survival lies in taking charge: building self-reliance, developing alternative political, economic, service and social structures, in which people can care for themselves . . . to promote nurturance of the earth and its peoples, rather than exploitation."[35] In the United States this challenge was taken up by a variety of women's groups using their gendered identity to promote peace.

On November 17, 1980, in what is frequently cited as the first public ecofeminist protest, two thousand women marched past Arlington cemetery to stage the highly publicized Women's Pentagon Action (WPA). Like the balloon release organized by the League Against Nuclear Dangers, the WPA used public theater to gain publicity. In a display that featured four huge paper-mache puppets, they encircled the Pentagon, put gravestones in the lawn, wove yarn across the entrances to symbolically reweave the web of life, and chanted, yelled, and banged on cans. They gathered "in fear for the life of this planet, our Earth, and the life of our children who are our human future," because, in the words of their unity statement, "We are in the hands of men whose power and wealth have separated them from the reality of daily life and from the imagination." Their demands included the end of all nuclear production, citing not only its potential to create death, but its contributions to the worldwide "vicious system of racist oppression and war." Protestors called for renewable energy for all people as well as healthy food, clean air and water, useful work, decent housing, childcare, reproductive freedom, and education "for children which tells the true history of our women's lives, which describes the earth as a home to be cherished, to be fed as well as harvested." They described all life as interconnected and stated their intention to live in "a healthy loving sensible way" among all peoples of the world.[36]

A New Lesbian Consciousness

Significantly, the WPA's unity statement also announced, "We want to be free to love whomever we chose . . . [and] we will not allow the oppression of lesbians."[37] Lesbians played a powerful role, not just in ecofeminism but also in contributing to

FIGURE 7.1 The Women's Pentagon Action. The names on the placards are of women martyred for their activism. (© Diana Mara Henry/www.dianamarahenry.com)

or leading some of the most innovative and intriguing possibilities for living in egalitarian and more environmentally sustainable ways. This role did not develop overnight. Even as feminism began to take hold in the late 1960s, women who identified as lesbian remained, for the most part, discreet about their sexuality. To do otherwise was to invite any number of repercussions including arrest, violence, loss of legal rights to one's children, dishonorable discharge from the military, involuntary hospitalization (which could involve electroshock treatments and other invasive therapies), and discrimination in hiring and housing. Lesbians nevertheless remained at the forefront of the burgeoning women's rights movement, despite the actions of Betty Friedan, who in 1969 characterized lesbians as divisive and sought to distance them from the National Organization for Women. Undeterred, lesbians continued to promote the rights of all women.[38]

Back to the Land

The gay rights movement contributed to a new lesbian presence in the Back to the Land movement. Aldo Leopold's *Sand County Almanac* (1949) and Helen Nearing and Scott Nearing's *Living the Good Life* (1954) helped lay the groundwork for the idea that there were alternative ways of living that were healthier for individuals and for the planet. During the 1960s, a trickle of people, mostly white and middle class, some of whom identified as ecofeminists, began moving to rural communities across the nation. Their numbers increased to more than a million in the following decade.[39] Experiments in organic farming and simple rural living were carried out by single families and by larger groups, often living and working communally. Shared values included independence and self-sufficiency, simplicity, efficiency, peace, and sustainability.

The proliferation of environmental problems and the ongoing and ecologically catastrophic war in Vietnam significantly contributed to a vision of an American nation in such deep trouble that only drastic measures, like separating entirely from mainstream values and practices, could reverse its course. Fueled by a desire to establish a meaningful connection with the earth, many in the modern Back to the Land movement believed themselves to be the nation's "hope for survival" by their transcendence of the violence, materialism, and environmental abuse afflicting mainstream society.[40] Although most Americans were not willing to commit such a radical lifestyle change, the movement's ideas had wide appeal. Alicia Bay Laurel's *Living on the Earth*, an illustrated survival guide, was a *New York Times* best seller in 1971.[41]

Many female "homesteaders" (as members of the Back to the Land movement often called themselves) took pride in protecting the environment by implementing sustainable practices such as growing and preparing their own food and clothing. They gleefully cast off boring, unchallenging jobs as well as make-up, girdles, bras, and other uncomfortable clothing, stopped shaving their body hair, and rejected practices they perceived to be stifling the freedom of most women in the mainstream. They gloried in their identities as "ultranatural earth mothers" who embraced the primitivism of a simplified lifestyle, and they argued that women's "nurturing, passive, receptive, and intuitive" energy was superior to the masculine attributes of "competition, aggression, acquisitiveness, and overreliance on reason."[42] Such women argued that the equality with men sought by feminists was actually a step down. The January 1976 issue of the *Green Revolution*, a publication for back-to-the-landers, asked, "Isn't woman's best chance of changing the world dependent on her having the freedom to exercise her superior, nurturing qualities? And could there be any place better for those qualities to blossom than in a truly functioning [communal] home?"[43] This vision of women's superiority emphasized the importance of female bonding and celebrated women's sisterhood and spiritual rebirth based on the earth's feminine energies.[44]

However, homesteading women were routinely disappointed to find that, even as women too often provided a disproportionate share of the labor, especially domestic labor, patriarchal beliefs and practices frequently remained intact. Some women homesteaders played with and manipulated gender stereotypes to their advantage, but most found them to be a barrier to true fulfillment.[45] Many women in the Back to the Land movement eventually found themselves worn down by the daily grind of activities crucial to the survival of themselves and their families, yet too often devalued by men (who placed a higher premium on their own contributions, such as building structures and working the land). Most homesteaders were disappointed that, despite their best efforts, they were unable to become totally self-sustaining. While their efforts created a lifestyle that was deeply satisfying in many ways, without the aid of modern technology (running water, refrigerators, washing machines, flush toilets, electric or gas stoves), it was also deeply exhausting. Middle-class people with little wilderness experience trying to survive exclusively off the land frequently found themselves living in poverty, sometimes even suffering from malnutrition and otherwise compromising the health of themselves and their children.[46] They found that "natural is not always better," especially when it came to treating serious injuries and illnesses. In the end, many returned to mainstream society, driven by frustration, exhaustion, discomfort, and lack of intellectual stimulation and drawn by modern health care and social and educational opportunities for their children.[47] A return to the mainstream rarely meant a rededication to consumerist values, however. Instead, most former homesteaders sought to incorporate into their new lives much of the environment-saving philosophies and practices of their old ones. Some returned to school in order to prepare for "green" careers that investigated or promoted sustainability and other forms of environmental awareness and protection.[48]

Among the liberating benefits of the Back to the Land movement for women were positive attitudes about menstruation, sex, natural childbirth, breastfeeding, and less restrictive childrearing practices. There was also a greater acceptance of sexual experimentation, including same-sex relations. Some women who initially came to the movement with husbands or boyfriends left them to form attachments with other women. Women who found their identities as lesbians on communes and wanted to live on the land in ways unencumbered by patriarchal divisions, beliefs, and practices were soon joined by urban women seeking to build a lesbian utopia.[49]

Some radical lesbian-feminists, convinced that the root causes of America's problems were male greed, egocentrism, and violence, believed that only a culture based on superior female values and women's love for each other could save the nation. Others embraced separatism for different reasons. In the early morning hours of June 28, 1969, patrons of the Stonewall Inn, a gay bar in New York City's Greenwich Village, resisted a routine police raid (gay bar patrons traditionally cooperated in hopes of a reduced charge). The three nights of rioting that ensued marked a turning point for many homosexuals, proud and relieved to at last be fighting oppression rather than ashamedly submitting to it. The Stonewall Riots, commemorated each June in gay pride parades in cities around the world, gave birth to the modern gay and lesbian movement, which many lesbians hoped would signal the beginning of a partnership in liberation between themselves and gay men. Instead they found that too often, by virtue of their sex, they remained second-class citizens even within this oppressed minority.

Some lesbians insisted that "women-only" spaces were the only way to ensure that lesbians' needs came first. Living in the country was considered superior to living in cities created and dominated by men because in urban centers both lesbian sexuality and efforts to transform society were constantly oppressed and diverted. Even the urban lesbian bar, "the gay equivalent of the country club, church social, and community center, all in one," came in for criticism.[50] In the years immediately following Stonewall, bar patrons were still vulnerable to raids and harassment. Moreover, the bar culture's emphasis on drinking discouraged political activism and contributed to increased rates of alcoholism. The separation from cities and suburbia offered by country life was considered crucial in the creation of models that would allow lesbians, many of whom had been discouraged from climbing trees or performing other "tomboy" outdoor activities as children, to improve their physical health and reclaim their sexual and environmental rights.[51] Moreover, these women, although often derided as "unnatural" by the straight community and therefore only suited to urban life, confidently took a holistic approach to society's problems by making nature central.[52]

In southern Oregon in 1972, the flow of women joining the Back to the Land movement became a wave of women immigrants.[53] The lesbians who settled in rural Oregon between Eugene and California's northern border were a far cry from the "gay ladies" of Cherry Grove. They sought not a temporary retreat into a kind of fantasy world but rather the creation of a new, viable, and sustainable alternative to patriarchy and capitalism.[54] These rural separatists viewed the land as a place where lesbians could restore their physical and spiritual health, away from corruption, oppression, and pollution.[55]

One resident recalls, "So much of the back to the land movement was about coming out, and coming into our power and identities as Lesbians. We intuitively knew we had to get out of the patriarchal cities, and redefine ourselves and our lives. We actually tried to build a new culture ... not [just] back to the land but back to ourselves."[56] This new culture included "a desire to live lightly on Mother Earth and in sympathy with nature."[57] Women erected (or adapted from existing shacks and cabins) small housing units that were easy to build and manage. These tiny residences (frequently less than ten by twelve feet, smaller than Thoreau's cabin on Walden Pond) represented safety, economy, and autonomy. Due to the conscious rejection of traditional women's roles these structures did not dedicate space to entertaining or child rearing. Instead of celebrating unbridled production, this new culture valued salvaging, recycling, and handcrafted materials over those industrially produced and store-bought. Sophisticated technology, heavy machinery, and animal products were eschewed by these women in favor of solar power, hand tools, and vegetarian organic foods in their desire to protect the environment as part of a larger effort to combat the evils of patriarchy and heterosexism. As one informational pamphlet from the Oregon Women's Land Trust put it, "We want to be stewards of the land, treating her not as a commodity but as a full partner and guide in this exploration of who we are."[58]

The women's plans to create a utopia on earth were never fully realized.[59] Despite the communities' desire to create an inclusive and diverse lesbian society, few women of color came to Oregon, and the mountainous terrain proved a barrier to women with disabilities and to the elderly. Because of relatively poor soil and chronic water shortages, the struggle for subsistence often overshadowed efforts to build a new kind of culture.[60] When there was time and energy to pursue less worldly concerns, residents were frequently divided over what constituted acceptable spiritual practices.[61] Yet none of the lesbians in residence termed their efforts a failure. They spoke of the empowerment they found in doing things for themselves and their recognition that nature is not an abstraction to be idealized, nor is it an "other" to be feared, tamed, subdued, or exploited, but rather "a friend, a sister, a lover (not to mention a workplace, a home, a refuge, and on some days a nuisance)."[62] These lesbians proved that there were ways of living that, however imperfect, did not hinge on profit or patriarchy and that instead allowed lesbians to live openly, freely, and consciously as partners with nature.[63]

The Pagoda

Just as the lesbians of Cherry Grove thrived due to their physical distance from mainstream society, rural lesbian communities of the Pacific Northwest enjoyed the privacy rendered by isolation. However, few lesbians were willing or able to live in such complete separation from the mainstream world. Even those who sought to create alternative communities were not necessarily drawn to rural life; others wished to pursue professions not valued or practicable in rural collective settings. In contrast to the Back to the Land lesbians of southern Oregon, a group of lesbians in Florida took a different approach to creating a lesbian environment. In 1977 Morgana MacVicar, a ritual performer and matriarchal belly dancer, combined resources with three other lesbians to buy the small cottage complex that would become the Pagoda, a womynspace in St. Augustine.

Like Cherry Grove, the Pagoda was less an effort to remake the world than an attempt to carve out a uniquely lesbian space within the existing one, but there were some important differences between the two. During its first four years the Pagoda served as a vacation destination for lesbians, then became increasingly residential. For both spiritual and tax reasons, the building at the community's center was officially declared a religious institution in 1979, strengthening the residents' communal identity and allowing the complex to exist legally as a woman-only space. Because lesbianism was not then an accepted lifestyle in St. Augustine and the Pagoda was not in a secluded location, the community did not publicly proclaim a lesbian identity. Residents were required to keep a "very low profile" in the outside community where, presumably, they worked. Unlike the lesbians in southern Oregon who, weather permitting, enjoyed music, nudity, and sexual activity out of doors, Pagoda residents were prohibited from appearing nude on the grounds and were urged to keep the volume of all voices and activities low, especially after dark. But beneath the imposed veneer of repression and orthodoxy there was a vibrant experiment occurring in lesbian community. In the center building called the Pagoda, the Temple of Love, residents were encouraged to participate in various activities and events celebrating women's culture and spirituality. "We maintain a very special energy here," noted Pagoda resident Elethia in 1982. "When I drive into the Pagoda, I feel like I've entered another space and time."[64] Emily Greene bought a cottage at the Pagoda in 1978. She found the Pagoda to be "life transforming" for herself and for others: "There was a real desire for egalitarianism, [and an] openness to diversity." The "rich, deep bonds we formed as we worked, played, and struggled to keep the Pagoda going" helped her to realize that she "wanted to always live in community, especially with Lesbians."[65]

According to Greene, the Pagoda "really did have the feel of an island of Lesbian paradise." Significantly, the Pagoda was a one-minute walk from the beach, offering immediate access to all the natural beauty and sense of timelessness, wonder, and freedom that the ocean evokes. The ocean had "invaded the soul" of Greene at an early age.[66] She recalls that "the beautiful setting of the Pagoda by the Sea was a big draw for so many women: as we traveled through this uncharted territory [of creating an egalitarian lesbian community], the ocean was such a comfort." Greene recounts walking on the beach "when life was almost overwhelming," then returning to "my little cottage with renewed strength to carry on."[67] Rituals and bonfires by the beach were common and sustained community identity.[68] The Pagoda's "sweet little" beach cottages "needed a lot of fixing up," but were relatively affordable in the community's early years.[69] Measuring six hundred square feet, they boasted "rustic-tacky charm." Lesbians at the Pagoda chose housing that offered convenience without extravagance, consciously living a relatively simple, environmentally friendly lifestyle. The dune grass that separated the structures from the beach, for example, remained undeveloped and without paved paths.

In this intimate environment, lesbians of the Pagoda were free to practice (or to reject) wide-ranging spiritualities, including goddess worship. Some residents who practiced Wicca participated in a variety of rituals including Moon Circles, which honor the various stages of the moon's appearance and draw from its power. Their beliefs, emphasizing the influence on humans caused by cycles in nonhuman nature, fostered environmental awareness. Practitioners sought to protect the earth they viewed as a holy mother.

The Pagoda's inhabitants were not as pointedly environmentally aware and active as their separatist sisters in Oregon, but certainly they were more conscious of the need for the conservation and preservation of resources than were the early residents of Cherry Grove. The way of life at the Pagoda represented an effort to live simply and in mutually beneficial harmony with nature. According to its promotional brochure, it offered paying guests "a place for womyn to renew our spirits and to recuperate from the onslaughts of patriarchy."[70] Residents and guests alike drew strength from the area's wild beauty, which they celebrated and protected. It helped them to create a supportive sisterhood of like-minded lesbians who could pursue their chosen spiritual practices and offer each other guidance and a sense of solidarity.

Six long-term community members made plans to expand the Pagoda in order to make room for the "Crone's Nest," a new kind of environment dedicated to the needs of aging lesbians.[71] It was not to be. The Pagoda's inviting physical environment ultimately contributed to its decline in the 1990s as a haven for lesbian living with an emphasis on simple, low impact ways of coexisting with nature. The dune grass providing easy access to the beach was replaced by condominiums, and a new bridge built directly in front of the property further diminished the peace and beauty that had played a large role in the community's success. Life at the Pagoda became more expensive and "things started to become difficult," according to a former resident. "Newer women [did not] want to continue struggling to work [out] our issues through our feelings, meetings, and consensus."[72] Beach erosion and the high price of local land contributed to the decision by three of the founding members to leave. Although heterosexuals bought some of the Pagoda properties, in the first decade of the twenty-first century the remaining lesbian residents retain aspects of lesbian community, albeit in reduced form.[73]

The Power of Women's Festivals

Most lesbians rejected the call to separatism and sought instead to find their rightful place in traditional society. But this often included a strong environmental element as well. Even before the riots at the Stonewall Inn, many closeted lesbians, emboldened by the women's rights movements that erupted in the 1960s, no longer felt compelled to live a lie and left their heterosexual marriages. Other women allowed themselves to examine honestly their sexuality for the first time. Free from the assumption that they must be heterosexual, they discovered, and celebrated, their same-sex desires. Many experienced this epiphany in a unique environment conducive to the empowerment of women: the women's music festival.

Women started performing in church basements and bookstores in the early 1960s but soon were gathering larger audiences in bigger venues. The impact on some women of hearing lesbian-themed music with an audience of other women was impossible to exaggerate. In the 1999 documentary *After Stonewall*, Torie Osborn highlights the large number of women who vividly remember attending their first women's music concert in the 1970s. Osborn certainly had not forgotten hers: "I can remember piling six . . . women . . . into my little baby blue Volkswagen and driving down from Burlington, Vermont, where there was no gay subculture, to see my first [women's music pioneer] Cris Williamson concert. One [of the six women] quit her job as a nurse so that she could form Coven Carpentry, so she could do lesbian carpentry. One left her husband—we're talkin' this concert literally changed people's lives. The empowerment had an ongoing impact. It was an extraordinary force."[74]

Women's concerts grew into festivals, described by Bonnie Morris as "a vibrant subculture" welcoming "the female outsider in search of an alternative community." Women's festivals rejected "the material objectification of women in violent U.S. media" and celebrated "the female sphere as a source of empowerment apart from men's gatherings."[75] Women were only nominally included in predominantly gay male groups like the Radical Faeries, whose Faerie Gatherings, beginning in the 1970s, offered "an extended retreat, usually in the woods, separated from the outside world" during which participants could explore spiritual dimensions of their sexuality. Women shared, however, the Faeries' recognition of the powerful role of isolated natural settings in fostering feelings of security, freedom, and self-acceptance.[76] "The only place we really feel safe," reflected one festival attendee in 1983, "is on the land, not in the city run by men. A lot of times we don't realize it until we leave and then we get slapped in the face by the contrast."[77] Women camped out and had the opportunity to buy and sell arts and crafts, carry out a variety of spiritual rituals and practices, go naked, and make social and sexual contacts. Perhaps the best known of women's annual music events that began during the 1970s and 1980s are Pennsylvania's Campfest and the Michigan Womyn's Music Festival.

Whether or not MWMF attendees participate in nature-centered spiritual rituals, all are required to clean up after themselves and respect their surroundings. The land is valued for its own sake, not merely for the special qualities it brings to the various events. Nonhuman rather than human nature takes precedence and is sustained. Instead of creating a permanent infrastructure on the land to facilitate the elaborate set-up procedures necessary to put on such a large event, after each festival much effort is expended to return the land as completely as possible to its natural ecological state. All nonorganic materials are removed. The electrical boxes that power the festival are buried so that no visible trace of human activity remains. "We reduce it all back to that meadow and ferns," notes organizer Sandy Ramsey. "If there is a very high impact deterioration happening somewhere . . . maybe we would do some mulching, seeding, landscaping. . . . We're very aware that we have to watch these things and do what needs to be done to make sure that we can continue to reuse them."[78]

Morris calls the MWMF "a wonderland of cultural anthropology," offering "a record spanning two generations or more of musicians, dancers, technicians, craftswomen, comedic emcees, workshop speakers, healthcare workers, kitchen chefs-for-8,000, and land stewards [that] represents the opportunity to examine the absolute best in cooperative community and what might be called an ongoing city of women (akin to 'Brigadoon,' appearing magically at yearly intervals)." A MWMF regular celebrates important differences rather than the similarities between the festival and Brigadoon: "Unlike Brigadoon, where everything is clean, the weather is perfect, and everyone is rich, white, heterosexual, able bodied, politically homogenous (i.e., unaware), Michigan brings together a largely lesbian sample of *everyone*."[79]

Although attendance is down to about half of the 8,000–9,000 reached during peak years, and Morris laments the passing of some of the early leaders and guiding lights of the festival, more than thirty years after the MWMF's debut, it lives on. Even in the second decade of the twenty-first century, the festival continues to invite women "into the familiar comfort of a time and space where we celebrate all things female." "The magic of Michigan" is described as "a city built up from the ground up by feminist values," where "healthy food, clean air, green woods, art and music will recharge batteries you didn't know were fading." "Make it to Michigan one time," organizers promise, "and it will call to you each and every August."[80]

Exclusively Lesbian Workshops and Meetings

Although lesbians flocked to Michigan, the festival has stayed open to all women-born-women, that is, those who were born and raised as girls and who identify as women, excluding transsexual and transgender women—one of several policies generating heated debate within the Queer community.[81] Many lesbians seek exclusively lesbian gatherings in which to meet, network, find strength, and create community. Lesbian meetings and political workshops, like Sisterspace, held annually in the Pocono Mountains beginning in 1975, and those organized in Gainesville, Florida, in the 1980s by LEAP (Lesbians for Empowerment, Action, and Politics), were frequently held at remote, outdoor sites that ensured privacy and encouraged "a passionate love for the natural world." LEAP organizers wanted lesbians to learn more about their connection to the earth. Seeking to heal both the environment and themselves, they set out to create a community that "will give us energy and power in our work of transforming the effects of the white man's patriarchy on this achingly beautiful planet."[82]

Two hundred and fifty lesbians gathered on October 19–21, 1984, at "a beautiful wooded private campground on the Suwannee River near Gainesville, Florida," where LEAP created its first "self-sufficient community by sharing . . . dreams, feelings, knowledge, skills, hopes, fears, art, spirituality, food, chores, tears, support and love." The result, according to the organizers of the following year's event, "has been enlightening, empowering, and is something we will carry through the rest of our lives."[83] Key to LEAP's success in "shar[ing] our actions . . . shar[ing] our ways of living lesbian lives" was its emphasis on partnership with the land, which was "beautiful with shaded oak groves, huge pines, open sunlit clearings, patches of deer moss, [and] sprinklings of zillions of kinds of Florida plant life." "We are here among the long leaf pines to find out more about ourselves and each other [and] more about our connection to the earth," LEAP reminded attendees, and urged them to "please enjoy the beauty of this land" but also to take care not to disrupt its delicate ecosystems.

Further evidence of LEAP's emphasis on nature as partner is apparent in much of its rhetoric: "We are . . . sharing this space with coral snakes, prickly pear cactus, and scorpions." "We are here," urged LEAP, "to help each other explore and discover the wisest, healthiest way to use the power that springs from our individual truths—for changing the world." A community vegetarian kitchen was partially dependant upon attendees' donations and "organized around the grand lesbian traditions of anarchy and chaos."[84] Because of the privacy the woods provided, nudity was "highly encouraged on the interior of the land," also contributing to the sense of being in a unique and accepting space.[85]

LEAP emphasized the power inherent in this alternative environment: "Coming into an all-lesbian space provides us with a particular kind of safety that is basically unknown in the world where most of us live. All of a sudden we are able to be ourselves in a truer way. The protective walls we keep up as we move through the patriarchal culture often come tumbling down—and sometimes real fast and dramatically. . . . Letting these feelings out is an important step toward our personal freedom and happiness and it also provides us with more energy for doing our political work."[86] Like the more inclusive women's festivals, LEAP was intended not as a temporary respite, but as a catalyst for creating permanent change. Attendees were warned that as they moved "back into patriarchal culture," they would likely find themselves "bombarded by the sickening truth of what . . . patriarchy is." LEAP organizers urged that attendees use the strength they gained through the LEAP experience: "Take a clearer look at our lives and figure out ways we can change things to better express our power; organize groups for political action and consciousness-raising; constantly validate ourselves and each other and the true incredible power of our presence in this world."[87]

Lesbian festival regular Retts Scauzillo "attended because it was crucial to my survival as a lesbian feminist. I needed to be with like-minded dykes, living and working together, to create our culture and practice our lesbianism. I went there hungry for love and sex and it was a place I could be all of me. I could take my shirt off, wear ripped or revealing clothes, flirt, be sexy, laugh, and talk lesbian feminist politics. We would agree on some things and disagree on others, but by the end of the night we were holding each other and dancing under the stars. I could be outrageous and radical, truly what they call high on life." Such liberation could take place only if one felt truly free. For Scauzillo and thousands of lesbians like her, the lesbian festival offered "the safest place on the planet. It made the outside world tolerable. . . . I grew up at these festivals and learned lessons that are with me today. Plus it was FUN!"[88]

The Decline of Women's Festivals

Many of the women's and/or lesbian festivals that flourished in the 1970s and 1980s disappeared in the 1990s due to poor attendance, in large part due to the success of feminism and of the gay and lesbian liberation movements. "The woman-identification of earlier festivals simply does not call out as spectacularly to young women who have grown up with more rights, with Title IX, with greater possibilities of becoming rabbis, lawyers, or politicians," notes Bonnie Morris.[89]

In the twenty-first century, Retts Scauzillo notes the festivals' intergenerational emphasis born out of desire to bring in younger women, such as Sisterspace's ODYQ (Old Dyke, Young Queer) forum, but doubts that it will bring in substantial numbers of young lesbians: "I try to think as a young queer woman now, would I need or want to go to women's music festivals? There are [lesbian-only] cruises and vacations and Dinah Shore parties. There are places like San Francisco and P-town Northampton and Key West." Although Scauzillo understands the draw of lesbian retreats more in the tradition of Cherry Grove than the MWMF, in her view, the "F-word [feminism] is missing in most of these places, vacations, and parties" and complains that, as "queer" replaces "dyke," "women are invisible in the new 'gender/studies' world." She recalls her recent positive experience at the Sisterspace festival, whose objectives focus exclusively on making women both visible and empowered. Features include education regarding issues of concern to the lesbian community and the fostering of a positive self-image for lesbians. Scauzillo also emphasizes the importance of the communal aspect of traditional women-only festivals like Sisterspace and Campfest, a feature conspicuously absent from the retreat activities enjoyed by younger lesbians. She singles out for praise "the unpaid workers, most of them lesbians who created these festivals and keep them going" and the sense of community that kind of participation produces. Also missing from most resort experiences (which frequently promote consumption rather than conservation) is the kind of environmental consciousness overtly cultivated and honored by Back to the Land lesbians and festivals like Michigan and LEAP. Despite her doubts, Scauzillo hopes that "festivals will start popping up and young dykes will start [re]claiming these institutions as their own."[90]

Lesbian alternative environments, both permanent and temporary, are by no means obsolete, but no longer do they provide the only safe space in which lesbians can enjoy the freedom to be themselves, find solidarity, and build community. Much of the effort to remake the world by creating environmental alternatives is a casualty of the rise of multiple new opportunities for lesbian sisterhood, including virtual lesbian communities on the Internet.[91]

Back to the Land Redux: Alapine Village

In 2009, Retts Scauzillo's desire for a revival of lesbian institutions was shared by former members of the Pagoda. After leaving Florida in 1997, three Pagoda cofounders relocated to northeastern Alabama where, over time, they acquired nearly 400 acres of rural land. They established a legal corporation and began developing about eighty acres into a lesbian community they named Alapine Village. On this land they carried out some of their original goals in far more isolation from the outside world than was ever possible at the Pagoda.[92] Former Pagoda member Emily Greene came to Alapine because she wanted "to be in nature" and to have lesbian neighbors. She was happy "to be back in community with people who want to live simply so that others may simply live." Her dreams were "to help us age the way we want without leaving our community, to be good care takers of this beautiful land and save some of it for our fellow creatures, [and] to have as low an impact as I possibly can on my environment."[93]

Environmental protection and sustainability have been paramount at Alapine. In 2009 Greene sought to negotiate an exchange of some Alapine land for an adjoining sixty-acre forest that was home to deer, coyote, squirrels, rabbits, and birds, in order to permanently protect that "pristine forest" from development.[94] Like most Alapine residents, she practices many environmentally friendly ways of living: she collects rain as her water source and uses a tankless hot water heater and a wood-burning stove. Vegetables grown in Alapine's community garden are another indication of the group's dedication to sustainability—and to eating as nutritiously as possible, good health being at a premium to the many members without medical insurance.

Alapine residents value the many additional benefits of their deep connections with the earth. Barbara Stoll shared many of Greene's dreams, and she too found in Alapine the opportunity to turn those dreams into reality. Stoll "just knew from an early age" that she was "meant to live in the woods." Finding "the consumerism and materialism of suburbia" to be "more than I could bear," she refers to Alapine as "my paradise." For Stoll, who had read "everything I could get my hands on regarding homesteading, alternative energy, sustainability, etc.," it is the place where she can "get back to the basics of life, the rawness of carving out a life that wasn't consumed by things and manmade ideologies." Life at Alapine allows her to find answers to life's most important questions: "How little could I live on, how much could I produce myself, how might I take life down to its simplest elements so that nature could flow through me without hindrance?"[95]

Like the lesbians in rural southern Oregon, Stoll designed her small, one-room cabin "for the highest efficiency," allowing her to "live lightly on the land" because "the most important aspect of living on the land for me was having as small a footprint as I possibly could." Although she describes herself as a hermit, Stoll rejoices in her ability to live in a like-minded community and has "visited other intentional communities in other states to learn about this much needed and wonderful way of life." Living at a "much calmer, more serene pace . . . surrounded by other like minded women," according to Stoll, allows for "stretching of the mind and new ideas to be considered and possibly implemented." She refuses, however, to romanticize her "very simple and frugal life," noting the psychological as well as the physical struggles at Alapine, where "the land brings emotion to the surface . . . [and] the woods do not let you hide from yourself," and where communal living in "a group of strong women with strong opinions" can be "very challenging."[96]

For all its difficulties, life at Alapine has profound meaning for Stoll: "I live an authentic life [because] I touch nature and the sacredness of life everyday." She vows to "heal the planet and its nonhuman inhabitants with my every action" and hopes that, even after she is "long dead," other women will "carry the torch and continue what we are trying to accomplish here, to preserve the beauty of life through nature and gentle, light-footed actions."[97]

The community worries about who those women bearing torches into the future might be. While at the Michigan festival in 2005, Alapine resident Emily Greene "became really aware and concerned about the lack of younger womyn attracted to living in community on the land."[98] By 2009, her worry became more acute. Although home to only twenty residents, Alapine was nevertheless one of the largest of the remaining "about 100 below-the-radar lesbian communities in North America." "We are really going to have to think about how we carry this on," notes Greene, or "in twenty to twenty-five years, we [lesbians in alternative communities] could be extinct."[99] At age sixty-three, Greene recognizes that younger lesbians are not eager to withdraw from heterosexual society because "the younger generation has not had to go through what we went through." Many Alapine residents had been "deeply scarred" by the discrimination and persecution they suffered at the hands of an openly homophobic society. They felt in the 1960s and 1970s "a real sense of the need to strongly identify as a woman and have women's space . . . the need to be apart" and grounded firmly in nature in order to draw on their own strength and empowerment. They recognize that "young feminists today recoil at the idea of identity politics."[100]

FIGURE 7.2 Members of the Alapine Village community enjoy sledding in 2011. (Courtesy of the Alapine Village community)

Although the members of Alapine Village live quietly and avoid publicity, as one way of reaching out to younger lesbians in their efforts to remain a viable and vibrant community they were willing to be the subject of a feature story in the *New York Times* in 2009. The web version of the story included a multimedia presentation featuring Alapine residents and their natural setting.[101] To help achieve its shared goal of "expanding into an intergenerational community, especially welcoming younger women," Alapine created a website.[102] In addition to celebrating the many social features of community living, much is made in the inviting web pages of the land's rolling hills, hardwood and pine forests, flowing mountain river, and hiking paths, as well as the ready availability of outdoor activities (bicycling, canoeing, kayaking, camping, and gardening). The website also features the community's use of wood heat, solar energy, propane, composting toilets, recycling, self-sufficiency, and its work toward making "homes, gardens and community buildings sustainable, with the ability to survive off grid."[103]

Nature remains an important component in this lesbian alternative environment and, its residents hope, one of the keys to attracting like-minded lesbians and assuring its future. The positive response generated by the *New York Times* was overwhelming. It "warmed the heart" of Greene and convinced her that "we are not a 'dying breed,' and that our form of community is very vibrant and alive." With "plenty of land [and] hard-working women," this former nursing-home care provider has renewed confidence that the women of Alapine can "create a new environment for Lesbians as we age."[104]

. . .

Empowered by the feminist movement that began in the 1960s and a growing awareness of environmental abuse as one strand in a broad web of injustice, some women sought to create alternative models of living. Through their efforts to transcend sexism, homophobia, and violence, lesbian communities made important contributions to environmental history and environmental justice movements. Some groups' recognition of nature as partner, emphasis on sustainability, and de-emphasis on materialism make them particularly valuable models of alternative, ecofriendly ways of living. Women's alternative communities, past and present, shed light not only on important aspects of the history of women, especially lesbians, but also on the power of place in fostering identity.[105] They provide as well thought-provoking examples of new ways of thinking, living, and valuing both human and nonhuman nature.

"The women must lead"

8   The Modern Environmental Justice Movement

"The Mother of the Environmental Justice Movement"

In 1945, the Chicago Housing Authority opened Altgeld Gardens, a residential community designed to provide low-income housing for the families of African-American World War II veterans and war workers, maintaining Jim Crow segregation.[1] Thirty-four years later, the community was in serious decline. Forty-four-year-old Altgeld Gardens resident Hazel Johnson founded the People for Community Recovery (PCR) to advocate for repairs to the row houses. She was especially concerned about dwellings with leaky ceilings and a water supply "coming from wells so contaminated you could see the film of chemicals floating on top."[2] Johnson's concerns were not about property values but rather the quality of life in the community. Her focus on home improvements, field trips for children, and community parties shifted in the early 1980s, however, following the cancer deaths of four little girls "whose bodies were so tiny they could fit in shoe boxes." Like most of the community's residents, Johnson had become "almost immune" to the number of local cancer deaths (her own husband, John Johnson, died of lung cancer at the age of forty-one), but the loss of such young children to cancer galvanized Johnson, herself a mother of seven. Without education or economic or political power, she set out to "find out what was going on."[3]

Hazel Johnson noted not only the high number of cancer deaths among her neighbors, but also the large variety of unexplained serious health problems in her community of about eight thousand residents. She linked her neighbors' health problems to environmental causes and, by personally handing out some twelve thousand forms, encouraged an onslaught of complaints to the Illinois Environmental Protection Agency. People for Community Recovery quickly became the first public housing environmental justice organization in the country. As a result of Hazel Johnson's initial survey, PCR demanded a comprehensive health study. The organization learned that the housing development had been built on land previously used as a sewage farm. That land had also seen service as an illegal dumping ground for PCBs (polychlorinated biphenyls, virtually banned by the EPA in 1979) by its landlord, the Chicago Housing Authority. Johnson dubbed the 190-acre Altgeld Gardens complex a "toxic doughnut" because it was surrounded by at least 50 landfills and more than 250 leaking storage tanks.[4] Under Johnson's calm, dignified, and tireless leadership, her community organization called for investigations into the polluting industries that were poisoning their neighborhood and demanded a permanent ban on waste facilities in the area.

Independent health surveyors confirmed the residents' suspicions: Fully half of the pregnancies studied in Altgeld Gardens resulted in miscarriage, stillbirth, premature birth, or birth defect. More than a quarter of the population surveyed suffered from asthma. Residents suffered myriad health problems, but 68 percent of former residents surveyed reported that their health problems disappeared when they left the community.[5] Leaving permanently was not a viable solution for most residents, however. Many could not afford to move, and Altgeld Gardens was also their home. In the words of Cheryl Johnson, Hazel Johnson's daughter, "I love the social fabric of the community. This is my sense of security. We should have a right to a clean environment. It shouldn't be a choice. So we've decided to stay here and fight for our community."[6]

PCR recognized that Altgeld Gardens and the surrounding communities were suffering not from the convergence of random isolated events, but from a pattern they identified as environmental racism. The organization called for an end to "artificially induced environmental destruction and the continued genocidal effects on people of color."[7]

Led by Johnson, the PCR lobbied for and won safe sewer and water lines. Other victories followed, and Johnson's reputation grew. In October 1991, she was dubbed the "Mother of the Environmental Justice Movement" at the First National People of Color Environmental Leadership Summit.[8] Three years later she witnessed President Bill Clinton sign Executive Order 12898, mandating that every federal agency make achieving environmental justice part of its mission.

FIGURE 8.1 A mother and her two children in their Altgeld Gardens kitchen in the development's early years. (Library of Congress, Prints and Photographs, LC-USZ62-51415)

Beginning in the 1970s, environmental awareness and concerns about sustainability increasingly permeated the national consciousness. The modern environmental justice movement emphasizes the right to safe and healthy ecological, physical, social, political, and economic environments for all people. Since those environments affect everyone, it might appear at first glance that there is nothing to be gained from an examination of the role of gender in the environmental justice movement. Yet gender, race and class are regularly addressed in environmental justice studies as factors that frequently subject people to injustice, but have also served to unify and mobilize those same people in their struggles against that injustice.[9] Not only has this strand of environmentalism invigorated those suffering from industrial toxins, but it has also energized others by connecting environmental protection to visions of social justice. Locally based organizations have emerged, with women in positions of leadership of groups made up primarily of working-class and nonwhite people.[10] Issues of sexuality in particular, especially as they relate to reproduction, have played a leading role in making women vulnerable to a variety of environmental injustices, but have also served as the key to their self-empowerment.

Love Canal

Hazel Johnson's title as the mother of the environmental justice movement was reaffirmed at the 2002 Environmental Justice Summit II, a meeting in Washington, D.C., of several thousand leaders from around the world. Johnson's name, however, is not as widely known in environmental circles as that of a white community activist, Lois Gibbs. Like Johnson, Gibbs was a homemaker who turned to activism to protect the health and safety of children from the contamination of her community.[11] The story of how Gibbs revealed the effects of the toxins buried under Love Canal, a neighborhood in Niagara Falls, New York, and how her leadership united her community in a spirited and ultimately successful crusade for accountability from the government and big business, broke in international headlines in the late 1970s.[12]

Beginning in the 1920s, the city of Niagara Falls began dumping its municipal refuse into an abandoned dry canal. The dumping accelerated in the 1940s when the U.S. Army also used the site for its refuse, including wastes from atomic weapon development in the Manhattan Project. A company eventually known as Hooker Chemical (later subsumed by Occidental Petroleum) was granted permission to dump its waste into the canal in the 1940s as well. The company ultimately bought the canal and the immediately surrounding land. The canal became the repository for 21,000 tons of chemicals, much of it in fifty-five-gallon barrels. Hooker Chemical covered the canal with dirt in 1953 and sold the land to the Niagara Falls School Board for one dollar. The deed states plainly that the land contained the waste products of chemical production, and absolved the company of any legal obligations in subsequent lawsuits. When the school built on top of the canal opened in 1955, new homes were built nearby to house the young working-class families it served. In this new neighborhood, the majority of the breadwinners worked in local industry, predominantly chemical manufacturing.

Heavy rains saturated the area in the 1970s, and some of the buried barrels began to rust through, releasing chemical sludge into the ground and its water. In 1977, journalist Michael Brown of the *Niagara Gazette* began reporting on the odors and inexplicable increase in birth defects and health problems in the area.[13] Lois Gibbs, who read Brown's stories, began to connect her son's health problems with his attendance at the school built on contaminated soil. Talking with her neighbors in 1978, she heard stories of miscarriages, birth defects, and mysterious childhood illnesses, as well as much unexplained property damage such as sinking foundations and prematurely rotting structures. In response to the complaints of Gibbs and others, including men who decried the decline of their property values, New York's health commissioner, Robert Whelen, declared on August 2, 1978, that the Love Canal landfill was a serious health threat. He urged the temporary relocation of pregnant residents and children under the age of two.[14]

Gibbs was furious rather than reassured. She asked Whelen directly, "If the dump will hurt pregnant women and children under two, what, for God's sake, is it going to do to the rest of us?!" In response she received what she called "engineering jargon and political answers that made no sense."[15] Five days later, President Jimmy Carter declared a Federal Health Emergency at Love Canal and provided funds to permanently relocate the 239 families living in the inner ring of homes circling the landfill site. People living outside that inner ring, including Lois Gibbs, were told they were not at risk.

The frightened and angry residents of the outer ring, primarily the women, formed the Love Canal Homeowners Association (LCHA), electing Gibbs president. With the help of Dr. Beverly Paigen, a cancer research scientist at Roswell Memorial Institute in Buffalo, New York, LCHA conducted a study similar to the one prompted by Hazel Johnson at Altgeld Gardens. It revealed rates of miscarriage, stillbirth, crib death, birth defects, and urinary disorders far above the national average.

Under Gibbs's leadership, local mothers staged marches to bring attention to the impact of the toxins on their children, attracting much media coverage. Proud to be full-time mothers and homemakers, they rejected as antifamily much of the burgeoning feminist movement, but they also embraced that movement's emphasis on women's right to speak freely and publicly and to demand government action.[16] Gibbs's early statements to the press included concerns about not being able to get a good price for her home when she first sought to put it on the market in June 1978. But her maternalistic rhetoric proved far more effective than discussion of economic concerns. Gibbs did not surrender entirely to traditional notions of womanhood, however. Steeped in the environmental awareness that had permeated the nation in the decade following the first Earth Day, she also cited her rights, as a taxpayer and as a citizen, to environmental health and safety. At the same time she rejected the tactics of Thomas Heisner, who emerged briefly as a local leader, but whose insistence that only those who lived closest to the canal were deserving of assistance quickly led to charges of selfishness and entitlement.

Like most men in the community, Heisner also sought to reinforce traditional gender roles. Some men resented the leadership roles exercised by women, perceiving them as an attack on the traditional male role of family leader, protector, and provider. Others, including Mayor Michael O'Laughlin, were worried as well about the impact that Gibbs's assertions could have on property values. O'Laughlin regularly described Lois Gibbs as "emotional" and charged that she "didn't want to deal with the facts."[17] Love Canal resident Charles Bryan also believed that, as a woman, Gibbs was a poor leader, but for different reasons. State health authorities revealed the negative potential of relying on the maternal stereotype as a protest tactic. They portrayed the full-time homemakers as well intentioned but essentially clueless about matters beyond domestic duties. They dismissed the health surveys generated by Gibbs and other members of the LCHA as "useless housewife data."[18] Bryan highlighted other perceived weaknesses of the authority-through-motherhood approach when he charged Gibbs with an inability to stand up to government officials: "Those government people are just walking all around that little girl. Lois is just this little skinny girl and she can't handle it. She's got to get up there and tell them to go screw themselves." That kind of assertiveness, he proclaimed, was "for a man to do."[19]

Bryan and other men remained focused on economics and were concerned that the desperate attempts to save the town's property values were leading to superficial and ineffectual remedies. They believed that union solidarity among men in the form of picketing might be the tool to stop the hasty remedial construction efforts because construction workers would not cross brother picketers at their site. Gender-based concerns caused other union men to respond quite differently: they refused to join picket lines because they feared being perceived as "chasing industry out," a position that invited union reprisals threatening to their livelihoods.[20]

Despite her male critics and their alternate approaches to the problem, Gibbs earned the lion's share of media attention—and the support of the majority of the community—because she emphasized the health and safety of children and was far more inclusive than Heisner in her calls for redress. Women's leadership was recognized as the most effective and therefore generally accepted during the crisis, but men pressed for a return to "normalcy" as soon as possible.[21]

As the community's most publicly recognized leader, Gibbs exhibited none of the racism of some of the white male members of LCHA and welcomed the appointment of an African-American minister to a state-established task force (ultimately named the Love Canal Revitalization Agency—LCRA). She and other women, including Barbara Quimby, were furious, however, when they were passed over for

LCRA membership and their credentials, as women and as full-time homemakers, ignored. Quimby asserted that the task force needed "women's point of view. . . . I'm not saying that men can't do the job. But I think that women have an advantage in this case. We can look, not only with our eyes, but with our hearts. We know whose children have birth defects. . . . We have so much more personal contact day-to-day than the people who are on the board now that we could do a superior job."[22] The all-male membership of LCRA nevertheless rejected plans to hastily revitalize the area in order to protect property values: "We can't be concerned with industry, tourism, and property. Our main concern has got to be people."[23]

Race and class as well as gender complicated the problem at Love Canal and its proposed solutions.[24] Businesswomen and single mothers alike raised some of the economic concerns usually voiced by men. Moreover, single mothers living at Griffon Manor, a federal housing project populated primarily by African Americans and located very near the canal, found that their race, and their status as renters rather than homeowners, resulted in their neglect by the media, politicians, bureaucrats, and many white members of the public.[25] White men in particular denigrated the concerns of renters, especially welfare recipients, claiming that only homeowners should receive financial restitution. Lois Gibbs rejected the argument that race or economic class enhanced or impeded the right of anyone to claim rightful restitution. She believed that the various efforts to dismiss the African-American community as undeserving were part of a conspiracy by state officials to keep those suffering from the toxic environment from creating a unified power base.[26] Since they were all mothers, she and most of the other white women in the LCHA shared maternalistic concerns and rhetoric with their black counterparts in Griffon Manor.

Racial tensions were nevertheless exacerbated by the Love Canal Homeowners Association, whose very name indicated that it privileged homeowners over renters. Gibbs and others immediately regretted that mistake and tried to remedy it by incorporating inclusive language into the organization's bylaws. Still feeling devalued by white homeowners, Griffon Manor residents looked to leaders in the African-American community and to civil rights organizations for support.[27] A group of Baptists, Methodists, Catholics, and Jews living near the Love Canal area founded the Ecumenical Task Force (ETF) of the Niagara Frontier to help their afflicted neighbors.[28] Its members were largely white, middle class, and educated. According to their pro tem chairman, Reverend Dr. Paul Moore, "In Niagara County, God's law has been broken, his eternal covenant violated, and we are reaping the bitter consequences: ecological disaster and human tragedy."[29] Like the women of LCHA and Griffon Manor, Moore used maternal rhetoric. He described the disaster at Love Canal as a violation of "our loving mother [earth] who gave us birth and faithfully sustains us [and who] as a vulnerable woman . . . is ravaged and raped by brutal exploiters . . . then discarded as worthless." According to Moore, "It is my duty to stoop to her weakness, bind her wounds, and heal her hurt."[30] Moore and other ETF members were concerned with more human, immediate matters as well, providing Love Canal residents funding and legal aid. And unlike those members of the LCHA who privileged white homeowners over African Americans, they expended considerable energy helping African-American renters find new homes.

The press was less moved by the more lofty and spiritually charged arguments of the ETF than it was by the forthright appeals of the LCHA, which continued to dominate headlines by using tactics like picketing at the canal every day for weeks (sometimes having their young daughters carry signs saying, "I want to be a mommy someday"), holding prayer vigils, and carrying symbolic coffins to the state capitol.[31] Not surprisingly, one of the biggest protests took place on Mother's Day in 1979 and included members of both LCHA and ETF, all carrying signs and banners expressing concern for either "mother earth" or the families of Love Canal.

In October 1979 Governor Hugh Carey announced that the state would purchase at fair market value the homes of the remaining residents who wanted to leave: a victory for homeowners but with no provisions for renters.[32] The state quickly bowed to pressure from the NAACP and included Griffon Manor in its relocation plan, but renters dependent upon federal assistance needed special government-issued certificates to gain access to safe, subsidized housing—certificates that were repeatedly delayed. As white outer ring homeowners celebrated the promised assistance, black families, the majority headed by single mothers, saw no escape from their toxic environments. Steeped in the civil rights and women's rights movements, African-American mothers noted that racial and class prejudice trumped even maternalism. They demanded equal protection: "We love our kids just like the home owners over there and . . . we should have the same protection that the home owners have."[33]

Despite ongoing plans and promises of relocation, homeowners were not receiving the support necessary to leave their homes. LCHA members felt increasingly marginalized as their frustrations mounted over the endless delays.[34] In May 1980, the EPA released the results of testing that revealed high levels of dioxin (polychlorinated dibenzodioxins, which are extremely toxic) in residents' blood. Thirty-three percent of residents tested showed chromosomal damage.[35] With the health of their own families at stake, panicked LCHA members turned to extralegal measures, even taking hostage two EPA officials on May 19, generating vast media attention. Only two days after the EPA officials were "detained," President Carter ordered the temporary relocation of 700 outer ring families.[36] In September, the

FIGURE 8.2 Lois Gibbs, her daughter Missy, and son Michael all carry signs at a Love Canal protest at the City Hall of Niagara Falls, New York, on October 16, 1978. (Photo by Penelope Ploughman © October 16, 1978, all rights reserved. Courtesy University Archives, State University of New York at Buffalo)

federal government agreed to assist New York in purchasing the remaining Love Canal homes.[37]

On December 11, 1980, Congress enacted the Comprehensive Environmental Response, Compensation, and Liability Act (CERCLA), commonly known as Superfund. According to the EPA, this law "created a tax on the chemical and petroleum industries and provided broad Federal authority to respond directly to releases or threatened releases of hazardous substances that may endanger public health or the environment."[38] This was the first time that federal emergency funds were used for something other than a natural disaster.

In 1983, more than thirteen hundred former Love Canal residents received a total of just under $20 million from Occidental, with an additional $1 million set aside for a Medical Trust Fund.[39] Despite the language in the original bill of sale absolving the sellers of all responsibility, in 1994 Hooker/Occidental was ruled negligent in its handling of waste and its sale of the land. It was sued by the EPA the following year and agreed to pay $129 million in restitution ($101 million to reimburse the government's superfund, plus $28 million in interest).[40] By 1997, virtually all of the 900 area families had moved and had received some financial restitution for their loss of health and property.[41]

Lois Gibbs became an internationally renowned activist devoted to promoting citizen involvement at the community level.[42] Toxic waste facilities, chemical emissions, and health risks from air and water pollution continued to disparately affect those considered to be powerless to prevent them: the impoverished, especially in communities of color.[43] All over the country women like Gibbs, untrained in environmental sciences but passionate about protecting their disadvantaged communities from corporate polluters, began action groups and protest movements deeply rooted in their local cultures, traditions, and gendered beliefs. African-American women in particular, frequently the heads of single-parent households, brought a legacy of assertiveness, leadership, and maternal concerns.[44] They played a prominent role in a number of community organizations, waging campaigns against environmental dangers in the workplace and the home, especially in areas known as "brown fields" because of their toxicity.[45]

In September 1996, about 100 of the roughly 2,000 residents of St. James Parish in southern Louisiana, formed St. James Citizens for Jobs and the Environment (SJCJE). The organization, whose members were mostly women, sought to lead its low-income, primarily African-American community in an effort to block the Shintech Corporation from constructing a polyvinyl chloride (PVC) plastics complex within a mile of a local school.[46] A founding SJCJE member was Emelda West, a seventy-two-year-old retired schoolteacher and mother of seven. Decades earlier West and her neighbors could pull shrimp from the Mississippi River and eat other local foods.[47] But their environment changed. They lived in the heart of what had become "Cancer Alley," an eighty-five-mile stretch along the Mississippi River between Baton Rouge and New Orleans, home to more than 160 industrial facilities accounting for 62 percent of all the toxic pollution produced in the state.[48]

Some local men agreed with Louisiana governor Mike Foster and with Ernest Johnson, president of the state's branch of the NAACP, that these facilities provided much-needed jobs.[49] SJCJE treasurer Gloria Roberts took the position of many women, countering claims that the proposed industrial complex would boost the local economy: "We've had forty years of 'economic development' in St. James Parish, yet people are [still] living in deplorable conditions." Emelda West argued that the industrial compounds actually intensified local poverty because the river had become so polluted its fish were not safe to eat, and residents were forced to buy bottled water. Fruit trees rooted in the toxic soil were "just rotting and falling," and vegetable gardens had to be abandoned. Worse still were the nosebleeds, rashes, and asthma plaguing local children because, observed West, "They're absorbing all this poison." West's oldest daughter had died at the age of thirty-six after six years of suffering with breast cancer that Emelda West blamed on exposure to environmental toxins at a young age.[50] This mother, grandmother, and great-grandmother was "very concerned about the children who attend our two elementary schools [because] both schools are within three miles of most of the largest industrial polluters in the parish."[51] The women of SJCJE held meetings in homes, schools, and churches to explain to their neighbors that the proposed plant would issue 600,000 pounds of toxins into the air each year, and discharge more than 6 million gallons of wastewater into the Mississippi River every day.[52]

SJCJE circulated petitions, wrote letters, and contacted local newspapers, gaining support from the Louisiana Environmental Action Network, Tulane University's Environmental Law Clinic, Xavier University's Deep South Center for Environmental Justice, and, ultimately, Greenpeace.[53] With the help of these institutions, the Shintech case became the litmus test of the executive order Hazel Johnson had witnessed President Bill Clinton sign into law in 1994.[54] During the long investigation by the EPA into the environmental impact of the proposed Shintech complex, West insisted on delivering her message to polluters personally. She was a member of the three-person uninvited delegation to Tokyo, where she spoke with Chihiro Kanagawa, president of Shintech's parent company, Shin-Etsu. West presented Kanagawa with a package of letters and petitions from more than a thousand St. James Parish residents opposing a Shintech plant in their community.[55] She accused the company of environmental racism and charged that the proposed industrial complex posed a health hazard "exacerbating damage to an already overly polluted region."[56] Before the EPA could render its final ruling, Shintech announced that it was dropping its plans to build in the area. West was honored for her role in the fight to prevent poor communities of color from becoming the dumping grounds for the nation's garbage, waste, and toxic industrial by-products.[57]

Texans Sylvia Herrera and Susana Almanza also took a more direct, personal approach to achieving environmental justice. When Herrera was seven years old her teacher in East Austin taped her mouth shut for speaking Spanish. At the age of five, while translating for her parents, who did not speak English, Almanza learned "what it's like when you're poor and you speak Spanish."[58] Herrera, who holds a Ph.D. in health education, noted that their community of primarily Latinos and some African Americans "has more than its fair share of zoning for industrial uses" and suffered from the "social, economic, and environmental problems . . . [that] disproportionately affect low-income people and people of color."[59] Although they supported the good work of mainstream environmental defense organizations like Greenpeace, Herrera and Almanza emphasized their culture's insistence that the human family remain central.[60] They viewed earth as a benevolent mother and believed that no one was "really looking at how the environment interlocks with humanity. Yes, we must take care of the water, but we also must take care of the people who're drinking it. . . . You can't just look at the environment without looking at the human beings who are being impacted."[61]

In May 1991 the two Chicanas cofounded People Organized in Defense of Earth and Her Resources (PODER), dedicated to looking at "all the environmental justice issues that affect communities of color."[62] When they learned later that year that Mobile Oil wanted to add an above-ground tank to the fifty-two-acre tank farm already polluting the surrounding predominantly residential community, they went door to door to gather information on existing pollution-related health problems. According to Herrera, as they gathered data on nosebleeds, asthma, and migraine headaches and shared that information, local residents "began to realize that the whole neighborhood was having similar problems and that it had something to do with air violations at the tank farm."[63] PODER held community meetings to share additional information with those affected, and its members also led a "toxic tour" for elected and school officials and the media. That tour generated considerable attention and sparked a series of investigations. PODER "kept after the regulatory agencies and elected officials—and in just eighteen months we were able to get them to relocate the tank farm."[64]

After challenging a series of discriminatory land use and economic development policies, PODER joined with the El Pueblo Network (EPN), an association of various East Austin community groups and neighborhood associations. In 1996, EPN met with members of the city of Austin's environmental board, planning commission, and city council to discuss zoning changes.[65] When the city council directed the city manager to begin a land use study, PODER worked with city officials. PODER and EPN scored what Almanza and Herrera called "a big victory" in 1997 when the city council declared that East Austin residents must be notified when industrial and commercial industries seek to locate in their community.[66]

Following their many successes, PODER faced a new problem: gentrification. The city of Austin designated East Austin, only five minutes away from down-town, a "desirable development zone, as if there were no people living here!" Unable to pay drastically increased property taxes, East Austin residents were evicted from their homes. Developers upgraded the homes for the more affluent, who were almost exclusively Anglos. According to Almanza, resettlement is "a battle that people of color, time and again, have lost: to urban renewal, to crime waves, to gentrification, and to the development or abandonment of property. This battle is not usually seen as an environmental issue, but in fact it is one of the most crucial of all urban environmental issues."[67] By encouraging local commu-nity leaders to participate aggressively in zoning meetings PODER succeeded in rezoning more than six hundred homes, allowing long-term residents to retain or regain possession.

Herrera and Almanza are confident that PODER can tackle newly emerging problems because of its insistence that local leadership prevail, even when working in conjunction with national environmental groups like Earth First! and the Audu-bon Society. According to Herrera, PODER tells these mainstream organizations, "We do want your support, but we don't want you to lead. . . . We, the grassroots people, speak for ourselves. . . . You're not the leader here, the community is."[68] People remain the focus in Herrera's spiritually informed view. The environment, she states firmly, "is not only about birds and salamanders, it is also about people," and asserts that "there was a time when we were all brothers and sisters, the night sky our ceiling, the earth our mother, the sun our father, and our parents the leaders, and justice their guide."[69]

The Role of Religion and Spirituality in Environmental Justice

The spiritual beliefs that guided the founders of PODER and the Ecumenical Task Force that served the African-American renters at Griffon Manor highlight the role of religious faith and spirituality in many environmental protection attitudes and efforts. For Christians, the exploitation of natural resources had long been justified by Scripture, such as Genesis 1:28 ("God blessed them, saying, 'Be fertile and multiply; fill the earth and subdue it. Have dominion over the fish of the sea, the birds of the air, and all the living things that move on the earth'"). With the growth of environmental awareness, however, people of many faiths reexamined their own sacred texts as well as the texts and environmental practices of others, including Native Americans, and reached quite different conclusions: they came to believe that people were designated by God to be the stewards of the earth, and it was their obligation to protect it as well as each other.[70] They were guided by biblical passages such as Isaiah 24:5 ("The earth is polluted because of its inhabitants"), Proverbs 29:18 ("Where there is no vision, the people perish"), and Numbers 35:34 ("Do not defile the land on which you live").[71] Their religious beliefs inspired them to remedy environmental problems as part of their dedication to carrying out God's word through social justice—helping, rather than just praying for, people in need.

As the modern environmental protection movement took hold, the religious values of traditional Native American environmental practices in particular were reexamined and held up as models for the present day. According to Northern Cheyenne environmental activist and tribal sociologist Gail Small, "Indian peoples believe in the spiritual nature of the environment. The federal agencies charged with helping us protect our physical surroundings cannot do so unless they understand the interdependence of environment, culture, and religion in the tribal way of life."[72] One celebrated concept was the Navajo tradition of *hozhó*, defined as "the harmonious balance between man and nature."[73] In accordance with terms expressed by Reverend Paul Moore at Love Canal, Navajos Esther Yazzie-Lewis and Jim Zion described traditional tribal religious beliefs as playing a large role in the community's modern environmental concerns: The earth "is important to [Navajos] as our Mother. As the Earth, she must not be disrespected or harmed in any way. There is an ancient Navajo belief that people should not dig into the earth—particularly with steel tools or machines."[74] Yet even as scholars, activists, and religious leaders began, in the 1970s and 1980s, to study such sustainable ways and philosophies, those indigenous populations were experiencing the results of unprecedented environmental abuses, particularly in the Southwest, home to some 350,000 Native Americans.[75]

The larger Four Corners area, which surrounds the point where Utah, Arizona, Colorado, and New Mexico meet, includes reservations of the Ute, Apache, Zuni, Hopi, and Navajo. The Navajo Nation, bounded by four sacred mountains, constituted the largest American Indian population in the United States, with 168,000 of its more than 255,000 enrolled members living in the Southwest. Navajos in Four Corners were suffering from extremely high rates of birth defects, cancer, and other radiation-associated illnesses stemming from mining practices dating back to the 1940s.[76] Reservations had long been off-limits to mining, but in 1919 the federal government passed an appropriations bill on behalf of the Bureau of Indian Affairs authorizing the secretary of the interior to allow leasing of Indian lands for prospecting and mining. The law favored mineral extraction, allowing prospectors to enter Indian lands without permission. Royalties to Indian tribes were minimal, and reclamation was not mentioned. The law was amended in 1938 to give tribal councils the authority to enter into lease agreements on reservation lands if approved by the secretary of the interior.[77]

Once the United States entered World War II, the government was especially intent on increasing its supplies of certain chemical elements. The all-male Navajo Tribal Council approved opening up its lands to development with the provision that Indian labor be used and disturbances to grazing lands be kept to a minimum. The council welcomed leasing opportunities, confident that mining would create jobs for tribal members and improve the area's standard of living.[78] From 1942 to 1986, the uranium mines located on Navajo lands provided ore, first for the Manhattan Project, and then for the huge weapons stockpile created during the arms race of the Cold War. The U.S. government was the sole customer for the private corporations that operated the mines and mills over this forty-four-year period.[79]

New Mexico and the Four Corners area became the largest uranium-producing region in the world, with a majority of the ore found on Indian lands, including a sizeable portion of the Navajo Nation. In New Mexico's Laguna Pueblo, Anaconda Uranium Corporation established in the early 1950s the Jackpile Mine, which quickly became the largest open-pit uranium mine in the United States. From 1949 to 1969, the Kerr-McGee Corporation operated several shallow mines in Cove, Arizona, and established a uranium-processing mill in Shiprock, New Mexico, the reservation's largest town. By 1976, the Department of Interior had leased close to 750,000 acres of Indian land for uranium exploration and development. Mount Taylor, or *Tsoodzil*, as the sacred mountain of the south is called in Navajo, became the site of the world's largest deep uranium mine.[80] From 1944 to 1986, nearly thirteen million tons of uranium ore were mined on Navajo Nation lands, an area of more than 27,000 square miles.[81]

While Indian women carried out domestic tasks and cared for livestock, approximately 15,000 Indian men worked in the mines. Navajo men, for generations poorly educated and doomed to lives of poverty, initially welcomed the steady employment they found there. In the 1940s and 1950s, miners used hand tools (picks, shovels, and wheelbarrows) to expose uranium ore. Later, tunnels were carved and blasted into cliff sides, with the most dangerous work relegated to the Navajos by their white supervisors. In 1979, former Navajo miner John H. Lee recalled his work of over twenty years: "The white men sat outside the mines and pushed us Navajos into those dusty mines right after dynamiting." Another former miner remembered: "They chased us in like we were slaves."[82] Their hard, racially segregated labor subjected the workers to cave-ins or, more commonly, being struck by smaller streams of dirt and/or stones loosened by the blasts. But the invisible risks turned out to be far more dangerous.

According to one former miner, "It used to be so dusty that we were always spitting up black stuff and . . . when we went home we all had headaches from breathing all that contamination."[83] Navajo Mae John, whose husband and two brothers died of lung cancer after working in the mines, emphasized that the miners were completely uninformed of the dangers of their environment and were not advised, let alone equipped by their employers, to take precautionary measures. The miners "even used to eat inside the mine, even drink the water dripping inside the mine. There were little puddles of water, and that's where they drank from. And they breathed in all that dust and smoke. That's what caused the people to die."[84]

Although miners were the most directly exposed to the deadly toxins and suffered the highest mortality rates, their wives and children also suffered due to inadvertent exposure. Mae John recalled that miners, including her husband, "used to wear their regular clothes and come home without changing them."[85] Fannie Yazzie reported that when her husband came from the mines, "he used to sleep in the clothes in the house." She hand-washed his work clothes along with the rest of the family's clothes. Two of their children died of unexplained causes at very young ages. Fannie Yazzie suffered from chronic chest pain and weak and aching joints, and three of her adult children had complicated and unexplained chronic ailments. She blamed the yellow dust to which they were all exposed over the years.[86]

Like the miners, millers were exposed to dangerous levels of dust and radon, an odorless, invisible radioactive gas. From 1959 to 1967, Navajo George Brown worked in the Tuba City uranium-processing mill. He explained, "I used to come out with yellowcake [mixed uranium oxide produced in the milling process] all over my clothes and shoes. My kids played with the shoes and shoelaces. My daughter got cancer of the lymph nodes. My sister used to do all the laundry. She came down with cancer and after treatment had a relapse."[87]

Miners were not just bringing home toxic dust for their families to breathe. Some, like Frank Nashcheenebetah, were building their houses out of radioactive materials provided by the mines. When blasted, uranium-bearing sandstone fractures into regular square and rectangular shapes. Mines were surrounded by heaps of the rock, left for the taking in open-air piles. Tailings, the radioactive residue after ore is ground up to extract the uranium, proved to be of excellent building quality when mixed with cement. Families across the Navajo Nation lived in homes constructed of radioactive stone and mortar. Although the U.S. government is, by law and by treaty, responsible for the Navajos' welfare, no effort was made to notify the residents of the toxic nature of their homes' very foundations of poured, contaminated cement. The government could hardly claim ignorance: the dangers of uranium mining and milling have been known to the scientific community since the early twentieth century, when the "mountain disease" suffered by miners in Germany and Czechoslovakia and identified as lung cancer in 1879 was recognized as the result of their exposure to radon, findings confirmed by Alice Hamilton's 1911–1921 investigations into radium and other industrial poisons in the United States.[88] It was not until 1967, after the uranium boom had peaked and more than 200 Navajo miners had died, that the government released the Public Health Service study, "Radiation Exposure of Uranium Miners," and required minimal safety standards.[89] Despite Navajos' lower-than-average use of tobacco, the uranium industry countered that the miners' lung cancer rates were due to excessive cigarette smoking. A long series of medical studies ensued.

The male-dominated Navajo tribal council pursued political remedies, passing resolution after resolution for Congress to provide compensation to the uranium miners and their survivors.[90] In 1979, former interior secretary Stewart Udall testified before a Senate Special Committee, "I have seen a lot of buck passing in the Government in my day, and I must say this is the most outrageous example that I have ever seen." Udall filed suit in federal court on behalf of ninety-five Navajo miners, widows, and their descendants against seven major mining corporations on the Navajo lands.[91] Udall charged that the corporations were aware of the radiation dangers but in a "breach of duty" did nothing to protect their workers.[92]

In the face of renewed environmental devastation, and as various appeals, including Udall's, to the legal and government system stalled, members of the Navajo community began to realize how deeply radiation had poisoned their entire web of life. Thousands of still-radioactive abandoned mines, tailings piles, water pits, and homes built from radioactive waste rock dotted the landscape. Like Altgeld Gardens and Griffon Manor, the area remained inhabited by those who, because of their race and class, had no alternative but to remain in their contaminated homes in the only community they had ever known.

In the Navajo Nation, women were the primary landholders. As Edward Plummer, Navajo superintendent of the Bureau of Indian Affairs (BIA) agency at Crown Point, New Mexico, explained, "The way it works is that the land usually belongs to the women. The society has been set up so that the land is passed from father to daughter and uncle to niece. When a Navaho [sic] man marries, he almost always goes to live on the woman's land. Most of these allotments were distributed between 1910 and 1930. At that time Indian-owned land was reduced from twenty-four million acres to two and a half million acres. Since the government didn't know about the uranium then, those who were given land were given both the surface and the subsurface rights to the land."[93]

As the land's owners, women bore the burdens of its loss. Hah-nah-ban Charley, an elderly woman living a traditional life, described in 1979 how she had recently lost the use, and with it the health, of her land: "One day a white man carrying papers came with an Indian and said, 'Mother, because all is well with you and you use your land well, and you have no problems with your neighbors or your allotment, we want you to put your thumbprint right here on this piece of paper.'" Only later did Charley realize that she had signed a contract giving a mining company access to her 160-acre allotment. "Later, I went to the BIA office and told them what had happened. But they just said, 'It is your fault. You signed the paper.'" Hundreds of holes were drilled on her land, and three large mud pits developed. Charley described the loss of her animals: "Some sheep drowned in the mud, others died—one right after the other, like they were poisoned." Altogether, three calves, sixteen sheep, eleven goats, and four horses died. "Now there is not enough food for the family because so many animals have died. A white man from the BIA came out to look at the dead sheep and said it's probably the water since the wells they dug have a runoff that goes right into the animals' stock pond."[94]

Navajo families not only noticed strange mutations in their surviving animals, but suffered themselves from birth defects, childhood sicknesses, and deaths. Rather than associating symptoms with the uranium radiation, researchers at the University of New Mexico published findings in 1976 describing "Navajo neuropathy," a lethal disease striking Navajo children in the same families. The disease was at first considered purely hereditary.[95]

Women of All Red Nations

What emerges from this deadly picture of health and environmental abuse in the Navajo Nation is a pattern of negligence by mining and milling corporations, as well as by the various branches of the state and U.S. governments involved in both atomic energy and the administration of Navajo Nation services. For many American Indians, negligence was much too mild a term. Members of Women of All Red Nations (WARN) chose instead to use the word genocide and mobilized American Indians, especially women, to fight against the threats to their health, lands, water, and reproduction.

Some Indian women had opposed uranium mining from the very beginning. Navajo Edith Hood was only a teenager when the exploratory drilling crew arrived in the 1960s. Although no one told her grandmother what was happening, Hood remembers her "running to stop them from making roads into the wooded areas. The stakes she drove into the ground did not keep them out." Hood's grandmother's instincts were soon confirmed: "There was no respect for people living there, and certainly no respect for Mother Earth."[96]

Other Indian women came to environmental activism less directly. In 1968, Native Americans, like African Americans, women, and other oppressed groups, began demanding civil and political rights. They formed the American Indian Movement (AIM). Stressing the connectedness of all Indian people, AIM was dedicated to restoring Indian lands and practices; its primary tactic was enforcing treaty obligations of the United States.[97] The 1973 occupation of Wounded Knee by members of AIM, during which two American Indian men were killed by federal forces, changed the perspectives of two American Indian women activists present, Madonna Thunder Hawk and Lorelei DeCora. They witnessed the subsequent loss of AIM momentum when many of its male leaders were imprisoned or fell victim to police abuse and FBI and CIA infiltration campaigns. These women joined with others, including many widows of uranium miners seeking compensation for their husbands' deaths, to found WARN in 1974. WARN focused its activism on the problems faced specifically by Indian women and their families, including poor nutrition, inadequate health care, forced or deceptive sterilization programs, and higher levels of domestic violence resulting from poverty, joblessness, substance abuse, and hopelessness.[98] In the case of Indian mining families, with so many men dead as a result of their work in the uranium industry, women were left to shoulder alone the burden of being the heads of their families. Without their husbands' income, they and their children slipped deeper into poverty.[99]

WARN members were motivated in their fight against the federal government by a variety of factors based on gender, including belief in women's spiritual power as well as the tradition of matrilineal inheritance of property. Rather than work in isolation, WARN gained strength from the liaisons it formed with a variety of feminist groups that recognized the unique concerns of minority women. Native American women were represented at the National Conference on Women in Houston in 1977 and worked closely with the National Organization for Women on a number of issues, including all forms of violence against Native American women.[100] WARN's cultural heritage as protectors of the earth was fortified by the newly found insistence that all women should exercise their political rights and demand equality.

According to the Council of Energy Resource Tribes, 75 to 80 percent of U.S. uranium reserves are on Indian land.[101] WARN disseminated information to help Indians understand legal rights concerning minerals, land, leases, and water.[102] It mapped from the southwest to the Northwest Territories of Canada an "Indian energy corridor, where major coal and uranium resources exist and are being exploited by the big corporations," calling it the "International Sacrifice Area." The organization reminded Indians, "This is an important time for our people to stand together.

If we let the corporations take away our land, we will no longer have a home. . . . [Your children] will have nothing if you don't keep the land to pass on from generation to generation. *We are the ancestors of those yet unborn.*"[103]

One of WARN's goals in the late 1970s and 1980s was to draw attention to health concerns involving reproduction, specifically the dramatic increases in sterilizations, as well as in miscarriages, birth defects, and childhood deaths due to cancer and other environmental diseases on Indian reservations, including the Navajo Nation. WARN urged its members to make the connection between indigenous people's lands and water sources being poisoned and the forced sterilizations of its women by the Indian Health Service (IHS) during the 1960s and 1970s.[104] WARN related the two practices to what its members understood to be a continuation of centuries of Euro-American genocide of Indian peoples.

At about the same time that Lois Gibbs and the Love Canal Homeowners Association were basing their authority in maternal concerns, WARN repeatedly emphasized the importance of reproduction in cultural self-determination. Its members saw their reproductive rights threatened by their contaminated surroundings as well as by more direct efforts to curtail their populations. Thousands of poor women and women of color were sterilized in the 1970s without informed consent. Many had given permission without completely understanding the physical and psychological ramifications. Because of tribal dependence on government agencies including the IHS, the Bureau of Indian Affairs, and the Department of Health, Education, and Welfare, Native American women suffered in particular from coerced sterilizations. Sterilization programs began increasing dramatically in the mid-1960s primarily because the largely white and male IHS medical personnel believed that smaller Indian families would reduce poverty rates and that fewer Indian births would reduce government welfare spending. Indian women and teenage girls were sterilized without their consent and often without their knowledge as an added step in another intervention, such as childbirth or an appendectomy. These women only discovered they could not have children when they sought medical assistance after being unable to become pregnant. In other cases, pregnant women were urged to sign sterilization consent forms during labor, when they were vulnerable and often medicated. The most persuasive and coercive technique, however, was to threaten mothers with the loss of their children to social welfare agencies if they did not agree to sterilization.[105]

WARN worked to inform Native American women of their rights and urged them to exercise those rights to resist this aggressive government-funded mass sterilization program: "The plan of sterilization is one way that the government has of weakening our nations . . . [because] to get control of our land it would be much easier if our numbers got smaller. We must think hard about keeping our right to bring life to the next generation."[106] In the words of physician Constance Redbird Pinkerton-Uri, "We [Indian women] have a new enemy and the enemy is the knife."[107]

WARN stressed the current generation's responsibility to right the wrongs of the past in order to benefit the future: "We must preserve our rights for the next generation to live the way we want to—SOVEREIGN!" Activists urged Native American women to "Control your own reproduction: not only just the control of the reproduction of yourselves . . . but control of the reproduction of your own food supplies, your own food systems" to rebuild traditional native cultures, religions, and ways of living with the earth.[108]

WARN held a five-day conference in June 1979, welcoming more than 1,200 people to "identify the many problems—the very threat to survival—currently facing the future generations of the Indigenous Nations of the Western Hemisphere."[109] Lakota activist Mary Crow Dog explained her reaction to the attacks on her people: "Like many other Native American women . . . I had an urge to procreate, as if driven by a feeling that I, personally, had to make up for the genocide suffered by our people in the past." Similarly, Mohawk Nation member Katsi Cook declared that "women are the base of the generations. Our reproductive power is sacred to us."[110] Conference leaders discussed a 1974 study that pointed out that only 100,000 "full-blood" Indian women of childbearing age lived in the United States. They argued that these women were the most likely victims of government sterilization programs.

While some Indian activists argued, "The sterilization campaign is nothing but an insidious scheme to get the Indians' land once and for all," WARN pointed out the wider ranging impact of these events: Tribal councils lost the respect of their communities when they were seen as unable to protect Indian women, weakening Indian political structures as well.[111] As WARN also noted, Navajo women's exceptionally high rate of birth defects compounded the damage caused to the Indian people by the intensive sterilization program. While the rate of birth defects per 100,000 births in the white U.S. population between 1973 and 1978 was 846.8, the rate among the Indian and Chicano population was 1,589 in Arizona and 2,114 in New Mexico. By 1981, the Navajo Health Authority found a dramatic increase in cancer of the ovaries and testicles among children, at least fifteen times the U.S. average, and bone cancers five times the U.S. average. These increases also coincided with the stepping up of uranium production.[112] The illnesses that affected miners and millers, along with their wives and children, were by the 1980s clearly affecting their grandchildren as well. Even sex ratios were affected. In Shiprock, New Mexico, for example, between 1963 and 1978, the ratio of male to female births was significantly altered. Public health officials confirmed that in Shiprock, "we're missing approximately 230 male births between 1963 and 1978." In all of the Four Corners states, but especially in New Mexico, the sex ratio began changing in 1953, precisely when uranium production accelerated in the area.[113]

At its 1979 conference, WARN identified five areas "that a People must have total control over, in their own lives, in order to call themselves sovereign": First, rebuild traditional government structures to resolve problems so that "our struggle as a total people won't be continually weakened." Second, revitalize traditional economies to "control the reproduction of . . . food supplies" on a strong land base. Third, control reproduction of people. "In terms of the children, in terms of guaranteeing the continuity of Our Peoples—the women must lead. The women must re-strengthen themselves. . . . This is coming back, along with the fight against sterilization programs." Fourth, provide education to perpetuate Indian "vision [and] belief in yourselves." Fifth, gain control over "that thing that creates your identity, creates your belief in yourself, creates your vision of the world."[114] WARN urged "learning and living the Indian way of life" as a means of defense against the constant attacks "from all sides in the struggle to survive as a race of Indian nations."[115]

The women of WARN, bolstered by the clout of the mainstream feminist movement, succeeded in their push for federal regulations to end sterilization abuse.[116] The poisoning of their land, however, persisted. Uranium was termed "the white man's new gold" by WARN. They cautioned, "The white man is always looking, looking for something that will be more precious than what they found before." Growing resistance to uranium mining involved reestablishing the traditional Indian relationship to land. WARN noted, "Indians are land possessed. Something of their very being is defined into the reverence of the soil, something of their very strength is the issue of the relationship to the land. . . . And before our time goes we must search out whatever it is that will re-establish that nature of relationships that will bring back the strengths of our people."[117]

Just weeks after WARN's 1979 conference gave hope and direction to many, the second largest radiation accident in the United States (surpassed only by Three Mile Island) occurred on the Navajo Nation in New Mexico. At the milling plants, every pound of uranium extracted resulted in the production of 500 to 1,000 pounds of tailings. This residue, in the form of thick, liquid sludge, was kept in extensive settling ponds until the water evaporated. On July 16, 1979, a retaining dam for liquid tailings at the United Uranium mill at Church Rock burst, sending 100 million gallons of radioactive waste down the Rio Puerco arroyo. It poured into the Little Colorado River, then the Colorado River, from which some flowed through the Grand Canyon, into irrigation water for California's Central Valley, and into the Gulf of Mexico. Nearly twenty years after the incident, Church Rock resident Larry King testified before the U.S. House of Representatives as to the immediate and ongoing repercussions of the flooding: "The contaminated fluids that escaped from the [United] uranium mill tailings pond ran through our property, into the Puerco River, where we watered our livestock. I remember the foul odor and yellowish color of the fluids. I remember that an elderly woman was burned on her feet from the acid in the fluid when she waded into the stream while herding her sheep. Many years later, when water lines were being installed in the bed of the Puerco, I noticed the same odor and color in a layer about eight feet below the stream bed. To this day, I don't believe that contamination from the spill has gone away."[118]

In the face of this new health threat, traditional beliefs and practices brought renewed strength to Navajo resistance to uranium development. As medicine man John Smith explained, Mount Taylor and other Navajo mountains "were embodied with a certain wealth, and we shouldn't begin to disturb them. Our elders have taught us that when you push nature, the balance changes, and she will fight back."[119] The work of WARN provided Navajo women and men the information they needed to mobilize further and fight to restore community control over their land. WARN's emphasis on women's health and reproduction helped to bring recognition to newly emerging consequences of old mining practices.

In the 1980s, one study based on the remains of 259 Navajo miners revealed lung cancer rates fifty-six times as high as those in the general population, and stomach cancer rates eighty-two times higher.[120] It also became increasingly clear that while less immediately exposed to the toxins of the mines, women, particularly the daughters and granddaughters of the original miners, were nonetheless suffering unique health effects. Studies in 1981 and 1982 revealed that teenage girls in the area developed breast, ovarian, and related cancers at seventeen times the national average; in two parts of the reservation filled with old mine pits near Shiprock, New Mexico, stomach cancer was two hundred times the average for women ages twenty to forty.[121]

As the full extent of the damage—and potential for future damage—became known, protests and outrage mounted. The Navajo nation benefited from the nascent environmental justice movement pioneered by women like Hazel Johnson, which demanded that the government no longer ignore the devastation caused to poor people, especially poor people whose neglect was compounded by racial discrimination. In the words of Gail Small, Americans and their government "would rather fight for a rain forest five thousand miles away than join the battle being waged by native peoples right here in their own backyards."[122] As a drawn-out (and still uncompleted) clean-up project began in the late 1980s, Department of Energy officials recognized the necessity of incorporating Navajo culture into efforts to heal the people and their land: They engaged Navajo medicine men, including Roger

Hathale, to perform blessing ceremonies. Yet Hathale remained pessimistic: "Mother Nature is getting angrier . . . digging through Mother Nature" was disturbing harmony, with death and disease the result. The only way to restore people's health was to restore *hózhó*.[123]

Spiritual remedies were slowly augmented by additional medical studies and legal reforms. In 1988, the suit Stewart Udall had filed on behalf of Navajo families in 1979 was finally concluded. The U.S. Supreme Court declined to rule, upholding the lower courts' decisions that the government was not liable. Udall battled on in court and in Congress and was instrumental in the writing and passage of the Radiation Exposure Compensation Act of 1990. As a result, Congress offered the families of Navajo miners who suffered ill effects from their labors an apology and compensation of $100,000 each.[124] Most of the miners' widows, however, had trouble collecting the money owed to them due to their inability to produce the required documentation, including detailed medical and occupational records as well as marriage licenses. This last proved a particular barrier, as they had married their husbands in the 1930s and 1940s in undocumented tribal ceremonies. Widows could petition tribal law judges to issue valid certifications of marriage, but this process required English proficiency. Mothers speaking only Navajo were forced to turn to their children to research and file on their behalf, which added to the already lengthy process.[125]

The slow-paced payouts did not end the ongoing effects of unleashing the toxins held in the earth. In 1995 the *American Journal of Public Health* published a study that determined, "Twenty-three years after their last exposure to radon progeny, these light-smoking Navajo miners continue to face excess mortality risks from lung cancer and pneumoconiosis and other respiratory diseases."[126] That same year lawyers attributed some of those problems directly to water poisoning. A lawsuit was filed in tribal court on behalf of victims' families against the operator of some of the uranium mines on the Navajo reservation near Cameron, Arizona. In 1996 and 1997, toxicologists sampled water from the pits, which were still in use. EPA regulations would soon limit the concentration of uranium in drinking water to fewer than thirty picocuries per liter, but in Cameron, uranium levels were as high as 139 picocuries in wells and up to 4,024 in many of the water pits located near the town and used by both people and animals.[127]

Lois Neztsosie was one of many women who regularly drank and bathed in that water. She had drunk about three liters per day from the pits while pregnant with two daughters, who would die of their illnesses.[128] Neztsosie also ate sheep that drank the water, and she used the water to make infant formula. The legal team's experts defined a clear pattern: "When these mothers drank uranium polluted water while pregnant, they bore children with Navajo neuropathy. When they were away from the old mines during their pregnancies, they bore healthy children."[129] The lawsuit dragged out to end in a cash payment to the families involved, but with no admission of guilt. In spite of growing environmental research, some medical researchers continue to emphasize genetics over environment, although their definition of Navajo neuropathy has been modified to include maternal exposure to uranium from contaminated water.[130]

In 2007, the U.S. House Committee on Oversight and Government Reform began an investigation into the consequences of uranium mining in view of its proposed renewal by mining corporations. In the same way that SJCJE treasurer Gloria Roberts responded to developers in Louisiana, Native American women countered the companies' promises of jobs and economic development with maternal concerns. Navajo Edith Hood testified about the cultural teachings she had learned from her grandfather, who was a medicine man and traditional leader: "He taught us to respect Mother Earth, for she gives us all the necessities of life. There is a Navajo concept called *Hózhó*. *Hózhó* is how we live our lives—it means balance, beauty and harmony between we, the Five-Finger People, and nature. When this balance is disturbed, our way of life, our health, and our wellbeing all suffer. The uranium contamination and mining waste at my home continues to disrupt *Hózhó*." Hood, however, presented her people as far from helpless.

Even as Hood recounted the ongoing tragedies caused by the residual effects of mining, she spoke of her personal role in surveying the ground as part of her community's involvement in monitoring the area for radiation. She spoke of what was required to restore the land and its people to physical and spiritual health, adding, "Today, there is talk of opening new mines. How can they open new mines when we haven't even addressed the health impacts and environmental damage of the old mines? Mining has already contaminated the water, the plants, and the air. People are sick and dying all around us." She urged the government to take responsibility and to act aggressively on behalf of her people, declaring, "We need your help to clean up the mess that the mining companies and the U.S. government have burdened us with. We need help to stop mining companies from coming in and making a new mess. We need to restore *Hózhó* so that we may live in balance and harmony with each other and nature as Navajo people, as *Diné*."[131]

Such appeals based in both science and spirituality appear to have had some effect on the efforts to resume, or start afresh, uranium mining in various places throughout the United States. Inspired by the victories of WARN and armed with both sophisticated scientific knowledge and the wisdom of the ancients, the Navajo Nation in particular continues to fight for the physical and spiritual health of the land and its people. In 2010, Colorado governor Bill Ritter approved a law requiring uranium mills to clean up existing contamination before expanding their operations.

Critics argued that the measure would cost jobs in both New Mexico and Colorado, because the Mount Taylor mine could not reopen without a place to mill its ore. Just as John Kennedy used women's maternal concerns to defend his decision to end above-ground nuclear testing, Ritter echoed WARN's rhetoric, "We should not think about ourselves, but about the generations to come. It's incumbent on us to turn this state over to the generation after us and the one that follows that in a better way than we found it."[132]

The Fight against Pesticides in the Fields

As Latinas contributed to environmental justice nationwide, their leadership affected a variety of landscapes.[133] Many Latinas emphasized their dual role as mothers and workers in combating environmental hazards in and around their homes as well as workplaces. In California, for example, aided by the organization Communities for a Better Environment, Latinas played a significant role in forcing the government to remove the mountain of concrete rubble, dubbed La Montaña, created by the freeway collapses during the 1994 Northridge earthquake and dumped in their community.[134] A workplace struggle of long standing is that of the United Farm Workers of America (UFW) against various pesticides, particularly those affecting reproduction.

Not all Latinas have couched their message in terms of maternalism, even as they practiced the Catholicism that often played an important role in their protests; significantly a female figure, Our Lady of Guadalupe, has been a major religious and cultural icon within the UFW. Dolores Huerta was a single mother of seven children in 1962 when she and Cesar Chavez cofounded the National Farm Workers Association, predecessor to the UFW. Huerta had long been a community activist in the San Joaquin Valley of central California, campaigning against the hunger, poverty, discrimination, language barriers, and police brutality suffered by many of the valley's residents, particularly its migrant farm workers. She flagrantly violated the gender rules of her culture when she rejected the notion that women best served efforts to better their community when they worked behind the scenes, supporting the male activists through clerical work as well as by performing all domestic tasks, including raising the children. Huerta instead followed traditionally male models of leadership, sometimes leaving her children in the care of others in order to give speeches and lead protests. She also clashed regularly and openly with Chavez. Instead of apologizing for disagreeing, then deferring to Chavez's authority as a man, Huerta viewed the airing of their differences as a necessary and healthy component of their partnership.[135]

FIGURE 8.3 Defying cultural stereotypes and gender expectations, Dolores Huerta speaks in support of United Farm Workers in the 1970s. (Walter P. Ruether Library, Wayne State University)

Although Huerta was eventually to have eleven children, she did not justify her leadership in terms of maternalism. Originally dismissive of the women's movement as a middle-class phenomenon, after meeting feminist leader Gloria Steinem in the late 1960s, she embraced its tenets and began consciously to challenge the gender discrimination that was rampant within the farm workers' movement. Even before meeting Steinem, however, Huerta had long argued that since men, women, and children all worked together in the fields, all were affected by the poor conditions, and all should seek empowerment and participate in efforts to remedy the injustices they faced. She took a holistic approach to the problems facing farm workers, seeking higher pay, collective bargaining rights, and access to benefits, including Aid to Families with Dependent Children and unemployment and disability insurance.[136]

Included in Huerta's demands on behalf of farm workers was the right to a safe working environment. While she noted that pesticides like DDT and Parathyon frequently used in commercial fields threatened the health of mothers, children, and possibly future generations of agricultural workers, she emphasized the damage they caused to all farm laborers as well as consumers in the here and now, and to the environment as a whole. She contributed to environmental achievements including agreements with growers to eliminate some of the most harmful pesticides and to notify workers about pesticides still being used. One ongoing concern is pesticide "drift," in which breezes cause pesticides dropped from crop dusting aircraft or sprayed in certain areas to land on workers in nearby fields.[137] Following the death of Chavez in 1993, Huerta became widely recognized as one of the movement's most energetic and successful activists.[138] Huerta's long activist career demonstrates that women can be effective leaders in environmental justice movements by challenging, rather than emphasizing, women's traditional values and priorities.

. . .

In 2000, the Academy Award–winner *Erin Brockovich* alerted millions of Americans to the reality of large, powerful businesses carelessly releasing contaminated wastes near unsuspecting poor or working-class neighborhoods. Based on a true story, the film traces how a working-class single mother without legal training used her rapport with local families affected by the toxins in their groundwater to help expose corporate wrongdoing, culminating in a $333 million settlement against Pacific Gas and Electric. Brockovich's story is well known, but it is only one chapter in the rich and varied history of the environmental justice movement. That movement turned conventional thinking about power upside down, as poor women, especially poor women of color living in the most toxic environments, asserted their personal authority against powerful corporations. These activists often braided gender into race, class, contemporary political movements—and sometimes religion—as they fought against the environmental exploitation of themselves and their neighbors. Some were influenced by aspects of feminism, others deferred to traditional gender roles, particularly an emphasis on maternalism. Some, like Lois Gibbs, incorporated both, but all strove to protect the physical, political, social, and economic health of their community's environment. The environmental justice movement put women on the frontlines of environmental protection and showcased their unique power and leadership.

"We told you so"

Epilogue: Women, Gender, and Environment in the Twenty-First Century

"Next time, listen to Mother"

On July 28, 1997, the National Cancer Institute (NCI) revealed that the release of radioactive iodine during the approximately ninety atmospheric nuclear bomb tests that took place in Nevada from 1951 to 1962 was larger than originally estimated. The combined total was at least ten times greater than that released during the 1986 explosion at the Chernobyl nuclear plant in the Ukraine. Protective actions, such as moving cows to shelter so that their milk might be less contaminated, could have been taken prior to the planned Nevada tests. Instead, tainted milk, the main pathway for radioactive iodine exposure, was widely consumed by children, leading to significant increases in childhood radiation absorbed dose (rad) rates. In 1997 the rad rate for the average American was 2, and the Public Health Service recommended medical monitoring for adults who had absorbed 10 rads or more as children. Adults living in the Nevada test site area site had received doses of from 5 to 16 rads, while children, because of their greater milk consumption, were subjected to from 50 to 160 rads, increasing their chances of developing thyroid cancer. According to Arjun Makhijani, president of the Washington, D.C.-based nonprofit Institute for Energy and Environmental Research, this risk was "especially tragic, because it could have been avoided. They knew when the tests were and chose not to warn the population, and they located the test site in the West, knowing there would be fallout over the whole country."[1]

In a 1997 article titled "Next Time, Listen to Mother," the *Los Angeles Times* noted that the NCI report "vindicates the women who warned in the 60s of radioactive fallout hazards." The Women's Strike for Peace received particular praise for its efforts to bring to national attention the dangers of radiation, especially in milk. Thirty-six years after the organization's first protest, *The Times* imagined those mothers, now elderly, "raging and fuming to America, 'We told you so, but you just wouldn't listen.'"[2]

As the twentieth century gave way to the twenty-first, more Americans were listening. Women's environmental efforts couched in maternalism continued to generate surprising respect from a society that increasingly emphasized egalitarianism. Environmental justice pioneers, despite their lack of formal education or scientific credentials, also continued to gain support. Hazel Johnson's crusade to remedy the environmental exploitation of the poor, especially those of color, extended far beyond the boundaries of Chicago's Altgeld Gardens. In 1992 she traveled to Rio de Janeiro as a delegate to the Earth Summit sponsored by the United Nations, speaking with the oppressed who lived in urban shantytowns as well as those in the endangered rainforests of the Amazon. She witnessed firsthand the environmental degradation suffered by black Africans under apartheid in South Africa and, in the early twenty-first century, participated in several world conferences dedicated to ending racism and developing environmental sustainability. Johnson's unofficial title of Mother of Environmental Justice was reinforced as she received numerous awards and honors for her efforts to educate people about the effects of environmental hazards faced disproportionately by low-income populations and people of color. Following Hazel Johnson's death in 2011, Cheryl Johnson continues her mother's work.[3]

Lois Gibbs, dubbed "Mother of the Superfund," continues to serve as executive director of the Center for Health, Environment, and Justice (CHEJ), the organization she founded in 1991. CHEJ is dedicated to "uniting community voices and facilitating collective action by building nationwide collaborative initiatives focused around specific environmental health issues."[4] CHEJ uses the same kind of demonstrations and other protest tactics that proved so successful in bringing media attention to Love Canal. With an annual budget of just over one million dollars, CHEJ receives about fifteen thousand requests for assistance each year from groups and communities. In 2011, CHEJ was one of many groups to point to the damaged nuclear reactors following the earthquake and tsunami in Japan (uniquely endangering pregnant women) as further evidence of the dangers of nuclear power and the ongoing need for citizen involvement in environmental protection efforts.[5] It is perhaps one indication of Gibbs's influence that a Gallup poll taken just months before the events in Japan revealed a "sharp" division in American attitudes toward nuclear power, with men far more supportive than women.[6] Gibbs has written several books and received a variety of honors, including environmental prizes and leadership awards. In its continuing efforts to mandate the cleanup of polluted school sites, CHEJ employs social media like Twitter to encourage discussions of school health.[7]

In recognition for her efforts to protect the health of her community, Lorelei DeCora, cofounder of Women of All Red Nations, was named a Robert Wood Johnson Foundation Community Health Leader in 1993 (an honor that included a $100,000 stipend). Her efforts to improve Native American health continue, as she encourages "people to return to traditional foods and food preparation."[8] Since 1996, DeCora has been running a wellness center in Winnebago, Nebraska, where she advocates "for Native-American health on a self-sufficiency model." She reported in 2002, "We've got 110 families back to planting again, because the Winnebago people are a planting people."[9] WARN cofounder Madonna Thunder Hawk also remains active on behalf of Native American causes. A popular public speaker in the twenty-first century, she raised funds for the Lakota People's Law Project in 2011.

In 2012, women own nearly half of the farmland in the United States. As they incorporate the lessons on soil erosion promulgated almost seventy years earlier by the We Say What We Think Club, they are ensuring that the land will remain productive for future generations. According to the Women, Land and Legacy report produced by Iowa State University and the U.S. Department of Agriculture, women show a "clear and strong consciousness about land-health issues and respect nature intrinsically—not for its productive value, but because it sustains life."[10] Because they care more about land preservation than about maximizing crop yields, women landowners are leading what is being hailed as a "farming revolution." The Women, Food and Agriculture Network, founded in 1997, is a particularly influential community of women involved in promoting this revolution in sustainable farming. Through the network, which is international in scope but especially strong in American midwestern agricultural states, women share the information, connections, and encouragement they need to "build food systems and communities that are healthy, just, sustainable, and that promote environmental integrity."[11]

Aspects of Women for a Peaceful Christmas's crusade have also been carried into the new millennium. First celebrated in 1992, International Buy Nothing Day (IBND) is a worldwide series of events and boycotts organized by Adbusters Media Foundation at the beginning of the major shopping season. IBND urges consideration of the major ecological and economic repercussions, including the perpetuation of sweatshop labor and waste of natural resources around the globe, created by women's consumption of consumer goods that is "most fevered" during the winter holiday season. Noting that the wealthiest 20 percent consume 80 percent of the world's resources, the IBND campaign, like that of its predecessor, offers women who have the resources to shop a variety of alternatives to rampant materialism, promoting a "shopping-frenzy-free" holiday season.[12] Consumers are asked to recognize and resist the oppression of women brought on by media insistence that women find ultimate power, joy, and fulfillment while bonding with each other in an endless round of spending sprees on nonessential, environmentally damaging goods.

Many nonessential services as well as goods are marketed primarily to women. The environmental problems posed by some of these services are gaining increased attention, as are the linkages between those hazards with race and class. While doing outreach in the Vietnamese community for Asian Health Services, Julia Liou learned of the health problems, including asthma, dermatitis, respiratory illness, miscarriages, and children with birth defects, that many nail-salon workers believed were the result of the chemicals they handled and breathed in on a daily basis. In 2005, Liou cofounded the California Healthy Nail Salon Collaborative (CHNSC) on behalf of the more than 114,000 manicurists in California, the majority of whom were Asian immigrants earning less than $18,200 a year, without health benefits or a clear understanding of the American legal or health care systems.[13] Citing the multitude of chemicals known to be harmful to human health that were nonetheless standard in nail salons, CHNSC set out to protect this almost exclusively female workforce facing a variety of cultural and linguistic barriers to building environmental awareness.[14] The collaborative worked to educate manicurists about their risks and rights and to raise public awareness, particularly in support of legislation to counteract the powerful cosmetics lobby that opposes chemical bans.[15]

The Role of Gender in Future Environmental Relations and Activism

While some environmental movements and organizations continue to pursue gender-based approaches to environmental problems, others eschew such tactics entirely. In the United States and around the world, women and men work together to protect the environment in governmental and nongovernmental agencies and organizations ranging from the Intergovernmental Panel on Climate Change and the United Nations Environment Programme, to the Sierra Club and the Nature Conservancy. Many find their goals, and their work, from clerical tasks to fundraising to science, to not be particularly gendered. They reject the notion that either sex is more suited to either causing or remedying toxic environments, contending that people of both sexes can be part of the problem as well as contributors to the solutions. Yet even some organizations that practice a gender-blind approach recognize that sex plays a role in the way people are affected by the environment and in response feature a division or branch devoted to women or women's issues, like the Women's Environmental Health Working Group of the Environment Funders Network. Others, such as the Women's Environmental and Development Organization, Women's Environmental Network, and many more at the local, state, national, and international levels, pointedly operate on feminist principles. Even as they support feminism's ongoing debunking of tired stereotypes, some of these groups emphasize the connections between the historical oppression of women and the ongoing oppression and exploitation of nonhuman nature.[16] While organizations and groups with a pervasive gender emphasis generally welcome all interested parties, frequently their membership is exclusively or predominantly female.

A variety of literature continues to bring feminine and feminist perspectives to nonhuman nature, including women's nature writing as well as a genre of science fiction that explores the limits and possibilities of ecofeminist theories.[17] One of the most important new trends is recognition of the importance of gender and environment across all times, cultures, and geographic boundaries.[18] In 2004, Kenyan activist Wangari Maathai became the first African woman to win the Nobel Peace Prize. This recognition of the Green Belt movement she began in the 1970s brought worldwide attention to the relationship between planting trees, environmental conservation, and women's rights. Following Maathai's death in 2011, her legacy of promoting hope and prosperity through self-sufficiency continues in Africa and other developing countries.

· · ·

Both the physiological differences between the sexes and the vast array of culturally created differences (that is, gender) have profoundly shaped environmental history. From pre-Columbian indigenous women working in the fields to woman-dominated environmental justice movements in the twenty-first century, American women have played a unique role in environmental history. Sometimes their involvement was for the worse, as when women's hat styles destroyed entire bird populations or when women carried out their role as homemakers by loading up their bathroom cabinets, kitchen cupboards, and laundry facilities with chemicals later found to be toxic. Sometimes their involvement was for the better, as when women's housekeeping imperative included cleaning up urban environments or when Rachel Carson alerted the world to the dangers of pesticides. In all cases, whether examining socially constructed differences or variances based in biology, much is to be gained from understanding gender, sex, and environment not as discreet entities, but as elements in a powerful synergy that significantly shapes American history.

Abbreviations

AGP	Amanda Gardner Papers, Sally Bingham Center, Rare Book, Manuscript, and Special Collections Library, Duke University, Durham, North Carolina.
AGRY	Alliance for Guidance for Rural Youth, Sally Bingham Center, Rare Book, Manuscript, and Special Collections Library, Duke University, Durham, North Carolina.
AHC	Adda Howie Collection, Wisconsin Historical Society, Madison, Wisconsin.
AHP	Alice Hamilton Papers, Schlesinger Library, Radcliffe Institute, Harvard University, Cambridge, Massachusetts.
ALFA	Atlanta Lesbian Feminist Alliance Collection, Sally Bingham Center, Rare Book, Manuscript, and Special Collections Library, Duke University, Durham, North Carolina.
CBC	Clara Burdette Collection, Huntington Library, San Marino, California.
CFWC	California Federation of Women's Clubs Collection, Huntington Library, San Marino, California.
CMMDY	Camp Moy Mo Da Yo Collection, Schlesinger Library, Radcliffe Institute, Harvard University, Cambridge, Massachusetts.
CPGC	Cambridge Plant and Garden Club Collection, Schlesinger Library, Radcliffe Institute, Harvard University, Cambridge, Massachusetts.
DFP	d'Autremont Family Papers, Schlesinger Library, Radcliffe Institute, Harvard University, Cambridge, Massachusetts.
DLC	Dolores Levy Collection, Wisconsin Historical Society, Madison, Wisconsin.

EHDP	Eleanor Hall Douglas Papers, Sally Bingham Center, Rare Book, Manuscript, and Special Collections Library, Duke University, Durham, North Carolina.
EL	Elwell Letters, Wisconsin Historical Society, Madison, Wisconsin.
EMP	Ellen Miller Papers, Wisconsin Historical Society, Madison, Wisconsin.
FWAP	Francis Walker Aglionby Papers, Sally Bingham Center, Rare Book, Manuscript, and Special Collections Library, Duke University, Durham, North Carolina.
HHC	Harriet Hardy Collection, Schlesinger Library, Radcliffe Institute, Harvard University, Cambridge, Massachusetts.
HNP	Haller Nutt Papers, Sally Bingham Center, Rare Book, Manuscript, and Special Collections Library, Duke University, Durham, North Carolina.
HSP	Harriet Strong Papers, Huntington Library, San Marino, California.
JC	Julia Carpenter notes, Bancroft Library, Berkeley, California.
LAND	League Against Nuclear Dangers Collection, Wisconsin Historical Society, Stevens Point, Wisconsin.
LAT	*Los Angeles Times*
LDM	Lyman Draper Manuscripts, Wisconsin Historical Society, Madison, Wisconsin.
LHD	Lorena Hays Diary, Bancroft Library, Berkeley, California.
LLD	Lucinda Lee Dalton autobiography, Bancroft Library, Berkeley, California.
MC	Mary Carr Diary, Sally Bingham Center, Rare Book, Manuscript, and Special Collections Library, Duke University, Durham, North Carolina.
MFP	McDonald Furman Papers, Sally Bingham Center, Rare Book, Manuscript, and Special Collections Library, Duke University, Durham, North Carolina.
NYT	*New York Times*
RJH	Richard James Hooker Collection, Schlesinger Library, Radcliffe Institute, Harvard University, Cambridge, Massachusetts.
VWH	Voices of Women Homemakers Collection, National Extension Home Makers Council Archives, Wisconsin Historical Society, Madison, Wisconsin.
WARN	Women of All Red Nations Collection, in League Against Nuclear Dangers Collection, Wisconsin Historical Society, Stevens Point, Wisconsin.
WFWC	Wisconsin Federation of Women's Clubs Records, Wisconsin Historical Society, Madison, Wisconsin.
WKL	William Kellogg Letter, Bancroft Library, Berkeley, California.
WPC	Women for a Peaceful Christmas, Wisconsin Historical Society, Madison, Wisconsin.
WSWWTC	We Say What We Think Club, Isabel Baumann Papers, Wisconsin Historical Society, Madison, Wisconsin.
ZFP	Zimmerman Family Papers, Huntington Library, San Marino, California.

Notes

INTRODUCTION

1. Lillian Schlissel, *Women's Diaries of the Westward Journey* (New York: Schocken, 1982), 28; 53.

2. Both Lillian Schlissel and John Mack Faragher assert that moving west was always a husband's decision. Julie Roy Jeffrey emphasizes that women were not entirely passive in that they at least participated in that decision making. See ibid.; John Mack Faragher, *Women and Men on the Overland Trail* (New Haven, Conn.: Yale University Press, 2001), 163; Julie Roy Jeffrey, *Frontier Women "Civilizing" the West? 1840–1880* (New York: Hill and Wang, 1998), 42.

3. Lydia Adams-Williams, "A Million Women for Conservation," *Conservation: Official Organ of the American Forestry Association* 15 (1909): 346–47.

4. George L. Knapp, "George L. Knapp Opposed Conservation," in *Major Problems in American Environmental History*, 2nd ed., ed. Carolyn Merchant (Boston: Houghton Mifflin, 2005), 322.

5. Carolyn Merchant, "Gender and Environmental History," *Journal of American History* 76, no. 4 (1990): 1117–21. Classic studies published prior to Merchant's plea include William Cronon, *Changes in the Land: Indians, Colonists, and the Ecology of New England* (New York: Hill and Wang, 1983, 2003); Roderick Nash, *American Environmentalism* (New York: McGraw-Hill, 1990) and *Wilderness and the American Mind* (New Haven, Conn.: Yale University Press, 1973, 1982); Donald Worster, *Rivers of Empire: Water, Aridity, and the Growth of the American West* (New York: Oxford University Press, 1985); and Mark Reisner, *Cadillac Desert: The American West and Its Disappearing Water* (New York: Penguin, 1986).

6. Studies that focused on women before Merchant's call include Walter O'Meara, *Daughters of the Country: The Women of the Fur Traders and the Mountain Men* (New York: Harcourt,

Brace, and World, 1968); Donald Rothblatt, Daniel Garr, and Jo Sprague, *The Suburban Environment and Women* (New York: Praeger, 1979); Joan M. Jensen, *With These Hands: Women Working on the Land* (New York: Feminist Press, 1981); Dona G. Gearheart, "Coal Mining Women in the West: The Realities of Difference in an Extreme Environment," *Journal of the West* 37, no. 1 (1988): 60–68; Glenda Riley, "'Wimmin Is Everywhere': Conserving and Feminizing Western Landscapes, 1870–1940," *Western Historical Quarterly* 29, no. 1 (Spring 1988): 5–23; Riley, *The Female Frontier: A Comparative View of Women on the Prairie and the Plains* (Lawrence: University Press of Kansas, 1988); and Riley, "Western Women and the Environment," *New Mexico Historical Review* 65 (April 1990): 267–75. Environmental studies that feature women and/or gender published since Merchant's call include Glenda Riley, *A Place to Grow: Women in the American West* (Arlington Heights, Ill.: Harlan Davidson, 1992); Maureen A. Flanagan, "The City Profitable, the City Livable: Environmental Policy, Gender, and Power in Chicago in the 1910s," *Journal of Urban History* 22, no. 2 (January 1996): 163–90; Priscilla Massmann, "A Neglected Partnership: The General Federation of Women's Clubs and the Conservation Movement, 1890–1920" (Ph.D. diss., University of Connecticut, 1997); Vera Mary Murphy, *Mining Cultures: Men, Women, and Leisure in Butte, 1914–1941* (Urbana: University of Illinois Press, 1997); Claudia Clark, *Radium Girls: Women and Industrial Health Reform, 1910–1935* (Chapel Hill: University of North Carolina Press, 1997); Mary Joy Breton, *Women Pioneers for the Environment* (Boston: Northeastern University Press, 1998, 2000); Virginia Scharff, ed., *Seeing Nature through Gender* (Lawrence: University Press of Kansas, 2003); Susan R. Schrepfer, *Nature's Altars: Mountains, Gender, and American Environmentalism* (Lawrence: University Press of Kansas, 2005); and several studies of Rachel Carson, including Linda Lear, *Rachel Carson: Witness for Nature* (New York: Owl Books, 1997).

7. See such works as Annabel Rodda, ed., *Women and the Environment* (London: Zed Books, 1993); Irwin Altman and Arza Churchman, eds., *Women and the Environment* (New York: Plenum, 1994); Janice Jiggins, *Changing the Boundaries: Women-Centered Perspectives on Population and the Environment* (Washington, D.C.: Island, 1994); Carolyn Merchant, *Earthcare: Women and the Environment* (New York: Routledge, 1995); Alison Ravetz, *Place of Home: English Domestic Environments, 1914–2000* (New York: Routledge, 1995); Brenda Martin and Penny Sparke, eds., *Women's Places: Architecture and Design, 1860–1960* (New York: Routledge, 2003); Sylvia Lorraine Bowerbank, *Speaking for Nature: Women and Ecologies of Early Modern England* (Baltimore: Johns Hopkins University Press, 2004); Barbara T. Gates, *Kindred Nature: Victorian and Edwardian Women Embrace the Living World* (Chicago: University of Chicago Press, 1999).

8. Carolyn Merchant argues for nature as partner in the web of life in *Reinventing Eden: The Fate of Nature in Western Culture* (New York: Routledge, 2003).

9. For examples of women naturalists see Tina Gianquitto, *Good Observers of Nature: American Women and the Scientific Study of the Natural World, 1820–1885* (Athens: University of Georgia Press, 2007); Harriet Kofalk, *No Woman Tenderfoot: Florence Merriam Bailey, Pioneer Naturalist* (College Station: Texas A&M University Press, 1989); Marcia Myers Bonta, *Women in the Field: America's Pioneering Women Naturalists* (College Station: Texas A&M University Press, 1991); Maxine Benson, *Martha Maxwell, Rocky Mountain Naturalist* (Winnipeg: Bison Books, 1999); Nancy J. Warner, *Taking to the Field: Women Naturalists in the Nineteenth-Century West* (master's thesis, Utah State University Press, 1995); Theodora C. Stanwell-Fletcher, *Driftwood Valley: A Woman Naturalist in the Northern Wilderness* (Corvallis: Oregon State University Press, 1999); Judith Reick Long, *Gene Stratton-Porter: Novelist and Naturalist* (Indianapolis:

Indiana Historical Society, 1990). For nature writers see Vera Norwood, *Made from This Earth: American Women and Nature* (Chapel Hill: University of North Carolina Press, 1993); Vera Norwood and Janice Monk, *The Desert Is No Lady: Southwestern Landscapes in Women's Writing and Art* (Tucson: University of Arizona Press, 1997); Lorraine Anderson, *Sisters of the Earth: Women's Prose and Poetry about Nature* (New York: Vintage, 2003); Susan Goodman and Carl Dawson, *Mary Austin and the American West* (Berkeley: University of California Press, 2009). For women environmentalists see Madelyn Holmes, *American Women Conservationists: Twelve Profiles* (Jefferson, N.C.: McFarland and Company, 2004); Kimberly Jarvis, *Franconia Notch and the Women Who Saved It* (Durham: University of New Hampshire Press, 2007); Glenda Riley, *Women and Nature: Saving the "Wild" West* (Lincoln: University of Nebraska Press, 1999).

10. See Merchant, *Major Problems*; Ted Steinberg, *Down to Earth: Nature's Role in American History* (New York: Oxford University Press, 2002); Carolyn Merchant, *American Environmental History, an Introduction* (New York: Columbia University Press, 2007); Kendall E. Bailes, ed., *Environmental History: Critical Issues in Comparative Perspective* (Lanham, Md.: Rowman and Littlefield, 1985).

11. Wendy Kaminer, "Crashing the Locker Room," *Atlantic*, July 1992. Available: http://www.theatlantic.com/issues/92jul/kaminer.htm (accessed June 5, 2009).

12. For evidence of the latter see Fawn Pattison, "Examining the Evidence on Pesticide Exposure and Birth Defects in Farmworkers: An Annotated Bibliography with Resources for Lay Readers" (Raleigh, N.C.: Agricultural Resources Center and Pesticide Education Project, 2006). Available: http://www.beyondpesticides.org/documents/Evidence_May06.pdf (accessed June 8, 2009).

13. Susan Badger Doyle, "Women's Experiences on the Westward Emigrant Trails," in *Encyclopedia of Women in the West*, ed. Gordon Bakken and Brenda Farrington (Thousand Oaks, Calif.: Sage, 2003), 313.

14. Allan G. Johnson, *The Gender Knot: Unraveling Our Patriarchal Legacy* (Philadelphia: Temple University Press, 1997).

15. Douglas Weiner, "A Death-Defying Attempt to Articulate a Coherent Definition of Environmental History," *Environmental History* 10, no. 3 (July 2005): 416. For other efforts to define environmental history see Jared Diamond, "The Evolution of Guns and Germs," in *Evolution: Society, Science, and the Universe*, ed. A. C. Fabian (Cambridge: Cambridge University Press, 1998), reprinted as "Predicting Environmental History," in Merchant, *Major Problems*, 2nd ed., 14; Donald Worster, "Doing Environmental History," in *The Ends of the Earth: Perspectives on Modern Environmental History*, ed. Worster (New York: Cambridge University Press, 1988), reprinted as "Doing Environmental History," in Merchant, *Major Problems*, 2nd ed., 4; Steinberg, *Down to Earth*, xii; Merchant, "What Is Environmental History?," *Major Problems*, 2nd ed., 1. See also Char Miller and Hal Rothman, *Out of the Woods: Essays in Environmental History* (Pittsburgh: University of Pittsburgh Press, 1997), 1.

16. J. R. McNeill, "Observations on the Nature and Culture of Environmental History," *History and Theory* 42, no. 4 (December 2003): 5–43.

17. For Hamilton see Wilma Ruth Slaight, "Alice Hamilton: First Lady of Industrial Medicine" (Ph.D. diss., Case Western Reserve University, 1974); Angela Nugent Young, "Interpreting the Dangerous Trades: Workers' Health in America and the Career of Alice Hamilton, 1910–1935" (Ph.D. diss., Brown University, 1982). A more recent biography is Barbara Sicherman, *Alice Hamilton: A Life in Letters* (Champaign: University of Illinois Press, 2003). For work on Dormon produced during the period see Donald M. Rawson, "Caroline Dormon: A Renaissance

Spirit of Twentieth Century Louisiana," *Louisiana History* 24, no. 2 (1983): 121–39; and Fran Holman Johnson, *The Gift of the Wild Things: The Life of Caroline Dormon* (Lafayette: University of Southwestern Louisiana Press, 1990). For Austin see Augusta Fink, *I-Mary: A Biography of Mary Austin* (Tucson: University of Arizona Press, 1983); and Esther F. Lanigan, *Mary Austin: Song of a Maverick* (Tucson: University of Arizona Press, 1997). A later biography is Heike Schaefer, *Mary Austin's Regionalism: Reflections on Gender, Genre, and Geography* (Charlottesville: University of Virginia Press, 2004). Studies of Carson produced in the 1970s and 1980s include Paul Brooks, *The House of Life: Rachel Carson at Work* (Boston: Houghton Mifflin, 1972) and *Speaking for Nature: How Literary Naturalists from Henry Thoreau to Rachel Carson Have Shaped America* (Boston: Houghton Mifflin, 1980). The best subsequent biography is Lear, *Rachel Carson*.

18. For more on areas receiving the most scholarly attention, see Elizabeth Blum, "Linking American Women's History and Environmental History: A Preliminary Historiography." Available: http://www.h-net.org/~environ/historiography/uswomen.htm (accessed January 15, 2009), as well as Carolyn Merchant, *The Columbia Guide to American Environmental History* (New York: Columbia University Press, 2002).

19. See Susan Schrepfer and Douglas Cazaux Sackman, "Gender," in *A Companion to American Environmental History*, ed. Sackman (Hoboken, N.J.: Wiley-Blackwell, 2010), 116–45. Environmental historians are increasingly investigating topics involving gender, women, and sexuality, but such research is still in its infancy. By 2007, the Forest History Society's *Environmental History Bibliography* included more than 40,000 entries. In 2012, the search term "forests" generated 4,746 entries, and "water" generated 2,599. The search term "women" generated 691 entries, "gender" generated 254, and "sexuality" only 18. Some entries appeared in more than one category, but even when these multiple listings are included in the final tally, the total for all three terms constitutes just over 2 percent of this database. Forest History Society, *Environmental History Bibliography*. Available: http://www.foresthistory.org/Research/biblio.html (accessed March 28, 2012).

20. For some of the ways in which gender caused women to see nature differently than did men, as well as some of the ways women's exposure to nature was limited by gender prescriptions, see Schrepfer, *Nature's Altars*. For essays focused specifically on the relationship between lesbianism and environment, see Catriona Mortimer-Sandilands and Bruce Erickson, eds., *Queer Ecologies: Sex, Nature, Biopolitics, and Desire* (Bloomington: University of Indiana Press, 2010). For the role of popular culture in shaping attitudes toward nonhuman nature, see Noel Sturgeon, *Environmentalism in Popular Culture: Gender, Race, Sexuality, and the Politics of the Natural* (Tucson: University of Arizona Press, 2008).

21. Links between gender, race, and environmentalism are explored in Linda Kalof et al., "Race, Gender, and Environmentalism: The Atypical Values and Beliefs of White Men," *Race, Gender, and Class* 9, no. 2 (2002): 1–19.

CHAPTER 1

1. Nathaniel Philbrick, *Mayflower* (New York: Penguin, 2006), 5; 108; 165.
2. Ibid., 165; 179; 206.
3. For an overview see Mary P. Ryan, "Where Have the Corn Mothers Gone? Americans Encounter the Europeans," in her *Mysteries of Sex: Tracing Women and Men through American History* (Chapel Hill: University of North Carolina Press, 2006). See also Melissa Leach and

Cathy Green, "Gender and Environmental History: From Representation of Women and Nature to Gender Analysis of Ecology and Politics," *Environment and History* 3 (1997): 352.

4. LHD, June 23, 1852, 16.

5. See Christopher Vecsey, "American Indian Environmental Religions," in *American Indian Environments: Ecological Issues in Native American History*, ed. Christopher Vecsey and Robert Venables (Syracuse, N.Y.: Syracuse University Press, 1980), 1–37.

6. See Hermien Soselisa, "The Significance of Gender in the Fishing Economy of the Goram Islands, Maluku," in *Old World Places, New World Problems: Exploring Issues of Resource Management in Eastern Indonesia*, ed. Sandra Pannell and Franz von Benda-Beckmann (Canberra: Australian National University, Centre for Resource and Environmental Studies, 1988), 321–35.

7. For an overview of pre-Columbian women see Carolyn Niethammer, *Daughters of the Earth: The Lives and Legends of American Indian Women* (New York: Simon and Schuster, 1977).

8. Virginia Scharff and Carolyn Brucken, *Home Lands: How Women Made the West* (Berkeley: University of California Press, 2010): 7–13.

9. Frances Watkins, "Southwestern Athapascan Women," *Southwestern Lore* 10, no. 33 (December 1944): 32; 34.

10. George Irving Quimby, *Indian Culture and European Trade Goods: The Archaeology of the Historic Period in the Western Great Lakes Region* (Madison: University of Wisconsin Press, 1966), 15.

11. Genevieve McBride, ed., *Women's Wisconsin: From Native Matriarchies to the New Millennium* (Madison: Wisconsin Historical Society Press, 2005), 3.

12. Robert Bieder, *Native American Communities in Wisconsin, 1600–1960* (Madison: University of Wisconsin Press, 1995), 26–28; 34; 41.

13. Colin G. Calloway, *First Peoples: A Documentary Survey of American Indian History* (Boston: Bedford/St. Martin's, 1999), 20.

14. See Morrill Marsten to the Reverend Jedediah Morse, November 1820, Thomas Forsyth Papers, Volume 1T, 65, LDM.

15. Merchant, *Earthcare*, 92–95.

16. Ryan, *Mysteries of Sex*, 28.

17. See James J. Rawls and Walton Bean, *California: An Interpretive History*, 7th ed. (New York: McGraw-Hill, 1998), 11–13; Richard Dasmann, *California's Changing Environment* (San Francisco: Boyd and Fraser, 1981), 1–8; George Phillips, *The Enduring Struggle: Indians in California History* (San Francisco: Boyd and Fraser, 1981), 4–12.

18. Ryan, *Mysteries of Sex*, 25.

19. See Riley, *A Place to Grow*, 15–33.

20. Samuel de Champlain, "The French Explorer Samuel de Champlain Describes the Lives of Huron Women and Men in the Great Lakes Region, 1616," in *Major Problems in American Women's History*, 2nd ed., ed. Mary Beth Norton and Ruth M. Alexander (Boston: Heath, 1996), 25.

21. George-Louis de Buffon, in Thomas Jefferson, *Notes on the State of Virginia* (New York: Norton, 1954), 59.

22. See Jensen, *With These Hands*, 4–5.

23. Thomas Forsyth, "Manners and Customs of the Sauk and Fox Nations of Indians," 1827, Thomas Forsyth Papers, Volume 9T, 21, LDM.

24. Paul le Jeune, S.J., to Bartholomew Jacquinot, S.J., 1633, in Reuben Gold Thwaites, ed., *The Jesuit Relations: and Allied Documents Travels and Explorations of the Jesuit Missionaries in New*

France 1610–1791, vol. 5 (Cleveland: Burrows Brothers, 1896–1901), 179. Available http://puffin. creighton.edu/jesuit/relations/relations_05.html (accessed September 1, 2011).

25. Ryan, *Mysteries of Sex*, 28, 38.

26. Cronon, *Changes in the Land*, 45, 151–52.

27. See Shepard Krech, *The Ecological Indian: History and Myth* (New York: Norton, 2000), as well as Michael E. Harkin and David Lewis, eds. *Native Americans and the Environment: Perspectives on the Ecological Indian* (Lincoln: University of Nebraska Press, 2007). See also William E. Tydeman, "No Passive Relationship: Idaho Native Americans in the Environment," *Idaho Yesterdays* 39, no. 2 (Summer 1995): 23–28; Guy Gugliotta, "Indians Hunted Carelessly, Study Says," *Seattle Times*, February 21, 2006.

28. Cronon, *Changes in the Land*, 80.

29. Forsyth, "Manners and Customs of the Sauk and Fox Nations of Indians," Forsyth Papers, 20, LDM.

30. Thomas Jefferson recorded that Indian women had a tradition of procuring abortions by the use of "some vegetable" whose efficacy extended "to prevent contraception for a considerable amount of time after." Jefferson, *Notes on the State of Virginia*, 60. Russell Thornton, *American Indian Holocaust and Survival: A Population History since 1492* (Norman: University of Oklahoma Press, 1987), 31. See also Demitri Shimkin, "Eastern Shoshone," in *Handbook of North American Indians (Great Basin)*, ed. Warren L. D'Azevedo (Washington D.C.: Smithsonian Institution, 1986), 330; Francis Riddell, "Maidu and Konkow," in *Handbook of North American Indians (California)*, ed. Robert F. Heizer (Washington D.C.: Smithsonian Institution, 1978), 381; T. N. Campbell, ed., "Coahuiltecans and Their Neighbors," in *Handbook of North American Indians (Southwest)*, ed. Alonso Ortiz (Washington D.C.: Smithsonian Institution, 1983), 352; O'Meara, *Daughters of the Country*, 84; John Demos, *The Tried and the True: Native American Women Confronting Colonization* (New York: Oxford University Press, 1995), 77; Ryan, *Mysteries of Sex*, 40; Niethammer, *Daughters of the Earth*, 19–21.

31. Thornton, *American Indian Holocaust and Survival*, 31; Adam Hochschild, *King Leopold's Ghost* (New York: First Mariner Books, 1999), 73. See also Liese M. Perrin, "Resisting Reproduction: Reconsidering Slave Contraception in the Old South," *Journal of American Studies* 35, no. 2 (2001): 258–59; 263; 266.

32. See Niethammer, *Daughters of the Earth*, 2–21.

33. Forsyth, "Manners and Customs," 15, Forsyth Papers, LDM.

34. Vecsey, "American Indian Environmental Religions."

35. See Nancy C. Unger, "Women, Sexuality and Environmental Justice in American History," in *New Perspectives on Environmental Justice: Gender, Sexuality, and Activism*, ed. Rachel Stein (New Brunswick, N.J.: Rutgers University Press, 2004), 48.

36. Philbrick, *Mayflower*, 186.

37. Chad Montrie, *Making a Living: Work and Environment in the United States* (Chapel Hill: University of North Carolina Press, 2008), 15.

38. Cronon, *Changes in the Land*, 167.

39. Ibid.

40. See Virginia Bouvier, *Women and the Conquest of California, 1542–1840: Codes of Silence* (Tucson: University of Arizona Press, 2001).

41. Philbrick, *Mayflower*, 188.

42. Susan Sleeper-Smith, *Indian Women and French Men: Rethinking Cultural Encounter in the Western Great Lakes* (Amherst: University of Massachusetts Press, 2001); Sylvia Van

Kirk, *Many Tender Ties: Women in Fur-Trade Society, 1670–1870* (Norman: University of Oklahoma Press, 1983; Winnipeg: Watson and Dwyer, 1996). See also Ryan, *Mysteries of Sex*, 46–47.

43. Ryan, *Mysteries of Sex*, 51.

44. Carol Devens, *Countering Colonization: Native American Women and Great Lakes Missions, 1630–1900* (Berkeley: University of California Press, 1992). See also Devens, "'If We Get the Girls, We Get the Race': Missionary Education of Native American Girls," *Journal of World History* 3, no. 2 (1992): 219–37, and "Separate Confrontations: Gender as a Factor in Indian Adaptation to European Colonization in New France," *American Quarterly* 38 (1986): 461–80; Karen Anderson, *Chain Her by One Foot: The Subjugation of Women in Seventeenth Century New France* (New York: Routledge, 1991); and Henry Bowden, *American Indians and Christian Missions: Studies in Cultural Conflict* (Chicago: University of Chicago Press, 1981).

45. For Spanish-speaking women's training of Indian girls and women in European domestic arts, see Miroslava Chavez-Garcia, *Negotiating Conquest: Gender and Power in California, 1770s–1880s* (Tucson: University of Arizona Press, 2004), 20–23.

46. See Rose Marie Beebe and Robert M. Senkewicz, eds., *Lands of Promise and Despair* (Berkeley: Santa Clara University and Heyday Books), 2001.

47. See Paul F. Starrs, "California's Grazed Ecosystems," in Carolyn Merchant, *Green Versus Gold: Sources in California's Environmental History* (Washington, D.C.: Island, 1998), 199–205.

48. Richard Rice, William Bullough, and Richard Orsi, eds., *The Elusive Eden: A New History of California*, 2nd ed. (New York: McGraw-Hill, 1996), 97–98. See also Dasmann, *California's Changing Environment.*

49. Kate Luckie, "Kate Luckie (Wintu) Deplores the Soreness of the Land," in Merchant, *Green Versus Gold*, 35.

50. Albert Hurtado, "Sexuality in California's Franciscan Missions: Cultural Perceptions and Sad Realities," *California History* 72 (Fall 1992): 370–85.

51. Jefferson, *Notes on the State of Virginia*, 60.

52. Joe Starita, *"I Am a Man": Chief Standing Bear's Journey for Justice* (New York: St. Martin's Press, 2007), 11.

53. See Robert J. Miller, *Native America, Discovered and Conquered: Thomas Jefferson, Lewis and Clark, and Manifest Destiny* (Lincoln, Neb.: Bison Books, 2008).

54. Ryan, *Mysteries of Sex*, 55.

55. Ira Berlin, *Generations of Captivity: A History of African-American Slaves* (Cambridge: Belknap Harvard, 2003), 56–57; 72. Another scholar claims "traders imported almost twice as many men as women," a smaller but still significant difference. Peter Kolchin, *American Slavery, 1619–1877* (New York: Hill and Wang, 1993), 23.

56. Deborah Gray White, *"Ar'n't I a Woman?" Female Slaves in the Plantation South* (1985; repr. New York: Norton, 1999), 67.

57. Berlin, *Generations of Captivity*, 83.

58. Liese Perrin, "Slave Women and Work in the American South" (Ph.D. diss., University of Birmingham, 1999), 89.

59. Carr Diary, February 22–29; July 3–7, 1861, MC.

60. Haller Nutt, "Cotton Picking," *Journal of the Araby Plantation, 1843–1850*, 118; 124–126, HNP.

61. Perrin, "Slave Women and Work in the American South," 110.

62. Ryan, *Mysteries of Sex*, 122. See also Daina Ramey, "'She Do a Heap of Work': Female Slave Labor on Glynn County Rice and Cotton Plantations," *Georgia Historical Quarterly* 82, no. 4 (Winter 1998): 707–34.

63. Philip D. Morgan, *Slave Counterpoint: Black Culture in the Eighteenth Century Chesapeake and the Low Country* (Chapel Hill: University of North Carolina Press, 1998), 207.

64. Ibid. See also Daina Berry, *Swing the Sickle for the Harvest Is Ripe* (Chicago: University of Illinois Press, 2007).

65. See Berry, *Swing the Sickle*, 17; Perrin, "Slave Women and Work in the American South," 19; S. Mintz, "Slavery Fact Sheet," *Digital History* (2007). Available: http://www.digitalhistory.uh.edu/historyonline/slav_fact.cfm (accessed February 16, 2008).

66. Haller Nutt, "Directions in the Treatment of the Sick," *Journal of the Araby Plantation, 1843–1850*, 194–95, HNP.

67. A. B. [John Billiller], "Rules for the Plantation," Sumterville, S.C., news clipping, November 3, 1847, MFP.

68. Nutt, "Directions in the Treatment." See also Perrin, "Slave Women and Work in the American South," 156–60.

69. Frederick Law Olmsted, *The Cotton Kingdom: A Traveler's Observations on Cotton and Slavery in the American Slave States* (New York: Knopf, 1953), 94.

70. Ibid., 170.

71. Ibid., 198–214. See also Mart Stewart, "Slavery and the Origins of African American Environmentalism," in *"To Love the Wind and the Rain": African Americans and Environmental History*, ed. Dianne Glave and Mark Stoll (Pittsburgh: University of Pittsburgh Press, 2006), 9–20; and Whitney Battle, "A Yard to Sweep: Race, Gender, and the Enslaved Landscape" (Ph.D. diss., University of Texas at Austin, 2004).

72. For historiography see Perrin, "Resisting Reproduction," 255–74. See also Sharla M. Fett, *Working Cures: Healing, Health, and Power on Southern Slave Plantations* (Chapel Hill: University of North Carolina Press, 2002), 65, 176–77.

73. Perrin, "Resisting Reproduction," 256.

74. Nutt, "Directions in Treatment of Sick," 194, HNP.

75. See Fett, *Working Cures*.

76. Scott Giltner, "Slave Hunting and Fishing in the Antebellum South," in Glave and Stoll, *"To Love the Wind and the Rain,"* 21–36.

77. Elizabeth Blum, "Power, Danger, and Control: Slave Women's Perception of Wilderness in the Nineteenth Century," *Women's Studies* 31, no. 2 (2002): 247–66. See also Stephanie M. H. Camp, "'I Could Not Stay There': Enslaved Women, Truancy and the Geography of Everyday Forms of Resistance in the Antebellum Plantation South," in *Women, Families, and Communities*, ed. Nancy Hewitt and Kirsten Delegard, vol. 1 (New York: Pearson Longman, 2008), 196–209.

CHAPTER 2

1. Frances Anne Kemble, *Journal of a Residence on a Georgian Plantation* (1863; repr. New York: Knopf, 1961), 87.

2. Ibid., 218.

3. Ibid., 6.

4. Ibid., 93.

5. Ibid., 114.

6. Ibid., 202–3. See also 378.

7. Mary Beth Norton, *Liberty's Daughters: The Revolutionary Experience of American Women, 1750–1800* (New York: Harper Collins, 1980), 169; 225.

8. Linda Kerber, "The Republican Mother: Women and the Enlightenment—An American Perspective (1976)," in *Toward an Intellectual History of Women*, ed. Kerber (Chapel Hill: University of North Carolina Press, 1997), 41–62.

9. Benjamin Rush, "Of the Mode of Education Proper in a Republic," in *Essays, Literary, Moral and Philosophical* (Samuel F. Bradford, 1798), 6–20. Reprinted in Frederick Rudolph, ed., *Essays on Education in the Early Republic* (Cambridge, Mass.: Belknap Press of Harvard University Press, 1965), 9–23.

10. Benjamin Rush, "Thoughts upon Female Education, Accommodated to the Present State of Society, Manners and Government, in the United States of America," in Rudolph, ed., *Essays on Education*, 77–78.

11. Julia Cowles, *The Diaries of Julia Cowles: A Connecticut Record, 1797–1803*, ed. Laura Moseley (New Haven, Conn.: Yale University Press, 1931), 76.

12. Hannah Buchanan to Thomas Buchanan, August 13, 1809, Box 1, Folder 5, RJH.

13. Eleanor Hall Douglas to Dear Sister, May 1, 1820, EHDP (emphasis in original).

14. "Learning Housekeeping Continued," (San Francisco) *Daily Evening Bulletin*, May 9, 1856, issue 28, col. A.

15. See Hannah Buchanan to Thomas Buchanan, September 11, 1809, Box 1, Folder 5, RJH.

16. Farmer's Cabinet, "Kitchen Garden," *New-Hampshire Statesman and State Journal*, April 6, 1839, E.

17. Clarissa Dillon, "'A Large, an [sic] Useful, and a Grateful Field': Eighteenth-Century Kitchen Gardens in Southeastern Pennsylvania, the Uses of Plants, and Their Place in Women's Work" (PhD diss., Bryn Mawr, 1986), 27.

18. Farmer's Cabinet, "Kitchen Garden," E.

19. American Farmer, "Kitchen Garden, for March," *Illinois Gazette*, March 28, 1829, D.

20. Dillon, "A Large Field," 77–79.

21. London Magazine, "Garden Operations for Ladies," *Lady's Book*, July 1831, 10.

22. Suellen Hoy, *Chasing Dirt: The American Pursuit of Cleanliness* (New York: Oxford University Press, 1995), 15.

23. Ibid., 7.

24. Ibid., 12.

25. See Dillon, "A Large Field."

26. Lydia Marie Child, *The American Frugal Housewife* (reissue ed., Carlisle, Mass.: Applewood, 1989). See also Sarah Hale, *Early American Cookery: "The Good Housekeeper" 1841* (New York: Dover, 1996); Mary Randolph, *The Virginia Housewife*, 4th ed. (Washington D.C.: Thompson, 1830).

27. For the important role that natural remedies played within slave culture, see Fett, *Working Cures*, especially chapter 3, "Sacred Plants," 60–83.

28. Marie d'Autremont to Mr. d'Autremont, September 1, 1807, file 1, 66, DFP; Charles d'Autremont to David Craft, December 28, 1897, file 1, 11, DFP.

29. Marie d'Autremont to Mr. d'Autremont, September 1, 1807, file 1, 68, DFP.

30. See Ivy Schweitzer, "Foster's 'Coquette': Resurrecting Friendship from the Tomb of Marriage," *Arizona Quarterly* 61, no. 2 (2005): 1–32. For fears that this separate female world could contribute to lesbianism in the early republic see Kristin M. Comment, "Charles Brockden Brown's 'Ormond' and Lesbian Possibility in the Early Republic," *Early American Literature* 40, no. 1 (2005): 57–78.

31. Deidre English and Barbara Ehrenreich, *"For Her Own Good": Two Centuries of the Experts' Advice to Women* (New York: Anchor, 2005), 157.

32. Mary Clavers, "Spring in the Woodlands," *Lady's Book* 31 (July 1845): 22.

33. "Garden Borders," *Godey's Lady's Book* 50 (May 1855): 477.

34. Barbara Welter, "The Cult of True Womanhood: 1820–1860." *American Quarterly* 18, no. 2, pt. 1 (1966): 151–74.

35. See E. Anthony Rotundo, *American Manhood* (New York: Basic Books, 1994), and Michael S. Kimmel, *Manhood in America: A Cultural History* (New York: Oxford University Press, 2005).

36. Welter, "Cult of True Womanhood."

37. Susan A. L., "Another Letter from the Western Wilds," *Godey's Lady's Book* 50 (February 1855): 130.

38. Nancy F. Cott, *The Bonds of Womanhood: "Woman's Sphere" in New England, 1780–1835* (New Haven, Conn.: Yale University Press, 1977).

39. Although few historians dispute what was being prescribed, there is controversy over to what degree it was internalized and by whom. While some historians defend Barbara Welter's conclusions about the era's women, others object to the negative connotations of the word "cult," and criticize Welter's definition of true womanhood as both simplistic and overly rigid. More recent additions to the controversy include Mary Kelley, "Beyond the Boundaries," *Journal of the Early Republic* 21, no. 1 (2001): 73–78; Carol Lasser, "Beyond Separate Spheres: The Power of Public Opinion," *Journal of the Early Republic* 21, no. 1 (2001): 115–23; Mary Cronin, "Redefining Woman's Sphere," *Journalism History* 25, no. 1 (1999): 13–25; Brian Gabrial, "A Woman's Place," *American Journalism* 25, no. 1 (2008): 7–29; Mary Louise Roberts, "True Womanhood Revisited," *Journal of Women's History* 14, no. 1 (2002): 150–155; Nancy A. Hewitt, "Taking the True Woman Hostage," *Journal of Women's History* 14, no. 1 (2002): 156–62.

40. LLD, Circle Valley Utah, 1865, 7.

41. Ibid.

42. LHD, April 6, 1852.

43. English and Ehrenreich, *"For Her Own Good,"* 157.

44. See English and Ehrenreich, "Microbes and the Manufacture of Housework," in English and Ehrenreich, *"For Her Own Good,"* 155–65; Ryan, *Mysteries of Sex*, 88–102.

45. Of the scholars who can at least agree on its existence, those who present the "woman's sphere" of the early to mid-nineteenth century as more negative than positive for women include Welter, "Cult of True Womanhood"; Gerda Lerner, "The Lady and the Mill Girl: Changes in the Status of Women in the Age of Jackson," *Mid-Continental American Studies Journal* 10 (Spring 1969): 5–15; and Mary P. Ryan, "Mothers of Civilization: The Common Woman, 1830–1860," in Ryan, *Womanhood in America* (New York: New Viewpoints, 1975), 137–81. A more positive interpretation of the "sphere" and its impact are discussed in Cott, *Bonds of Womanhood*; Kathryn Kish Sklar, *Catherine Beecher: A Study in American Domesticity* (New Haven, Conn.: Yale University Press, 1973) and *Florence Kelley and the Nation's Work: The Rise of Women's Political Culture, 1830–1900* (New Haven, Conn.: Yale University Press, 1995); Carroll Smith-Rosenberg,

"The Female World of Love and Ritual: Relations between Women in Nineteenth-Century America," *Signs: Journal of Women in Culture and Society* 1 (1975): 1–30; and Daniel Scott Smith, "Family Limitation, Sexual Control, and Domestic Feminism in Victorian American," in *Clio's Consciousness Raised*, ed. Mary Hartman and Lois Banner (New York: Harper & Row, 1974): 119–36. There is also extensive literature debating the "sphere's" relationship between women and organized religion, westward migration, slavery, and a host of other topics.

46. LLD, 7.

47. See Katherine Clinton, "Pioneer Women in Chicago, 1833–1837," *Journal of the West* 12, no. 2 (April 1973): 317–24.

48. Patricia Anne Carter, *"Everybody's Paid but the Teacher": The Teaching Profession and the Women's Movement* (New York: Teachers College Press, 2002), 34.

49. Ibid., 36.

50. Ibid., 38.

51. Ibid., 20.

52. Nancy Woloch, *Women and the American Experience* (New York: McGraw-Hill, 1984), 129.

53. Mary Irving, "A Peep at the Prairie," *National Era* 5, no. 212 (January 23, 1851): 13.

54. Clavers, "Spring in the Woodlands," 22.

55. "Landscape Gardening. Operating with Wood," *Godey's Lady's Book* 46 (March 1853): 252.

56. Norwood, *Made from This Earth*, 20–21.

57. Judith H. Dobrzynski, "The Grand Women Artists of the Hudson River School," Smithsonian.com, July 21, 2010. Available: http://www.smithsonianmag.com/arts-culture/The-Grand-Women-Artists-of-the-Hudson-River-School.html?c=y&;page=2# (accessed April 28, 2011).

58. Norwood, *Made from This Earth*, 25.

59. Susan Fenimore Cooper, *Rural Hours* (New York: Gregory P. Putnam, 1850), 161–62.

60. Norwood, *Made from This Earth*, 33.

61. Ibid., 34; Cooper, *Rural Hours*, 91.

62. Norwood, *Made from This Earth*, 37.

63. Ibid., 39.

64. Cooper, *Rural Hours*, 217.

65. Norwood, *Made from This Earth*, 34–35; Cooper, *Rural Hours*, 277.

66. Norwood, *Made from This Earth*, 4.

67. Rebecca Harding Davis, *Life in the Iron Mills and Other Stories* (New York: The Feminist Press at CUNY, 1985), 12; 23. Davis is described as having "protoecofeminist sensibilities" in Michele L. Mock, "'A Message to Be Given': The Spiritual Activism of Rebecca Harding Davis," *NWSA Journal* 12, no. 1 (Spring 2000): 44–67.

68. See Daina Ramey Berry, "'I Had to Work Hard, Plow, and Go and Split Wood Jus' Like a Man': Skill, Gender, and Productivity," in *Swing the Sickle*, 13–34.

69. David Christy, *Cotton Is King; or, The Culture of Cotton, and Its Relation to Agriculture, Manufactures, and Commerce, to the Free Colored People, and to Those Who Hold That Slavery Is in Itself Sinful*, 2nd ed. (Cincinnati: Moore, Wilstach, Keys & Co., 1855).

70. Kemble, *Journal of a Residence*, 202.

71. Ibid., 202.

72. Ibid., 219.

73. Ibid.

74. Eugene Genovese, "Cotton, Slavery, and Soil Exhaustion in the Old South," *Cotton History Review* 2, no. 1 (1961), 5.

75. Charles Schuler, "A Louisiana Convention Declares War on the Boll Weevil, 1903," in Merchant, *Major Problems in Environmental History*, 2nd ed., 218.

76. James C. Giesen, "'The Truth about the Boll Weevil': The Nature of Planter Power in the Mississippi Delta," *Environmental History* 14, no 4 (October 2009): 692.

77. Ibid., 689–90.

78. Morgan, *Slave Counterpoint*, 184.

79. Scholars Daniel Littlefield and Judith Carney show that women's agricultural expertise in rice, indigo, corn, and cotton production stemmed from specialized knowledge and hand-tool experience garnered from working crops native to their homelands and passed down through the generations. Daniel Littlefield, *Rice and Slaves: Ethnicity and the Slave Trade in Colonial South Carolina* (Urbana: University of Illinois Press, 1991); Judith Carney, *Black Rice* (Cambridge, Mass.: Harvard University Press, 2001), 162. For more on women slaves in rice production see "Slavery," in *A Hard Fight for We: Women's Transition from Slavery to Freedom in South Carolina*, ed. Leslie A. Schwalm (Urbana: University of Illinois Press, 1997), 19–72.

80. Morgan, *Slave Counterpoint*, 193.

81. Judith Carney's conclusion "that knowledge of rice cultivation enabled slaves arriving in South Carolina to enjoy greater autonomy from their owners than was possible for other crops" is questioned by David Eltis, Philip Morgan, and David Richardson in "Agency and Diaspora in Atlantic History: Reassessing the African Contribution to Rice Cultivation in the Americas," *American Historical Review* 112, no. 5 (December 2007): 1328–58. The controversy was revisited in "*AHR* Exchange: The Question of 'Black Rice,'" *American Historical Review* 115, no. 1 (February 2010): 123–71.

82. Genovese, "Cotton, Slavery," 6.

83. Ibid.

84. Ibid., 8.

85. Ibid., 7.

86. Ibid., 9.

87. Another solution to replenishing exhausted soils is crop rotation, which helps counteract the leaching and erosion of soil dedicated too long to supporting a single crop. It was also not routinely practiced, as planters were unwilling to take land away from their proven cash crop. Ibid., 10.

88. See E. N. Elliott, ed., *Cotton Is King, and Pro-Slavery Arguments; Comprising the Writings of Hammond, Harper, Christy, Stringfellow, Hodge, Bledsoe, and Cartwright, on This Important Subject* (1860; repr., New York: Negro Universities Press, 1969).

89. Olmsted, *Cotton Kingdom*, 410.

90. Ibid., 411.

91. A similar, although more flagrant refusal to terrace white-owned hillsides occurred in Kenya during 1948–1949. Known as "The Revolt of the Women," local workers undermined colonists' efforts to stem soil erosion. See Fiona Mackenzie, "Political Economy of the Environment, Gender and Resistance under Colonialism: Murang'a District, Kenya, 1910–1950," *Canadian Journal of African Studies* 25, no. 2 (1991): 226–56.

92. See Judith Ann Giesberg, *Civil War Sisterhood: The U.S. Sanitary Commission and Women's Politics in Transition* (Boston: Northeastern University Press, 2006); Justin Martin, *Genius of Place: The Life of Frederick Law Olmsted* (Cambridge, Mass.: Da Capo Press, 2011), 178–229.

CHAPTER 3

1. Jessie May, "My Little Neighbor," *Godey's Lady's Book* 64 (April 1862): 356.

2. M. C. P., "Letter to the Publisher," *Godey's Lady's Book* 49 (December 1854): 538.

3. Mrs. E. F. Ellett, "The Pioneer Mothers of Michigan," *Godey's Lady's Book* 44 (April 1852): 266.

4. Carolyn Kirkland, *A New Home—Who'll Follow?* (New York: C. S. Frances, 1839), 64.

5. Ibid., 14, 38.

6. Ibid., 246.

7. Annette Kolodny, *The Land Before Her: Fantasy and Experience of the American Frontiers, 1630–1860* (Chapel Hill: University of North Carolina Press, 1984), 148–49.

8. Mary Irving, "The Romance of Society," *National Era* 5, no. 213 (January 30, 1851): 17.

9. The degree of influence gender prescriptions had on the activities and attitudes of the newcomers both on the trail and once in the West remains the subject of lively debate. Some scholars offer compelling case studies that demonstrate that the two sexes saw the overland trek through different eyes, and that the white middle-class women who constituted most of the female presence in the westward movement clung almost possessively to their traditional roles. See Schlissel, *Women's Diaries*, 3–4. Sandra Myres and Julie Jeffrey, on the other hand, emphasize a full spectrum of women's attitudes and experiences in the West. Sandra Myres, *Westering Women and the Frontier Experience, 1800–1915* (Albuquerque: University of New Mexico Press, 1992) and *Ho for California! Women's Overland Diaries from the Huntington Library* (San Marino, Calif.: Huntington Library Press, 2007); Jeffrey, *Frontier Women*. Others, including John Mack Faragher, highlight the similarities between men and women on the trail and in the West. See Faragher, *Women and Men on the Overland Trail*. Melody Miyamota asserts that pioneers clung to gendered ideology, but ultimately made significant adjustments. Melody Miyamoto, "No Home for Domesticity? Gender and Society on the Overland Trails" (Ph.D. diss., Arizona State University, 2006). A variety of books focus on pioneering women in general, including Linda Peavy and Ursula Smith's *Pioneer Women: The Lives of Women on the Frontier* (Norman: University of Oklahoma Press, 1998); Ruth Moynihan, Susan Armitage, and Christian Fischer Dichamp's edited collection *So Much to Be Done: Women Settlers on the Mining and Ranching Frontiers* 2nd ed (Lincoln: University of Nebraska Press, 1998); and Susan Roberson's edited collection *Women, America, and Movement: Narratives of Relocation* (Columbia: University of Missouri Press, 1998). Others take a more regional approach, such as Terri Baker and Connie Henshaw's edited collection, *Women Who Pioneered Oklahoma: Stories from the WPA Narratives* (Norman: University of Oklahoma Press, 2007); JoAnn Levy's *They Saw the Elephant: Women in the California Gold Rush* (Norman: University of Oklahoma Press, 1992); Glenda Riley's *Female Frontier* and "Women on the Wisconsin Frontier, 1836–1848," in McBride's edited collection *Women's Wisconsin*. For single women homesteaders see Paula Mae Bauman, "Single Women Homesteaders in Wyoming, 1880–1930," *Annals of Wyoming* 58:1 (1986): 39–53.

These various accounts are filled with women gathering fuel along westward trails; building sod houses; clearing land; planting, harvesting, and preserving crops; herding animals; milking cows; and tending chickens.

10. Schlissel, *Women's Diaries*, 10–11, 27.

11. Laura Ingalls Wilder, *Little House on the Prairie* (New York: Harper and Row, 1971), 1–2.

12. Robert Chadwell Williams, *Horace Greeley: Champion of American Freedom* (New York: New York University Press, 2006), 42–43.

13. Edwin H. Grant, "To the Unemployed of Our Eastern Cities," (Washington, D.C.) *National Era* 11, no. 572 (December 17, 1857): 201.

14. "Free Soil for Free Men," (Belleville, Ill.) *Advocate*, in *National Era* 11, no. 549 (July 9, 1857): 112.

15. Schlissel, *Women's Diaries*, 21.

16. Anders Stephanson, *Manifest Destiny: American Expansion and the Empire of Right* (New York: Hill and Wang, 1996), 42.

17. See Richard W. Etulain, *Does the Frontier Experience Make America Exceptional?* (Boston: Bedford/St. Martins, 1999).

18. Raymond W. Settle and Mary Settle, eds., *Overland Days to Montana* (Glendale, Calif.: Arthur Clark Company, 1971), 15.

19. Schlissel, *Women's Diaries*, 150.

20. Moynihan, Armitage, and Dichamp, eds., *So Much to Be Done*, 5.

21. Schlissel, *Women's Diaries*, 10, 95.

22. For an account of girls' experiences, see Mary Barmeyer O'Brien, *Toward the Setting Sun: Pioneer Girls Traveling the Overland Trails* (Helena, Mont.: TwoDot, 1999); for boys' experiences, see O'Brien, *Into the Western Winds: Pioneer Boys Traveling the Overland Trails* (Helena, Mont.: TwoDot, 2002).

23. Mollie Dorsey Sanford, *Mollie: The Journal of Mollie Dorsey Sanford in Nebraska and Colorado Territories, 1857–1866* (Lincoln: University of Nebraska Press, 1959), 3.

24. Schlissel, *Women's Diaries*, 83.

25. Ibid., 84.

26. Ibid., 23–24.

27. See Cathy Luchetti and Carol Olwell, *Women of the West* (New York: Orion, 1982), 27.

28. See Schlissel, *Women's Diaries*, 98–99.

29. Ibid., 106.

30. Settle and Settle, *Overland Days*, 51.

31. Schlissel, *Women's Diaries*, 106.

32. LHD, 4.

33. Ibid., 5.

34. Sanford, *Mollie*, 2.

35. Settle and Settle, *Overland Days*, 43.

36. Ibid., 44–45.

37. Kenneth Holmes, ed., *Covered Wagon Women: Diaries and Letters from the Western Trails*, vol. 2 (Glendale, Calif.: Arthur C. Clark, 1983), 237.

38. Harriet Fish Backus, *Tomboy Bride: A Woman's Personal Account of Life in Mining Camps of the West* (Boulder, Colo.: Pruett, 1969).

39. Margaret Frink, *Journal of the Adventures of a Party of California Gold Seekers* (Fairfield, Wash.: Ye Galleon Press, 1987), 10.

40. Joanna Stratton, *Pioneer Women: Voices from the Kansas Frontier* (New York: Simon and Schuster, 1981), 34.

41. Wilder, *Prairie*, 3.

42. Schlissel, *Women's Diaries*, 28.

43. Holmes, *Covered Wagon Women*, vol. 2, 235.

44. Stratton, *Pioneer Women*, 44–45.

45. Frink, *Journal*, 14.

46. Ibid., 92.

47. WKL, 66.

48. Rice, Bullough, and Orsi, *Elusive Eden*, 2nd ed., 171.

49. Frink, *Journal*, 84.

50. Ibid., 52.

51. Schlissel, *Women's Diaries*, 106; Frink, *Journal*, 84.

52. Frink, *Journal*, 92.

53. See Robert J. Willoughby, ed., *The Great Western Migration to the Gold Fields of California, 1849–1850* (Jefferson, N.C.: McFarland and Company, 2003); LeRoy R. Hafen and Ann W. Hafen, eds., *Journals of Forty-Niners, Salt Lake to Los Angeles* (Winnipeg: Bison Books, 1998).

54. Frink, *Journal*, 30.

55. Settle and Settle, *Overland Days*, 43–44.

56. Ibid., 54, 58.

57. Ibid., 48.

58. Schlissel, *Women's Diaries*, 85.

59. Settle and Settle, *Overland Days*, 51–52.

60. Schlissel, *Women's Diaries*, 84.

61. Faragher, *Women and Men*, 109.

62. Myres, *Westering Women*, 141–42.

63. Sarah Hepburn Hayes, "Peter Allan's Panther Chase," *Godey's Lady's Book* 40 (February 1850): 134.

64. Luchetti and Olwell, *Women of the West*, 29.

65. Charles Dana Wilbur, *The Great Valleys and Prairies of Nebraska and the Northwest* (Omaha, Neb.: Daily Republican Print, 1881).

66. David Laskin, *The Children's Blizzard* (New York: HarperCollins, 2004), 57–58.

67. Ibid., 185.

68. Jensen, *With These Hands*, 108.

69. Luchetti and Olwell, *Women of the West*, 29; Jensen, *With These Hands*, 108–9; Kolodny, *The Land Before Her*, 140.

70. Moynihan, Armitage, and Dichamp, *So Much to Be Done*, 125; 128.

71. Alice B. Neal, "Life at a Post," *Godey's Lady's Book* 46 (January 1853): 58.

72. Ellett, "Pioneering Mothers of Michigan," 226.

73. See Myres, *Westering Women*.

74. Ada Colvin, in Holmes, *Covered Wagon Women*, vol. 11, 52.

75. Rice, Bullough, and Orsi, *Elusive Eden*, 184.

76. Mary Beth Norton and Ruth Alexander, eds., *Major Problems in American Women's History*, 4th ed. (Boston: Houghton Mifflin, 2007), 232–36. Magoffin's diary is published in Stella M. Drumm, ed., *Down the Santa Fe Trail and into Mexico* (Lincoln: University of Nebraska Press, 1982).

77. Moynihan, Armitage, and Dichamp, *So Much to Be Done*, 127.

78. In Worster, *Rivers of Empire*, 97. Emphasis in original.

79. Schlissel, *Women's Diaries*, 148. For the evolution of the relationship between gender and women's roles during the transition from homesteading to settlement, see Chad Montrie, "'Men Alone Cannot Settle a Country': Domesticating Nature in the Kansas-Nebraska Grasslands," *Great Plains Quarterly* 25, no. 4 (2005): 245–58. See also Cynthia Prescott, "'Why She Didn't Marry Him': Love, Power, and Marital Choice on the Far Western Frontier," *Western Historical Quarterly* 38, no. 1 (2007): 25–45.

80. See Stratton, *Pioneer Women*, 145.

81. WKL, 44 (emphasis in original).

82. WKL, 66, 70–71, 97, 108.

83. LHD, 16.

84. Ibid., 30.

85. Ibid., 27–28, 30. See also Sarah Royce, *A Frontier Lady: Recollections of the Gold Rush and Early California* (Lincoln: University of Nebraska Press, 1960), 109–110.

86. LHD, 37.

87. Ibid., 43.

88. Moynihan, Armitage, and Dichamp, *So Much to Be Done*, 117.

89. Adrienne Caughfield, *True Women and Westward Expansion* (College Station: Texas A&M University Press, 2005), 101.

90. See Jacqueline Barnhart, *The Fair but Frail: Prostitution in San Francisco, 1849–1900* (Reno: University of Nevada Press, 1986).

91. Sylvia Van Kirk, "The Role of Native Women in the Creation of Fur Trade Society in Western Canada, 1670–1830," *Frontiers* 7, no. 3 (1984): 11. See also Theda Perdue, *Cherokee Women: Gender and Culture Change, 1700–1835* (Winnipeg: Bison Books, 1999), and Bruce White, "The Woman Who Married a Beaver: Trade Patterns and Gender Roles in the Ojibwa Fur Trade," *Ethnohistory* 46, no. 1 (1999): 109–47.

92. Beebe and Senkewicz, *Lands of Promise and Despair*, 172; 289.

93. Chavez-Garcia, *Negotiating Conquest*, 57.

94. Albert Hurtado, *Indian Survival on the California Frontier* (New Haven, Conn.: Yale University Press, 1990), 161.

95. Chavez-Garcia, *Negotiating Conquest*, 153.

96. Ibid., 165. See also Deborah Kanter, "Native Female Land Tenure and Its Decline in Mexico, 1750–1900," *Ethnohistory* 42 (Fall 1995): 607–16. For more on how race and ethnicity interacted with gender in the West, see Vicki Ruiz, "Shaping Public Space/Enunciating Gender: A Multiracial Historiography of the Women's West, 1995–2000," *Frontiers* 22, no. 3 (2001): 22–25; S. J. Kleinberg, "Race, Region, and Gender in American History," *Journal of American Studies* 33, no. 1 (1999): 85–88; Paula Nelson, "Women and the American West: A Review Essay," *Annals of Iowa* 50, nos. 2–3 (1989): 269–273; Katherine Benton-Cohen, "Common Purposes, Worlds Apart: Mexican-American, Mormon, and Midwestern Women Homesteaders in Cochise County, Arizona," *Western Historical Quarterly* 36 (Winter 2005): 429–52.

97. Chavez-Garcia, *Negotiating Conquest*, 155.

98. In Caughfield, *True Women and Westward Expansion*, 26.

99. Benton-Cohen, "Common Purposes, Worlds Apart," 439–40.

100. The proportion of Mexican-American women homesteaders doubled after American homesteading laws were liberalized in 1900. Ibid., 442.

101. Deena Gonzalez, *Refusing the Favor: The Spanish-Mexican Women of Santa Fe, 1820–1880* (New York: Oxford University Press, 1999), 103.

102. Chavez-Garcia, *Negotiating Conquest*, 123.

103. Gonzalez, *Refusing the Favor*, 83, 85, 87.

104. Ibid., 10. Chavez-Garcia agrees with Gonzalez's conclusions, but emphasizes that American takeover also created opportunities for women to contest power relationships in marriage and family. Chavez-Garcia, *Negotiating Conquest*, xvi.

105. Gonzalez, *Refusing the Favor*, 43.

106. Chavez-Garcia, *Negotiating Conquest*, 149.

107. Ibid., 156.

108. Ibid., 149; Gonzalez, *Refusing the Favor*, 49.

109. Gonzalez, *Refusing the Favor*, 104.

110. John Downey to Mrs. Erastus Burr, January 14, 1857, ZFP.

111. For more on the evolution of gender spheres see Nancy C. Unger, "The Two Worlds of Belle La Follette," *Wisconsin Magazine of History* 83 (Winter 1999–2000): 82–110.

112. Chauncey Elwell to Mother, January 1, 1854, EL.

113. Catherine Elwell to Mother, October 21, 1852; December 31, 1853; to Sister Sabra, June 9, 1891, EL.

114. See Moynihan, Armitage, and Dichamp, *So Much to Be Done*, 111–12; Riley, *Women and Nature*.

115. Julia Carpenter, "Notes on Western Travels, 1882–1887," 3; II4; II17; 18, JC.

116. For African American women in the West, see Tricia Martineau Wagner, *African American Women of the Old West* (Helena, Mont.: TwoDot, 2007); Luchetti and Olwell, *Women of the West*, 43, 46–47; Schlissel, *Women's Diaries*, 136–38; Caughfield, *True Women and Western Expansion*, 97–116. For the aesthetics and cultural significance of traditional gardening practices of rural African American women, see Richard Westmacott, *African-American Gardens and Yards in the Rural South* (Knoxville: University of Tennessee Press, 1992); Dianne Glave, "'A Garden So Brilliant with Colors, So Original in Its Design': Rural African American Women, Gardening, Progressive Reform, and the Foundation of an African American Environmental Perspective," *Environmental History* 8, no. 3 (2003): 395–411.

117. Sally Spaulding, "The Letters of Ellen Spaulding Miller," 8, EMP.

118. Ellen Miller to family, May 15, 1876, EMP.

119. Sally Spaulding, "The Letters of Ellen Spaulding Miller," 32, EMP.

120. Ida McMechen to Mary Post Zimmerman, June 1, 1932, Box 2, File 41, ZFP.

121. LHD, 6.

122. Settle and Settle, *Overland Days*, 60.

123. Mary Dodge Woodward, "The Diary of Mary Dodge Woodward," in *The Checkered Years*, ed. Mary Cowdrey (Caldwell, Idaho: Caxton, 1937), 38.

124. Ibid., 81.

125. Jane Haigh and Claire Murphy, *Gold Rush Women* (Anchorage: Alaska Northwest Books, 1997), 20.

126. Ibid., 48–49.

127. Lael Morgan, *The Good Time Girls of the Alaska-Yukon Gold Rush* (Fairbanks, Alaska: Epicenter, 1998); Haigh and Murphy, *Gold Rush Women*, 12.

128. Mary McCarty to Minnie Patterson, December 9, 1900, Box 2, File 32, ZFP.

129. Mary McCarty to Mary Post Zimmerman, May 25, 1903; July 31, 1906, Box 2, File 32, ZFP.

130. Ella Chase to Mary Post Zimmerman, October 11, 1900, Box 1, File 15, ZFP.

131. Ella Chase to Mary Post Zimmerman, December 17, 1900. Box 1, File 15, ZFP.

132. See Mary McCarty to Mary Post Zimmerman, February 17, 1913, Box 2, File 32, ZFP.

CHAPTER 4

1. Adda F. Howie, "Women in Agriculture," 2–3; 8, Folder 1, AHC.

2. "Mrs. Howie's Dairy Farm," *Farm Sentinel*, August 14, 1902, news clipping, Folder 3, AHC.

3. George H. Dacy, "America's Outstanding Woman Farmer," *The Forecast*, April 1925, clipping, 252, Folder 3, AHC. See also Larry F. Graber, "America's Outstanding Woman Farmer," 1932, Folder 3, AHC.

4. Dacy, "America's Outstanding Woman Farmer."

5. Undated statement by Adda Howie, 3, Folder 1, AHC.

6. See Elizabeth Blum, "Women, Environmental Rationale, and Activism during the Progressive Era," in Glave and Stoll, "*To Love the Wind and the Rain*," 77–92.

7. Nancy C. Unger, *Fighting Bob La Follette: The Righteous Reformer*, rev. ed. (Madison: Wisconsin Historical Society, 2008), 86–87.

8. See Suellen Hoy, *Chasing Dirt*, 59–86; Flanagan, "The City Profitable"; Carolyn Merchant, "Women of the Progressive Conservation Movement, 1900–1916," *Environmental Review* 8, no. 1 (Spring 1984): 55–85.

9. Mary Austin, *Land of Little Rain* (1903; repr., Albuquerque: University of New Mexico Press, 1974).

10. For considerably more nuance, see Marc C. Carnes and Clyde Griffen, eds., *Meanings of Manhood* (Chicago: University of Chicago Press, 1990).

11. Carolyn Merchant, "George Bird Grinnell's Audubon Society: Bridging the Gender Divide in Conservation," *Environmental History* 15 (January 2010): 4–5.

12. See Cindy Aron, *Working at Play: A History of Vacations in the United States* (New York: Oxford University Press, 1999), 158.

13. Ibid., 174–75, 231.

14. Ibid., 173.

15. William James, "The Moral Equivalent of War," *McClure's Magazine* 35 (August 1910): 467.

16. David McCullough, *The Johnstown Flood: The Incredible Story behind One of the Most Devastating "Natural" Disasters America Has Ever Known* (New York: Touchstone, 1968), 253.

17. Ibid., 268.

18. See Andrew Isenberg, *The Destruction of the Bison: An Environmental History, 1850–1920* (New York: Cambridge University Press, 2001).

19. For the gender divide, see Kathryn Kish Sklar, "Two Political Cultures in the Progressive Era: The National Consumers' League and the American Association for Labor Legislation," in *U.S. History as Women's History: New Feminist Essays*, ed. Linda K. Kerber, Alice Kessler-Harris, and Kathryn Kish Sklar (Chapel Hill: University of North Carolina Press, 1995), 36–62.

20. John F. Reiger, *American Sportsmen and the Origins of Conservation*, 3rd rev. ed., (Corvallis: Oregon State University Press, 2000); Daniel Justin Herman, *Hunting and the American Imagination* (Washington, D.C.: Smithsonian Institution Press, 2001).

21. Isenberg, *Destruction of the Bison*, 169.

22. Sarah Watts, *Rough Rider in the White House: Theodore Roosevelt and the Politics of Desire* (Chicago: University of Chicago Press, 2003); Peter Bayers, "Frederick Cook, Mountaineering in the Alaskan Wilderness and the Regeneration of Progressive Era Masculinity," *Western American Literature* 38, no. 2 (Summer 2003): 170–93; Tina Loo, "Of Moose and Men: Hunting for Masculinities in British Columbia, 1880–1939," *Western Historical Quarterly* 32, no. 3 (Autumn 2001): 296–320; Gail Bederman, *Manliness and Civilization: A Cultural History of Gender and Race in the United States, 1880–1917* (Chicago: University of Chicago Press, 1996).

23. Andrea Smalley, "'Our Lady Sportsmen': Gender, Class, and Conservation in Sport Hunting Magazines, 1873–1920," *Journal of the Gilded Age and Progressive Era* 4, no. 4 (October 2005): 359, 364.

24. See Allison Hepler, *Women in Labor: Mothers, Medicine, and Occupational Health in the United States, 1890–1980* (Columbus: Ohio State University Press, 2000).

25. Susan Fitzgerald, "Women in the Home," in *One Half the People: The Fight for Woman Suffrage*, ed. Anne Firor Scott and Andrew MacKay Scott (Champaign: University of Illinois Press, 1982), 114–15. Jane Addams offered an expanded version of these arguments in the pamphlet *Why Women Should Vote*, in *Modern History Sourcebook*, 1999. Available: http://www.fordham.edu/halsall/mod/1915janeadams-vote.html (accessed June 23, 2010).

26. Wendy Keefover-Ring, "Municipal Housekeeping, Domestic Science, Animal Protection, and Conservation: Women's Political and Environmental Activism in Denver, Colorado, 1894–1912" (M.A. thesis, University of Colorado, 2002), 105.

27. Clara Bradley Burdette, "The College Woman and Citizenship," *The Syracusan* (June 15, 1917): 3, Box 125, file 1, CBC.

28. Rheta Childe Dorr, *What Eight Million Women Want* (Boston: Small, Maynard, and Co., 1910), 327.

29. Fitzgerald, "Women in the Home," 115.

30. "What Is Meant by Conservation?" *Ladies' Home Journal* 28 (November 1911): 23, 95, in *Conservation in the Progressive Era: Classic Texts*, ed. David Stradling (Seattle: University of Washington Press, 2004), 33.

31. Massmann, "A Neglected Partnership," unnumbered page, 3, 5.

32. Mrs. Overton Ellis, in Merchant, "Women of the Progressive," 73–74.

33. Mary I. Wood, *The History of the General Federation of Women's Clubs for the First Twenty-Two Years of Its Organization* (Norwood, Mass.: Norwood, 1912), 147.

34. Elizabeth M. Howe, "Ellen Swallow Richards, 1882," 4, Alumnae Department, Vassar College Library, enclosure from Don Scherer to author (author's collection).

35. Darlene Stille, *Extraordinary Women Scientists* (Chicago: Children's Press, 1995), 168.

36. Howe, "Ellen Swallow Richards," 6.

37. Keefover-Ring, "Municipal Housekeeping," 66, 168.

38. See Mary Joy Breton, "First Lady of Environmental Science," in her *Women Pioneers for the Environment*, 47–63; Robert Clarke, *Ellen Swallow: The Woman Who Founded Ecology* (New York: Follett, 1973); and John McBrewster, Frederic P. Miller, and Agnes F. Vandome, eds., *Ellen Swallow Richards* (Mauritius: Alphascript, 2010).

39. Burdette, "The College Woman and Citizenship," 4, CBC.

40. Martin Melosi, "Environmental Justice, Political Agenda Setting, and the Myths of History," *Journal of Policy History* 12, no. 1 (2000): 53.

41. Addams, *Why Women Should Vote.*

42. Martha Conine in Keefover-Ring, "Municipal Housekeeping," 4–5.

43. Melosi, "Environmental Justice," 53. See, for example, Clark, *Radium Girls*; Hepler, *Women in Labor*; Angela Gugliotta, "Class, Gender, and Coal Smoke: Gender Ideology and Environmental Injustice in Pittsburgh, 1868–1914," *Environmental History* 5 (April 2000): 165–193; Mark Speltz, "'An Interest in Health and Happiness as Yet Untold': The Woman's Club of Madison, 1893–1917," *Wisconsin Magazine of History* 89 (Spring 2006): 2–15.

44. Melosi, "Environmental Justice," 54.

45. Jane Addams, *20 Years at Hull-House* (1910; repr., Charleston, S.C.: BookSurge Classics, 2004), 242.

46. Keefover-Ring, "Municipal Housekeeping," 116, 119, 124.

47. Sarah Huftalen, "The Use of the Hand Book in Rural Schools," *Midland Schools: A Journal of Education* 24, no. 10 (July 1910): 300.

48. Keefover-Ring, "Municipal Housekeeping," 83.

49. Jennifer Koslow, *Cultivating Health, Los Angeles Women and Public Health Reform* (Piscataway, N.J.: Rutgers University Press, 2009), 159.

50. See Jennifer Koslow, "Putting It to a Vote: The Provision of Pure Milk in Progressive Era Los Angeles," *Journal of the Gilded Age and Progressive Era* 3, no. 2 (April 2004): 111–44; Meredith Eliassen, "Got Pure Milk? Dr. Adelaide Brown's Crusade for San Francisco's Safe Milk Supply," *Argonaut* 18, no. 1 (January 2007): 36–51. While recognizing the role of women, Daniel Block puts greater emphasis on the role of male public health officers in "Saving Milk through Masculinity: Public Health Officers and Pure Milk, 1880–1930," *Food and Foodways: History and Culture of Human Nourishment* 15, no. ½ (January–June 2005): 115–34.

51. (St. Paul, Minn.) *Pioneer Press*, June 5, 1905, in Keefover-Ring, "Municipal Housekeeping," 87.

52. Barbara Sicherman, "Working It Out: Gender, Profession, and Reform in the Career of Alice Hamilton," in *Gender, Class, Race, and Reform in the Progressive Era*, ed. Noralee Frankel and Nancy S. Dye (Lexington: University Press of Kentucky, 1991), 127.

53. Ibid., 128.

54. Alice Hamilton, *Do Women in Industry Need Special Health Legislation?* Pamphlet No. 12, Consumers League of Connecticut, Box 2, Folder 28, AHP.

55. Alice Hamilton, "Industrial Diseases, Charities and the Commons, Vol. XX," Box 2, file 29, typescript, 655–59, AHP.

56. See Edna E. Raphael, "Sociological Strategies in the Investigation of Women's Occupational Health" and "Working Women—a Population at Risk," Box 7, folder 126, HHC.

57. Alice Hamilton, "Women Workers and Industrial Poisons," Box 2, Folder 28, 2–4, AHP. See also Hamilton, *Exploring the Dangerous Trades* (1943; repr., Miller, S.D.: Miller Press, 2008).

58. See Bill Kovarik and Mark Neuzil, "The Radium Girls," in Neuzil and Kovarik, *Mass Media and Environmental Conflict: America's Green Crusades* (Thousand Oaks, Calif.: Sage, 1996), 33–52; Clark, *Radium Girls*; and Deborah Blum, *The Poisoner's Handbook* (New York: Penguin, 2010).

59. Hamilton, "Industrial Diseases, Charities and the Commons, Vol. XX."

60. Elizabeth Beardsley Butler, *Women and the Trades: Pittsburgh, 1907–1908* (New York: Russell Sage Foundation, 1909), 95–96.

61. Esther Katz, "Sanger, Margaret." *American National Biography Online*, February 2000. Available: http://www.anb.articles//15-00598.html (accessed September 28, 2002).

62. Although Sanger never advocated forcibly limiting reproduction based on race, class, or ethnicity, her beliefs in "scientific eugenics" and assertions that some people carried genetic material that should not be passed down to future generations contribute to a common criticism that many of the early urban environmentalists were essentially tainted by cultural constraints and values. See Jean H. Baker, *Margaret Sanger: A Life of Passion* (New York: Hill and Wang, 2011).

63. Judy Yung, *Unbound Feet: A Social History of Chinese Women in San Francisco* (Berkeley: University of California Press, 1995), 35.

64. Ibid., 37. See Peggy Pascoe, *Relations of Rescue: The Search for Female Moral Authority in the American West, 1874–1939* (New York: Oxford University Press, 1993).

65. See Frankel and Dye, *Gender, Class, Race*.

66. Victoria Wong, "Square and Circle Club: Women in the Public Sphere," *Chinese America: History and Perspectives* (1995): 127–53.

67. Yung, *Unbound Feet*, 155.

68. Ibid., 152.

69. Blum, "Women, Environmental Rationale, and Activism," 86.

70. Elizabeth Blum, *Love Canal Revisited: Race, Class, and Gender in Environmental Activism* (Lawrence: University Press of Kansas, 2008), 133.

71. Colin Fisher, "African Americans, Outdoor Recreation, and the 1919 Chicago Race Riot," in Glave and Stoll, *"To Love the Wind and the Rain,"* 73.

72. Mary Church Terrell, "Club Work of Colored Women, 1901," and "Lynching from a Negro's Point of View," in Norton and Alexander, *Major Problems in American Women's History*, 4th ed., 294–97. See also Eileen Boris, "The Power of Motherhood: Black and White Activist Women Redefine the 'Political,'" *Yale Journal of Law and Feminism* 2, no.1 (Fall 1989): 25–49.

73. Blum, "Women, Environmental Rationale, and Activism," 91. See also Nancy Dye, "Introduction," to Frankel and Dye, *Gender, Class, Race*, 7.

74. Melosi, "Environmental Justice," 55.

75. Notes, Box 1, file 1, CPGC.

76. Ibid., Box 1, File 3, CPGC.

77. Centennial Timeline, Box 1, File 9, 7; 9, CPGC.

78. See Rae Eighmey, *Food Will Win the War: Minnesota Crops, Cooks, and Conservation during World War I* (St. Paul: Minnesota Historical Society Press, 2010).

79. Centennial Timeline, 9, CPGC.

80. Lydia Adams-Williams, "A Million Women for Conservation," 346–47.

81. Massmann, "A Neglected Partnership," 90. See, for example, Jack E. Davis, *An Everglades Providence: Marjory Stoneman Douglas and the American Environmental Century* (Athens: University of Georgia Press, 2009); Jarvis, *Franconia Notch*.

82. Newell Searle, "Minnesota National Forest: The Politics of Compromise, 1898–1908," *Minnesota History* 42, no. 7 (1971): 250.

83. Keefover-Ring, "Municipal Housekeeping," 128.

84. Kevin Armitage, *The Nature Study Movement: The Forgotten Populizer of America's Conservation Ethic* (Lawrence: University Press of Kansas, 2009), 52.

85. Ibid., 115.

86. Blum, "Women, Environmental Rationale, and Activism," 85.

87. "Mrs. Marion Crocker on the Conservation Imperative, 1912," in Merchant, *Major Problems in American Environmental History*, 2nd ed., 325.

88. Armitage, *Nature Study*, 117.

89. Ibid., 124.

90. Kevin C. Armitage, "Bird Day for Kids: Progressive Conservation in Theory and Practice," *Environmental History* 13, no. 3 (July 2007): 529.

91. Armitage, *Nature Study*, 128.

92. Ibid., 134.

93. Ibid.

94. Merchant, "Women of Progressive," 59.

95. Adams-Williams, "A Million Women for Conservation," 346–47. Adams-Williams's estimate is confirmed in Merchant, "Women of Progressive," 65.

96. Merchant, "Women of Progressive," 63.

97. Joseph Kastner, "Long before Furs, It Was Feathers That Stirred Reformist Ire," *Smithsonian* 25, no. 4 (July 1994): 100.

98. Sarah Orne Jewett, *A White Heron* (Memphis: General Books, 2009).

99. Kastner, "Long before Furs," 97.

100. Bill Bryson, *A Walk in the Woods: Rediscovering America on the Appalachian Trail* (New York: Broadway, 1998), 204–5.

101. Massmann, "Neglected Partnership," 168.

102. Merchant, "Women of Progressive," 64–65.

103. Kastner, "Long before Furs," 103.

104. "Ill Advised Sarcasm," *La Follette's Magazine* 5, no. 35 (August 30, 1913): 3. Such widespread anthropocentric and jingoistic views left many women believing that, in the words of environmental historian Carolyn Merchant, "Man the moneymaker had left it to woman the moneysaver to preserve resources." Merchant, "Women of the Progressive," 65.

105. Porter practiced what scholar Amy Green has termed a form of "muscular womanhood" grafted "onto the true womanhood ideal." Amy Green, "She Touched Fifty Million Lives," in Scharff, *Seeing Nature through Gender*, 235.

106. Armitage, *Nature Study*, 165–67.

107. Keefover-Ring, "Municipal Housekeeping," 146.

108. Robin W. Doughty, *Feather Fashions and Bird Preservation: A Study in Nature Protection* (Berkeley: University of California Press, 1975), 63.

109. Ibid., 72.

110. "Mrs. Marion Crocker," 324.

111. Merchant, "Women of Progressive," 68.

112. Act of October 3, 1913, c. 16, 38 Stat. 114 (148); Migratory Bird Treaty Act, c. 128, § 2, 40 Stat. 755 (1918).

113. Armitage, *Nature Study*, 202.

114. Adam Rome, "'Political Hermaphrodites': Gender and Environmental Reform in Progressive America," *Environmental History* 11, no. 3 (July 2006): 440–463. "Sweeping Back the Flood," *San Francisco Call*, December 13, 1909. Available: http://chroniclingamerica.loc.gov/lccn/sn85066387/1909-12-13/ed-1/seq-1/;words=Sweeping+Flood+Back?date1=1836&rows=20&se

archType=basic&state=&date2=1922&proxtext=sweeping+back+the+flood&dateFilterType=yearRange&;index=0 (accessed June 14, 2011).

115. See Robert V. Hine and John Mack Faragher, *Frontiers: A Short History of the American West* (New Haven, Conn.: Yale University Press, 2000), 176–90.

116. "The Pros and Cons of the Great Hetch Hetchy Dam Debate, 1913," in *Major Problems in the Gilded Age and Progressive Era*, ed. Leon Fink, 2nd ed. (Boston: Houghton Mifflin, 2001), 422–23.

117. Merchant, "Women of Progressive," 78.

118. Theodore Roosevelt, "President Theodore Roosevelt's Conservation Message, 1907," in Fink, *Major Problems in the Gilded Age*, 413.

119. Watts, *Rough Rider in the White House.*

120. George L. Knapp, "George L. Knapp Opposed Conservation," in Merchant, *Major Problems in American Environmental History*, 2nd ed., 322.

121. Blum, *Love Canal Revisited*, 28.

122. Ibid.

123. Susan Albertine, "The Life Writings of Harriet Strong," *Biography* 17, no. 2 (Spring 1994): 161, 165, 175.

124. Harriet Strong, "Water Power Hearings before the Committee on Water Power of the House of Representatives," 65th Cong., 2d sess., 14–27 (1918), part 4 (Washington D.C.: Government Printing Office), in HSP.

125. Clara Burdette, Box 139, File 3, CBC.

126. Merchant, "Grinnell's Audubon Society," 8.

127. Merchant, "Women of Progressive," 76.

128. Maureen Flanagan, *America Reformed: Progressives and Progressivisms, 1890s–1920s* (New York: Oxford University Press, 2007), 178.

129. Rome, "'Political Hermaphrodites'"; Gugliotta, "Class, Gender, and Coal Smoke"; Merchant, "Women of Progressive," 57–85.

130. Rome, "'Political Hermaphrodites,'" 456. See Kevin P. Murphy, *Political Manhood: Red Bloods, Mollycoddles, and the Politics of Progressive Reform* (New York: Columbia University Press, 2008).

131. Rome, "'Political Hermaphrodites,'" 455–56; Merchant, "Women of Progressive," 77.

132. Schrepfer, *Nature's Altars.*

133. Unsigned, undated, to Food Administrators and Fellow Workers, Box 137, File 5, CBC. See also Eighmey, *Food Will Win the War.*

134. See Elaine F. Weiss, *Fruits of Victory: The Woman's Land Army of America in the Great War* (Washington, D.C.: Potomac, 2008).

135. Clara Burdette, undated, "Neighborly Chat," Box 139, File 5, CBC.

136. Sicherman, "Working It Out," 142–43.

CHAPTER 5

1. Mary K. Sherman to Clara Burdette, December 1, 1922, Box 44, File 4, CBC.

2. John James Tigert to Mary K. Sherman, November 9, 1922, Box 44, File 4, CBC.

3. "Report on Garden Week," Biennial Convention of General Federation, June 3–13, 1924, Box 47, File 1, CBC.

4. Ibid.

5. Undated, unsigned plan to conduct a National Garden Week in the Spring, Box 44, File 4, CBC. William Levitt, founder of the housing community Levittown, similarly proclaimed, "No man who owns his own house and lot can be a communist. He has too much to do." David Halberstam, *The Fifties* (New York: Ballantine, 1984), 130.

6. Herbert Hoover to Mrs. Thomas O. Winter, January 17, 1924, Box 46, File 1, CBC.

7. Mrs. John D. Sherman, "The Home Makers of America," October 21, 1926, Box 136, File 11, CBC.

8. Mrs. W. L. Lawton, Official Statement of Policy, May 15, 1924, Box 46, File 5, CBC.

9. In Scharff and Brucken, *Home Lands*, 113.

10. Rebecca Conard, *Places of Quiet Beauty: Parks, Preserves, and Environmentalism* (Iowa City: University of Iowa Press, 1997), 81.

11. Ibid.

12. Ibid., 83.

13. Merchant, "Women of Progressive," 79.

14. Riley, *Women and Nature*, 131. See also Kimmel, *Manhood in America*, 117–56.

15. See Susan A. Miller, *Growing Girls: The Natural Origins of Girls' Organizations in America* (Piscataway, N.J.: Rutgers University Press, 2007), and Leslie Paris, *Children's Nature: The Rise of the American Summer Camp* (New York: New York University Press, 2008).

16. Schrepfer, *Nature's Altars*, 153–54.

17. Ibid., 155.

18. *Camp Fire Girls* (1912; repr., Carlisle, Mass.: Applewood, n.d.), 25.

19. Ibid., 23.

20. Ibid., 8, 12, 14.

21. Ibid., 8, 16.

22. Ibid., 16.

23. W. J. Hoxie, *How Girls Can Help Their Country: Handbook for Girl Scouts* (1913; repr., Carlisle, Mass.: Applewood, n.d.), vii.

24. Ibid., 12. Emphasis in original.

25. Ibid., 2.

26. Ibid., 108–9.

27. Ibid., 16.

28. Ibid., 14.

29. Schrepfer, *Nature's Altars*, 157.

30. *Camp Fire Girls*, frontispiece.

31. Hoxie, *How Girls Can Help*, 110, 80.

32. Ibid., 63.

33. Ben Jordan, "'Conservation of Boyhood': Boy Scouting's Modest Manliness and Natural Resource Conservation, 1910–1930," *Environmental History* 15, no. 4 (October 2010): 614, 617, 628.

34. Ibid., 629.

35. *Camp Fire Girls*, 16–17.

36. Schrepfer, *Nature's Altars*, 157.

37. Ibid., 158.

38. "History," *Campfire USA*. Available: http://www.campfireusaia.org/about-us-history.php (accessed November 1, 2011); "Our Eleventh Birthday," *Everygirls' Magazine* 10, no. 6 (February 1923): back cover.

39. History—Girl Scouts Timeline 1912–1919; 1920s, *Girl Scouts*. Available: http://www. girlscouts.org/who_we_are/history/timeline/1920s.asp (accessed November 1, 2011).

40. "Lou Henry Hoover: A Biographical Sketch," *The Herbert Hoover Presidential Library and Museum*. Available: http://hoover.archives.gov/education/louhenrybio.html (accessed September 1, 2010).

41. Paul C. Mishler, review of Paris, *Children's Nature*, in *Journal of American History* 96, no. 2 (September 2009): 590.

42. Riley, *Women and Nature*, 131.

43. Ibid., 132.

44. Camp Moy Mo Da Yo brochures, Box 1, Files 1–3, CMMDY.

45. "Camp Moy Mo Da Yo to Medical School," Box 1, File 1, CMMDY.

46. Jane Newburder Kauders, "The Husbands' Lament," in *Splash*, 20th anniversary edition, 1916–1935, Walden Camp Papers, Box 1, Schlesinger Library, Cambridge, Mass.

47. Interview by author, Virginia Remy Whitebread, June 4, 2011.

48. "Celebrating This Year Its Twenty Fifth Anniversary of Service," Box 17, File "Statement of Purpose," AGRY.

49. "Report of Visit of Miss Cox to a Virginia School," Box 17, File "Statement of Purpose," AGRY.

50. Field Notes, May 14, 1925, Box 17, File "Statement of Purpose," AGRY.

51. Ibid., January 1925.

52. Ibid., November 10, 1924.

53. Ibid., December 2, 1924. See also Melissa Walker, *All We Knew Was to Farm: Rural Women in the Upcountry South, 1919–1941* (Baltimore: Johns Hopkins University Press, 2000); Rebecca Sharpless, *Fertile Ground, Narrow Choices: Women on Texas Cotton Farms, 1900–1940* (Chapel Hill: University of North Carolina Press, 1999).

54. Dianne Glave, "Rural African American Women, Gardening, and Progressive Reform in the South," in Glave and Stoll, "*To Love the Wind and the Rain*," 43.

55. Ibid., 45.

56. "Celebrating This Year Its Twenty Fifth Anniversary of Service."

57. Montrie, *Making a Living*, 14.

58. Transcript, "Young Workers Who Make Our Clothes No. 1," *American School of the Air*, January 7, 1937, Box 46, File "Radio Broadcast Scripts, 1925–1937," AGRY.

59. Bryant Simon, "'New Men in Body and Soul': The Civilian Conservation Corps and the Transformation of Male Bodies and the Male Politic," in Scharff, *Seeing Nature through Gender*, 80–102.

60. Heather Van Wormer, "A New Deal for Gender: The Landscapes of the 1930s," in *Shared Spaces and Divided Places: Material Dimensions of Gender Relations and the American Historical Landscape*, ed. Deborah L. Rotman and Ellen-Rose Savulis (Knoxville: University of Tennessee Press, 2003): 219.

61. Blanche Wiesen Cook, *Eleanor Roosevelt, Vol. 2, 1933–1938* (New York: Viking, 1999), 88–91.

62. Simon, "New Men in Body and Soul," 82–83, 87–90, 95–97.

63. Aron, *Working at Play*, 253.

64. See Timothy Egan, *The Worst Hard Time: The Untold Story of Those Who Survived the Great American Dust Bowl* (New York: Mariner, 2006); Donald Worster, *Dust Bowl: The Southern Plains in the 1930s* (New York: Oxford University Press, 1979). For a woman's experiences of the Dust Bowl, see Ann Marie Low, *Dust Bowl Diary* (Lincoln: University of Nebraska Press, 1984).

65. Mrs. John F. Duvall, Bulletin, October 1936, Los Angeles District, Box 30, File 3, CFWC.

66. Violet Cotrell, Transcript, Washington, 9, VWH.

67. Dorothy Tolley, Transcript, Arkansas, 10, VWH.

68. Mary Raymond, Transcript, Wyoming, 7, VWH.

69. Norwood, *Made from This Earth*; Marcia Myers Bonta, *American Women Afield: Writings by Pioneering Women Naturalists* (College Station: Texas A&M University Press, 1995); Lorraine Anderson and Thomas S. Edwards, eds., *At Home on This Earth: Two Centuries of U.S. Women's Nature Writing* (Lebanon, N.J.: University Press of New England, 2002); Rachel Stein, *Shifting the Ground: American Women Writers' Revisions of Nature, Gender, and Race* (Charlottesville: University of Virginia Press, 1997).

70. See Mary Leffler Cochran, *Fulfilling the Dream: The Story of National Garden Clubs, Inc., 1929–2004* (St. Louis: National Garden Clubs, Inc., 2004); Cameron Binkley, "'No Better Heritage Than Living Trees': Women's Clubs and Early Conservation in Humboldt County," *Western Historical Quarterly* 33 (Summer 2002): 179–203.

71. Unsigned, undated, Genevieve Branstad to Dear Chairman, Box 2, File C, WFWC.

72. Genevieve Branstad, "Report on the Division of Conservation of the WFWC," February 1938, 1, Box 2, Folder C, WFWC.

73. Genevieve Branstad, "Report on the Division of Conservation of the WFWC," 1936, Box 2, Folder C, 2–3, WFWC.

74. Ibid.; Unsigned, undated, Genevieve Branstad to Dear Chairman, Box 2, File C, WFWC; Branstad, "Report on the Division of Conservation of the WFWC," February 1938, Box 2, Folder C, 1, WFWC.

75. Gigi La Budde, "Intro Information," enclosed in Gigi La Budde to Gail Lamberty, forwarded to author, August 6, 2002. In 1990 Wilhelmine La Budde became the first woman to be inducted into the Wisconsin Conservation Hall of Fame.

76. Transcript of Wilhelmine La Budde, Letter to Editor, *Milwaukee Journal*, enclosed in La Budde to Lamberty, forwarded to author, August 6, 2002.

77. Transcripts enclosed in La Budde to Lamberty, forwarded to author, August 6, 2002.

78. La Budde was the first woman to serve in the Wisconsin Conservation Congress and was a leader in making conservation education a priority within the WFWC.

79. Rafe Gibbs, "What Became of the Shelter Belt?," *Popular Mechanics* 99, no. 5 (May 1953): 106.

80. Delores Levy, manuscript, "Branching Out," 90, DLC.

81. Tressa Waters, Transcript, Tennessee, 7, VWH.

82. Wisconsin Development Authority, "Wisconsin Gets a Power Program" (Madison: WDA, 1938), 7. Available: http://www.wisconsinhistory.org/turningpoints/search.asp?id=1007 (accessed January 20, 2006).

83. Script, 25 May 1937, 4; 7, Box 1, Folder 1, WSWWTC.

84. Helen Matheson, "'We Say What We Think' Club Is 13 Years Old," *Wisconsin State Journal*, June 4, 1950, sec. 2, 1, Box 1, Folder 14, WSWWTC.

85. Ibid.

86. The broadcast day and time varied during the program's long run.

87. Matheson, "Club Is 13 Years Old," 2.

88. Script, June 25, 1941, 1, Box 1, Folder 5, WSWWTC.

89. See Merchant, "Women of Progressive," 57–85; Merchant, *The Death of Nature: Women, Ecology, and the Scientific Revolution* (repr., San Francisco: HarperSanFrancisco, 1990); Nancy

C. Unger, "Gendered Approaches to Environmental Justice," in *Echoes from the Poisoned Well*, ed. Heather Goodall, Paul C. Roiser, and Sylvia Washington (Lanham, Md.: Lexington, 2006), 17–34.

90. Script, July 26, 1937, Box 1, Folder 1, WSWWTC.

91. Script, September 6, 1937, Box 1, Folder 1, WSWWTC.

92. Ibid.

93. Ibid.

94. Ibid.

95. Ibid., 6; Script, April 25, 1938, 4, Box 1, Folder 2, WSWWTC.

96. See for example, Nancy C. Unger, "The We Say What We Think Club," *Wisconsin Magazine of History* 90, no. 1 (Autumn 2006): 16–27; Unger, "Women for a Peaceful Christmas: Wisconsin Homemakers Seek to Remake American Culture," *Wisconsin Magazine of History* 93, no. 2 (Winter 2009–2010): 2–15; Ted Moore, "'Democratizing the Air': The Salt Lake Women's Chamber of Commerce and Air Pollution, 1936–1945," *Environmental History* 12 (January 2007): 80–106.

97. Stephanie A. Carpenter, *On the Farm Front: The Women's Land Army in World War II* (DeKalb: Northern Illinois University Press, 2003).

98. See Polly Welts Kaufman, *National Parks and the Woman's Voice: A History* (1996; updated, Albuquerque: University of New Mexico Press, 2006).

99. Jan C. Dawson, "'Lady Lookouts' in a 'Man's World' During World War II: A Reconsideration of American Women and Nature," *Journal of Women's History* 8 (Fall 1996): 99–113; Sarah Smith, "They Sawed up a Storm," *Northern Logger and Timber Processor* 48 (April 2000): 8–10, 61–62; Roxane S. Palone, "Women in Forestry—Past and Present," *Pennsylvania Forests* 92 (Spring 2001): 7–10.

100. With so many women entering heavy industry, claims during the Progressive Era that women were more prone to suffer from the effects of industrial toxins prompted a comprehensive critical review of the literature on women's occupational health problems. The report, *Women in Industry*, sponsored by the U.S. Army, concluded that "no good scientific evidence existed to indicate that non-pregnant female workers were any more susceptible to toxic substances than male workers." Andrea Hricko and Cora Marrett, "Women's Occupational Health," Box 7, Folder 126, 6–7, HHC.

101. Emma Harris, "Emma the Welder," *San Jose Mercury News*, February 24, 2000, 1E, 9E.

102. Sonya Jason, "From Gunpowder Girl to Working Woman," *Newsweek*, February 23, 2004, 20.

103. Evie Foster, Transcript, Alaska, 1, VWH.

104. Script, June 24, 1942, 1–8, Box 1, Folder 6, WSWWTC.

105. Script, March 31, 1943, 4, Box 1, Folder 7, WSWWTC.

106. Mrs. Albert Toepfer, "Conservation of Natural Resources: 'Glimpses Ahead,'" May 25, 1949, Box 2, Folder C, WFWC.

107. Script, June 30, 1943, 1, Box 1, Folder 7, WSWWTC.

108. Ibid., 3.

109. Ibid., 4.

110. Ibid., 5.

111. Script, July 13, 1943, 2, Box 1, Folder 7, WSWWTC.

112. Ibid., 2, 5.

113. Script, June 30, 1943, 6, Box 1, Folder 7, WSWWTC.

114. Ibid., 3, 5 (emphasis in original).

115. Douglas Blackmon, *Slavery by Another Name: The Re-Enslavement of Black Americans From the Civil War to World War II* (New York: Doubleday, 2008), 375.

116. Fisher, "African Americans," in Glave and Stoll, *"To Love the Wind and the Rain,"* 74. For a history of African American resort vacationers, see Aron, *Working at Play*, 212–18.

117. Judy Yung, *Unbound Voices: A Documentary History of Chinese Women in San Francisco* (Berkeley: University of California Press, 1999), 378.

118. Connie Chiang, "Imprisoned Nature: Toward an Environmental History of the World War II Japanese American Incarceration," *Environmental History* 15 (April 2010): 239.

119. John Armor and Peter Wright, *Manzanar* (New York: Times Books, 1988), 105.

120. Jeanne Wakatsuki Houston and James D. Houston, *Farewell to Manzanar* (New York: Bantam, 1973), 72; Armor and Wright, *Manzanar*, 91–92.

121. Armor and Wright, *Manzanar*, 94.

122. Valerie Matsumoto, "Japanese Women during World War II," *Frontiers: A Journal of Women Studies* 8, no. 1 (1984): 9.

123. Houston and Houston, *Farewell to Manzanar*, 72.

124. See Samuel O. Regalado, "Incarcerated Sport: Nisei Women's Softball and Athletics during Japanese American Internment," *Journal of Sport History* 27, no. 3 (Fall 2000): 431–44.

125. See Thomas Y. Fujita-Rony, "Remaking the 'Home Front' in World War II: Japanese American Women's Work and the Colorado River Relocation Center," *Southern California Quarterly* 88, no. 2 (June 2006): 161–204.

126. In Susan McKay, *The Courage Our Stories Tell: The Daily Lives and Maternal Child Health Care of Japanese American Women at Heart Mountain* (Powell, Wy.: Western History Publications, 2002), 97.

127. In Linda Trinh Vo and Marian Sciachitano, eds., *Asian American Women: The Frontiers Reader* (Lincoln: University of Nebraska Press, 2003), 48.

128. Matsumoto, "Japanese Women," 10–11.

CHAPTER 6

1. Virginia Kemp Fish, "We Stopped the Monster: LAND in Retrospect," manuscript presented at the Midwest Sociological Society, March 10–13, 1994, 8, Box 1, Folder 1, LAND.

2. Ibid., 5; Virginia Kemp Fish, "Widening the Spectrum: The Oral History Technique and Its Use with LAND, a Grass-Roots Group," *Sociological Imagination*, 31, no. 2 (1994): 106.

3. Fish, "Monster," 12; Gertrude Dixon, "How We Have Organized and Acted," in *A Primer on Nuclear Power*, ed. Jack Miller (Melville, Minn.: Anvil, 1981), 44.

4. Elaine Carey, "Honors Piling up for the Rebel Nun," January 25, 1987, *Toronto Star*, in Box 1, File 4, LAND.

5. Catherine Gudis, *Buyways: Billboards, Automobiles, and the American Landscape* (New York: Routledge, 2005), 77.

6. Another typical series read, "Brother Speeders/Let's Rehearse/All Together/'Good Morning, Nurse.'" Curtis Kant, "Warning Signs," *Reminisce* (May/June 2006): 66.

7. Fish, "Spectrum," 106.

8. Betty MacDonald, *Anybody Can Do Anything* (Philadelphia: Lippincott, 1950), 252.

9. Betty MacDonald, *The Egg and I* (Philadelphia: Lippincott, 1945), 37.

10. MacDonald, *Anybody*, 35.

11. Paula Becker, "Betty MacDonald's *The Egg and I* Is Published on October 3, 1945," *History Link: The Free Online Encyclopedia of Washington State History*, August 14, 2007. Available: http://www.historylink.org/index.cfm?DisplayPage=output.cfm&;file_id=8261 (accessed May 17, 2010); MacDonald, *Egg and I*, 171.

12. MacDonald, *Egg and I*, 59.

13. Ibid., 90.

14. Ibid., 113.

15. Paula Becker, "Libel Trial against Betty MacDonald of *Egg and I* Fame Opens in Seattle on February 5, 1951," *HistoryLink: The Free Online Encyclopedia of Washington State History*, August 31, 2007. Available: http://www.historylink.org/index.cfm?DisplayPage=pf_output.cfm&;file_id=8270 (accessed May 17, 2010). See also Beth Kraig, "Betty and the Bishops: Was *The Egg and I* Libelous?" *Columbia Magazine* 12, no. 1 (Spring 1998): 17–22, and Paula Becker, "Seattle Jury Finds for the Defendants in Libel Suit against *Egg and I* Author Betty MacDonald on February 20, 1951," *HistoryLink: The Free Online Encyclopedia of Washington State History*, September 5, 2007. Available: http://www.historylink.org/index.cfm?DisplayPage=output.cfm&;file_id=8271 (accessed May 30, 2011).

16. MacDonald, *Egg and I*, 192. Emphasis in original.

17. Schrepfer, *Nature's Altars*, 197–98.

18. Ibid., 198.

19. Mike Taugher, "Her Spoiled View Inspired a Clearer Vision for the Bay," *San Jose Mercury News*, November 7, 2011, A1, A9.

20. See Elaine May, *Homeward Bound: American Families in the Cold War Era* (New York: Basic Books, 1988).

21. See Alice Kessler-Harris, *In Pursuit of Equity: Women, Men, and the Quest for Economic Citizenship in 20th-Century America* (New York: Oxford University Press, 2003), 203.

22. Quoted in Schrepfer, *Nature's Altars*, 203.

23. Schrepfer, *Nature's Altars*, 188–89.

24. Ibid., 207.

25. See Christopher Sellers, *Crabgrass Crucible: Suburban Nature and the Rise of Environmentalism in Twentieth-Century America* (Chapel Hill: University of North Carolina Press, 2012).

26. Merle Kinsley, "Women's Fight," LAT, October 20, 1954, A4.

27. This marked one of the first anti-smog marches by Los Angeles women, but not their first protest against increased air pollution. See Bess M. Wilson, "C. of C. Women's Division Urged to Back Smog Bill," LAT, March 27, 1947, A5; Joan Martin, "Women Marshaled in Battle on Smog," LAT, March 19, 1950, A4.

28. "Grand Jurors Quiz Mayor in Smog Inquiry," LAT, October 21, 1954, 1–2, 27.

29. See Dorothy Marshall, "Women's Appeal," LAT, September 28, 1955, A4.

30. See Ted Steinberg, *American Green: The Obsessive Quest for the Perfect Lawn* (New York: Norton: 2006).

31. See Adam Rome, "Building on the Land: Toward an Environmental History of Residential Development in American Cities and Suburbs, 1870–1990," *Journal of Urban History* 20, no. 3 (May 1994): 419.

32. For ways that chemicals have been changing the internal ecosystems of humans, livestock, and wildlife, see Nancy Langston, *Toxic Bodies: Hormone Disruptors and the Legacy of DES* (New Haven, Conn.: Yale University Press, 2011).

33. See Dee Garrison, *Bracing for Armageddon: Why Civil Defense Never Worked* (New York: Oxford University Press, 2006).

34. See B. L. Larson and K. E. Ebner, "Significance of Strontium-90 in Milk: A Review," *Journal of Dairy Science* 41, no. 12 (1995): 1647–62.

35. W. K. Wyat, "50,000 Baby Teeth," *The Nation*, June 13, 1959, 535–37.

36. Ethel Barol Taylor, *We Made a Difference: My Personal Journey with Women Strike for Peace* (Philadelphia: Camino, 1998), ix.

37. Blum, *Love Canal Revisited*, 143. See also Katha Pollitt, "Phallic Balloons against the War," *The Nation*, March 24, 2003, 9; Amy Swerdlow, *Women Strike for Peace: Traditional Motherhood and Radical Politics in the 1960s* (Chicago: University of Chicago Press, 1993).

38. Blum, *Love Canal Revisited*, 145.

39. Ibid., 146.

40. Taylor, *We Made a Difference*, xi.

41. Lear, *Rachel Carson*, 259–60.

42. The original broadcast in October was followed up by a program dedicated to reviews of the book. The readings were rebroadcast in the 7:00 p.m. slot in December. Randall Davidson to author, e-mail, January 29, 2009.

43. Rachel Carson, *Silent Spring*, 25th anniversary ed. (New York: Houghton Mifflin, 1987), 8.

44. Lear, *Rachel Carson*, 423.

45. See Michael B. Smith, "'Silence, Miss Carson!' Science, Gender, and the Reception of *Silent Spring*," *Feminist Studies* 27, no. 3 (Fall 2001): 733–52.

46. Julia Corbett, "Women, Scientists, Agitators: Magazine Portrayal of Rachel Carson and Theo Colborn," *Journal of Communication* 51 (2001): 728.

47. Frank Graham, *Since Silent Spring* (Boston: Houghton Mifflin, 1970), 60; Stewart Udall, "How the Wilderness Was Won," *American Heritage* (February–March 2000): 105.

48. Martha Freeman, ed., *Always, Rachel: The Letters of Rachel Carson and Dorothy Freeman, 1952–1964* (Boston: Beacon, 1995), xvi.

49. Corbett, "Women, Scientists, Agitators," 729.

50. Lear, *Rachel Carson*, 409.

51. Vera Norwood, "Rachel Carson," in *The American Radical*, ed. Mari Jo Buhle (New York: Routledge, 1994), 318. See also Maril Hazlett, "Voices from the Spring: *Silent Spring* and the Ecological Turn in American Health," in Scharff, *Seeing Nature through Gender*, 103–28.

52. "Rachel Carson Dies of Cancer, 'Silent Spring' Author Was 56," April 15, 1964, NYT, 1, 25.

53. Carson's first book, *The Sea Around Us*, had been published in thirty languages and garnered a number of prestigious prizes, including the National Book Award.

54. The attacks continue, as do the rebuttals. See Priscilla Coit Murphy, *See What a Book Can Do: The Publication and Reception of Silent Spring* (Boston: University of Massachusetts Press, 2005); Reed Karaim, "Not So Fast with the DDT: Rachel Carson's Warnings Still Apply," *American Scholar* 74, no. 3 (Summer 2005): 53–59; John Quiggin and Tim Lambert, "Rehabilitating Carson," *Prospect Magazine* 146 (May 24, 2008). Available: http://www.prospectmagazine.co.uk/2008/05/rehabilitatingcarson/ (accessed November 23, 2008); Naomi Oreskes and Erik M. Conway, *Merchants of Doubt: How a Handful of Scientists Obscured the Truth on Issues from Tobacco Smoke to Global Warming* (New York: Bloomsbury, 2010); Thomas Dunlap, ed., *DDT, Silent Spring, and the Rise of Environmentalism: Classic Texts* (Seattle: University of Washington Press, 2008).

55. See Udall, "How Wilderness Was Won," 98–105; Norwood, "Carson," 313–15; and Mark Hamilton Lytle, *The Gentle Subversive: Rachel Carson,* Silent Spring, *and the Rise of the Environmental Movement* (New York: Oxford University Press, 2007).

56. Adam Rome, "'Give Earth a Chance': The Environmental Movement and the Sixties," *Journal of American History* 90, no. 2 (September 2003): 536.

57. Thomas Dunlap, *In the Field, Among the Feathered: A History of Birders and Their Guides* (New York: Oxford University Press, 2011), 142–43.

58. Rob Zaleski, "Soaring Triumph: Banning DDT Brought Eagles Back, and It Started in Wisconsin," *Capital Times,* July 14, 2007, A1; see also Elizabeth Kolbert, "Turf War," *New Yorker,* July 21, 2008; "Lorrie Otto: 'Godmother of Natural Landscaping,'" Milwaukee Public Television. Available: http://www.mptv.org/garden_paths/guests.php (accessed January 26, 2009).

59. In David J. Webber, "Senator Gaylord Nelson, Founder of Earth Day," January 1996. Available: http://web.missouri.edu/~polidjw/Nelson.html (accessed December 29, 2005).

60. Lady Bird Johnson and Carlton B. Lees, *Wildflowers across America* (New York: Abbeville, 1988), 11.

61. Ibid., 12.

62. Ibid., 15.

63. Ronald Bailey, "Earth Day, Then and Now," *Reason* (May 2000). Available: http://www.reason.com/issues/show/346.html (available May 10, 2009).

64. Rome, "Give Earth a Chance," 537.

65. Ibid., 540.

66. See Todd Gitlin, *The Sixties* (New York: Bantam, 1993); T. V. Reed, *The Art of Protest: Culture and Activism from the Civil Rights Movement to the Streets of Seattle* (St. Paul: University of Minnesota Press, 2005).

67. The complete slogan was "Better Things for Better Living . . . through Chemistry."

68. Donna Warnock, pamphlet, *What Growthmania Does to Women and the Environment,* circa 1985, Syracuse: Feminist Resources on Energy and Ecology, Box 14, File 6, ALFA.

69. Ibid.

70. Branstad, report, Box 2, Folder C, WFWC.

71. See Unger, "Women for a Peaceful Christmas, 2–15.

72. See "Women Plan Boycott for Peace on June 21," *Des Moines Register,* June 3, 1971, 11, Folder 8, WPC.

73. William S. Becker, "Women Push for Peaceful Christmas," clipping from unidentified Canton, Illinois, newspaper, Folder 8, WPC.

74. William S. Becker, "'Money Talks,' Women Say; Boycott Shopping in War, Waste Production," unidentified clipping, Folder 8, WPC.

75. Dorothy Link, "Christmas Can Be Saved for Future Generations," *Catholic Herald Citizen,* November 21, 1971, Folder 8, WPC.

76. Whitney Gold, "Women for Peaceful Christmas Shun Opulence," *Capital Times,* November 14, 1971, 7, Folder 8, WPC.

77. Nan Cheney to Rosalie Corson, October 30, 1971, Folder 6, WPC.

78. Becker, "Money Talks."

79. News Release, November 1973, Folder 8, WPC.

80. WPC Press Release, November 1973, Folder 8, WPC.

81. Gold, "Shun Opulence"; Nancy Hambrecht, "Women Promote Peaceful Presents," *Daily Cardinal*, December 4, 1974, Folder 8, WPC.

82. 1974 Press Release, Folder 8, WPC.

83. "Fair Displays Ideas for Non-Commercial Holiday," *Capital Times*, November 27, 1974, Folder 8, WPC.

84. Ibid.

85. Link, "Christmas Can Be Saved."

86. Hambrecht, "Women Promote Peaceful Presents"; Press Release, 1974, Folder 8, WPC.

87. Dorothy Bartol, minutes 1972, in Centennial Timeline, Box 1, File 9, CPGC.

88. "Making the Switch," pamphlet, *Clean Water Action Project Toxic and Non-Toxic Household Products*, Box 15, Folder 153, CPGC; Bill Premo, "Household Waste Day," *Cambridge Tab*, April 23, 1988, Box 16, Folder 164, CPGC.

89. Annette B. Cottrell, "A Few Selected Memories," Box 2, File 1, 17, CPGC.

90. Unsigned, "My Recollections about the Environment, the Plant and Garden Club, and Nuclear Technology," Box 2, CPBC.

91. Ibid.

92. Joanne J. Turnball to President Matina S. Horner, October 2, 1986, Box 16, Folder 167, CPGC.

CHAPTER 7

1. In the documents of many organizations' efforts to create lesbian alternative communities, "woman" is spelled variously as "womyn," "womon," and "wimmin." Those original spellings are maintained in this chapter.

2. Bonnie J. Morris, "Valuing Woman-Only Spaces," *Feminist Studies* 31, no. 3 (Fall 2005): 623.

3. "General Festival Information," *Michigan Womyn's Music Festival*. Available: http://www.michfest.com/festival/index.htm (accessed April 7, 2007).

4. Ibid.

5. Conversation between Susan Wiseheart et al., "Michigan," in *Lesbian Land*, ed. Joyce Cheney (Minneapolis: Word Weavers, 1985), 97.

6. Morris, "Valuing Woman-Only Spaces," 622.

7. *After Stonewall*, produced by John Scagliotti, Janet Baus, and Dan Hunt, directed by John Scagliotti (First Run Features, 1999).

8. Bonnie J. Morris, *Eden Built by Eves: The Culture of Women's Music Festivals* (New York: Alyson, 1999), xiii. Emphasis in original.

9. Ibid., 328.

10. Angela Y. Davis, *Blues Legacies and Black Feminism* (New York: Pantheon, 1998), 133; 137.

11. Mabel Hampton in *Before Stonewall*, produced by Robert Rosenberg, John Scagliotti, and Greta Schiller, directed by Greta Schiller (First Run Features, 1985).

12. Joan Nestle, "Excerpts from the Oral History of Mabel Hampton," *Signs* 18, no. 4 (Summer 1993): 934.

13. Eric Garber, "A Spectacle in Color: The Lesbian and Gay Subculture of Jazz Age Harlem." Available: http://xroads.virginia.edu/~UG97/blues/garber.html (accessed January 10, 2009).

14. Nestle, "Excerpts from Oral History," 935.

15. Ibid.

16. Elizabeth Kennedy, *Boots of Leather, Slippers of Gold: The History of a Lesbian Community* (New York: Routledge, 1993). See also George Chauncey, *Gay New York* (New York: Basic Books, 1994), and Gary Atkins, *Gay Seattle* (Seattle: University of Washington Press, 2003).

17. See Esther Newton, "The 'Fun Gay Ladies': Lesbians in Cherry Grove, 1936–1960," in *Creating a Place for Ourselves: Lesbian, Gay, and Bisexual Community Histories*, ed. Brett Beemyn (New York: Routledge, 1997).

18. Ibid., 145.

19. Ibid., 147.

20. Ibid., 147–48.

21. Ibid., 149.

22. Ibid., 150.

23. Ibid.

24. Ibid., 149.

25. Ibid., 156.

26. Ibid.

27. Esther Newton, *Cherry Grove, Fire Island: Sixty Years in America's First Gay and Lesbian Town* (Boston: Beacon, 1995), 186.

28. Ibid., 185.

29. Newton, "'Fun Ladies,'" 157.

30. See Robert W. Collin, *The Environmental Protection Agency: Cleaning Up America's Act* (New York: Greenwood Press, 2005).

31. See Benjamin Kline, *First along the River: A Brief History of the U.S. Environmental Movement*, 3rd ed. (Lanham, Md.: Rowman and Littlefield, 2007).

32. See Noel Sturgeon, *Ecofeminist Natures: Race, Gender, Feminist Theory, and Political Action* (New York: Routledge, 1997); Karen Warren, ed., *Ecofeminism* (Bloomington: Indiana University Press, 1997); Mary Heather MacKinnon and Moni McIntyre, eds., *Readings in Ecology and Feminist Theology* (Kansas City: Sheed and Ward, 1995).

33. Carolyn Merchant, *Radical Ecology* (New York: Routledge, 1992), 205.

34. Suzanne Braun Levine and Mary Thom, *Bella Abzug* (New York: Farrar, Straus and Giroux, 2007), 61.

35. Warnock, *What Growthmania Does to Women and the Environment*.

36. "The Women's Pentagon Action," *Women and Life on Earth*. Available: http://www.wloe.org/women-s-pentagon-action.77.0.html (accessed April 27, 2009). See also Sturgeon, *Ecofeminist Natures*.

37. "The Women's Pentagon Action."

38. Lillian Faderman, *Odd Girls and Twilight Lovers* (New York: Columbia University Press, 1991), 201.

39. See Eleanor Agnew, *Back from the Land* (Chicago: Ivan R. Dee, 2004), and Jeffrey Jacob, *New Pioneers: The Back-to-the-Land Movement and the Search for a Sustainable Future* (Philadelphia: Pennsylvania University Press, 1997), 5.

40. See Agnew, *Back from the Land*, and Jacob, *New Pioneers*. See also Ryan H. Edgington, "'Be Receptive to the Good Earth': Health, Nature, and Labor in Countercultural Back-to-the-Land Settlements," *Agricultural History* 82, no. 3 (Summer 2008): 279–308; Joan Gross, "Capitalism and Its Discontents: Back-to-the-Lander and Freegan Foodways in Rural Oregon," *Food and Foodways: History and Culture of Human Nourishment* 17, no. 2 (April–June 2009): 57–79.

41. Gretchen Lemke-Santangelo, *Daughters of Aquarius: Women of the Sixties Counterculture* (Lawrence: University Press of Kansas, 2009), 106–7.

42. Ibid., 37, 56, 61.

43. Ibid., 161.

44. Ibid., 162. For accounts of male expectations and experiences, see Tim Hodgdon, *Manhood in the Age of Aquarius: Masculinity in Two Countercultural Communities, 1965–83* (New York: Columbia University Press, 2008).

45. Lemke-Santangelo, *Daughters of Aquarius*, 3–4.

46. Ibid., 82.

47. Agnew, *Back from the Land*, 169. See also 194–212.

48. See ibid., 213–30.

49. See Catherine B. Kleiner, "'Doin' It for Themselves': Lesbian Land Communities in Southern Oregon, 1970–1995" (Ph.D diss., University of New Mexico, 2003).

50. Danielle Hain, *Stopping Stereotypes: Problem Drinking and Alcoholism in the LGBT Community*, catalog number 110 (Tempe, Ariz.: Do It Now Foundation, 2007). Available: http://doitnow.org/pages/110.html (available May 11, 2009).

51. Faderman, *Odd Girls*, 216. See also Myra Lilliane Splitrock, "Traveling in Dykeland," in Cheney, *Lesbian Land*, 167–68.

52. See Cheney, *Lesbian Land*.

53. Tee Corinne, "Little Houses on Women's Lands" (1998; print-out of a link no longer available [http://www-lib.usc.edu/~retter/teehouses.html]; accessed June 1, 2002). See Tee Corinne, *The Little Houses on Women's Lands: Feminism, Photography, and Vernacular Architecture* (Wolf Creek, Ore.: Pearlchild, 2002).

54. Catriona Sandilands, "Lesbian Separatist Communities and the Experience of Nature," *Organization and Environment* 15 no. 2 (June 2002): 131–61.

55. Ibid., 138.

56. Corinne, "Little Houses."

57. Ibid.

58. Sandilands, "Lesbian Separatist," 139.

59. Ibid., 140.

60. Ibid.

61. See Catherine Kleiner, "Nature's Lovers: The Erotics of Lesbian Land Communities in Oregon, 1974–1984," in Scharff, *Seeing Nature through Gender*, 242–62.

62. Sandilands, "Lesbian Separatist," 146.

63. See Jennifer Marie Almquist, "Incredible Lives: An Ethnography of Southern Oregon's Women's Lands" (M.A. thesis, Oregon State University, 2004).

64. Morgana and Elethia, "The Pagoda," in Cheney, *Lesbian Land*, 111–15.

65. Emily Greene e-mail to author, February 11, 2009.

66. Ibid.

67. Emily Greene e-mail to author, February 12, 2009.

68. Joan S. Rabin and Barbara Slater, "Lesbian Communities across the United States: Pockets of Resistance and Resilience," *Journal of Lesbian Studies* 9, no. 1/2 (2005): 175.

69. Greene e-mail to author, February 11, 2009.

70. Scan of brochure, circa 1990s, e-mail attachment, Emily Greene to author, February 11, 2009.

71. Greene e-mail to author, February 11, 2009.

72. Ibid.

73. Rabin and Slater, "Lesbian Communities," 175.

74. *After Stonewall.*

75. Morris, "Valuing Women-Only Spaces," 620–21.

76. "Who Are the Radical Faeries?" radfae.org. Available: http://www.radfae.org/ (accessed November 10, 2011).

77. Conversation between Wiseheart et al., "Michigan," 97. Festivals also continue to be places where media stereotyping of lesbians is visibly rejected. See Ann Ciasullo, "Making Her (In)Visible: Cultural Representations of Lesbianism and the Lesbian Body in the 1990s," *Feminist Studies* 27, no. 3 (2001): 577–608.

78. Malinda Lo, "Behind the Scenes at the Michigan Womyn's Festival," April 20, 2005. Available: http://www.afterellen.com/archive/ellen/Music/2005/4/michigan3.html (accessed May 17, 2007).

79. Morris, "Valuing Women-Only Spaces," 627; Therese Edell, "Michigan," in Cheney, *Lesbian Land*, 100. Emphasis in original.

80. "General Festival Information," *Michigan Womyn's Music Festival.*

81. See Bev Jo, "Lesbian Community: From Sisterhood to Segregation," in *Lesbian Communities: Festivals, RVs, and the Internet*, ed. Esther Rothblum and Penny Sablove (New York: Harrington Park, 2005), 135–43; and Morris, *Eden*, 147–77.

82. Newsletter, October 25–27, 1985, "The Next Southern Leap," ALFA.

83. Ibid.

84. Ibid., cover page–1.

85. Ibid., 2.

86. Ibid., 6.

87. Ibid.

88. Retts Scauzillo, "Women-Only Festivals: Is There a Future?" *About: Lesbian Life.* Available: http://lesbianlife.about.com/od/musicreviews/a/WomenOnlyFest.htm (accessed April 7, 2007). Emphasis in original.

89. Morris, "Valuing Women-Only Spaces," 625.

90. Scauzillo, "Women-Only Festivals."

91. See Rothblum and Sablove, *Lesbian Communities.*

92. Sarah Kershaw, "My Sister's Keeper," NYT, February 1, 2009, style section, 1.

93. Emily Greene e-mail to author, February 11, 2009.

94. Ibid.

95. Barbara Stoll e-mail to author, February 20, 2009.

96. Ibid.

97. Ibid.

98. Greene e-mail to author, February 25, 2009.

99. Kershaw, "Sister's Keeper."

100. Ibid.

101. Article available:http://www.nytimes.com/2009/02/01/fashion/01womyn.html?_r=2&scp=2&scp=sarah%20kershaw&;st=cse. Multimedia presentation at http://www.nytimes.com/interactive/2009/02/01/style/20090201-women-feature/index.html (accessed February 3, 2009).

102. "Alapine Community Association, Inc." Available: http://alapine.com/ (accessed February 3, 2009).

103. Ibid.

104. Greene e-mail to author, February 11, 2009.

105. See Scott Herring, *Another Country: Queer Anti-Urbanism* (New York: New York University Press, 2010).

CHAPTER 8

1. Robert Bullard, "Environmental Justice Movement Loses Southside Chicago Icon Hazel Johnson," *OpEdNews.com*, January 14, 2011. Available: http://www.opednews.com/articles/Environmental-Justice-Move-by-Robert-Bullard-110114-323.html (accessed February 7, 2011).

2. Dawn Turner Trice, "Far South Side Environmental Activist Hazel Johnson and Her Daughter 'Decided to Stay Here and Fight,'" *Chicago Tribune*, March 1, 2010. Available: http://articles.chicagotribune.com/2010-03-01/news/ct-met-trice-altgeld-0229-20100228_1_sewer-line-hazel-johnson-cancer (accessed April 4, 2011).

3. Ibid.

4. Ibid.

5. Sylvia Hood Washington, *Packing Them In: An Archeology of Environmental Racism in Chicago, 1865–1954* (Lanham, Md.: Lexington, 2005), 199.

6. Trice, "Far South Side."

7. Washington, *Packing Them In*, 200.

8. Bullard, "Environmental Justice Movement Loses."

9. Much scholarship on environmental justice is being carried out by sociologists, political scientists, and legal experts, but researchers are increasingly accentuating crucial historical and gendered viewpoints. Stein's *New Perspectives on Environmental Justice* offers a wide-ranging and particularly valuable introduction to the history of efforts to achieve environmental justice in communities suffering from factors including poverty, racism, ethnocentrism, sexism, and homophobia. See also Robert D. Bullard and Damu Smith, "Women Warriors of Color on the Front Line," in *The Quest for Environmental Justice: Human Rights and the Politics of Pollution*, ed. Robert D. Bullard (San Francisco: Sierra Club Books, 2005), 62–84; Andrea Simpson, "'Who Hears Their Cry?' African American Women and the Fight for Environmental Justice in Memphis, Tennessee," in *The Environmental Justice Reader: Politics, Poetics, and Pedagogy*, ed. Joni Anderson, Mei Mei Evans, and Rachel Stein (Tucson: University of Arizona Press, 2002), 82–104; Giovanna Di Chiro, "Environmental Justice from the Grassroots: Reflections on History, Gender, and Expertise," in *The Struggle for Ecological Democracy: Environmental Justice Movements in the United States*, ed. Daniel Faber (New York: Guilford, 1998), 104–36; Barbara Epstein, "Grassroots Environmental Activism: The Toxics Movement and Directions for Social Change," in *Earth, Air, Fire, Water: Humanistic Studies of the Environment*, ed. Jill Ker Conway, Kenneth Kiniston, and Leo Marx (Amherst: University of Massachusetts Press, 1999), 170–83. Case studies of environmental racism abound. See Washington, *Packing Them In*; Melissa Checker, *Polluted Promises: Environmental Racism and the Search for Justice in a Southern Town* (New York: New York University Press, 2005); Eileen McGurty, *Transforming Environmentalism: Warren County; PCBs, and the Origins of Environmental Justice* (Piscataway: Rutgers University Press, 2007).

10. Karen Brodkin, *Power Politics: Environmental Activism in South Los Angeles* (Piscataway, N.J.: Rutgers University Press, 2009), 14.

11. Blum, *Love Canal Revisited*, 1.

12. For background to the crisis see Craig E. Colten and Peter N. Skinner, *The Road to Love Canal: Managing Industrial Waste before EPA* (Austin: University of Texas Press, 1995). Overviews include Peter Neushul, "Love Canal: A Historical Review," *Mid America* 69, no. 3 (1987): 125–38; Thomas H. Fletcher, *From Love Canal to Environmental Justice: The Politics of Hazardous Waste on the Canada-U.S. Border* (Peterborough, Ont.: Broadview, 2003); Adeline Levine, *Love Canal: Science, Politics, and People* (Lexington, Mass.: Lexington, 1982).

13. For the role of the press in publicizing the crisis at Love Canal see Penelope Denise Ploughman, "The Creation of Newsworthy Events: An Analysis of Newspaper Coverage of the Man-Made Disaster at Love Canal" (Ph.D. diss, State University of New York, Buffalo, 1984).

14. Blum, *Love Canal Revisited*, 26–27.

15. Ibid., 27.

16. Ibid., 35. See also Amy Hay, "Recipe for Disaster: Motherhood and Citizenship at Love Canal," *Journal of Women's History* 21, no. 1 (Spring 2009): 111–34, and "Recipe for Disaster: Chemical Wastes, Community Activists, and Public Health at Love Canal, 1945–2000" (Ph.D. diss., Michigan State University, 2005).

17. Blum, *Love Canal Revisited*, 54.

18. Lois Gibbs, "The Start of a Movement," in Boston University School of Public Health, ed., *Lessons from Love Canal: A Public Health Resource*, 2003. Available: http://www.bu.edu/lovecanal/canal/date.html (accessed April 14, 2011).

19. Blum, *Love Canal Revisited*, 59.

20. Ibid., 60.

21. Ibid., 62.

22. Ibid., 43.

23. Ibid.

24. On class, see Ken Geiser, "Toxic Times and Class Politics," *Radical America* 17, no. 2 (1983): 39–50.

25. For environmental racism beyond Love Canal, see Bullard, *The Quest for Environmental Justice*; Julie Sze, *Noxious New York: The Racial Politics of Urban Health and Environmental Justice* (Cambridge, Mass.: MIT Press, 2006); Washington, *Packing Them In*; Robert D. Bullard, ed., *Confronting Environmental Racism: Voices from the Grassroots* (Cambridge: South End Press, 1999); Luke Cole and Sheila Foster, *From the Ground Up: Environmental Racism and the Rise of the Environmental Justice Movement* (New York: New York University Press, 2000); Checker, *Polluted Promises*; Robert Bullard and Beverly Hendris, "The Politics of Pollution: Implications for the Black Community," *Phylon* 47, no. 1 (1986): 71–78.

26. Blum, *Love Canal Revisited*, 68.

27. Ibid., 73.

28. Richard Newman, "'Where There Is No Vision, the People Perish,' Comparative Religious Responses to Hurricane Katrina and Love Canal," *Journal of Southern Religion* 12 (2010): 3. Available: http://jsr.fsu.edu (accessed April 12, 2011). See also Amy M. Hay, "A New Earthly Vision: Religious Community Activism in the Love Canal Chemical Disaster," *Environmental History* 14, no. 3 (2009): 502–27.

29. Blum, *Love Canal Revisited*, 92.

30. Ibid., 95.

31. Ibid., 97; Gibbs, "The Start of a Movement."

32. Blum, *Love Canal Revisited*, 76–77.

33. Ibid., 85. See also Gregory Roberts, "Environmental Justice and Community Empowerment: Learning from the Civil Rights Movement," *American University Law Review* 48, no. 1 (October 1998): 229–67.

34. Newman, "'Where There Is No Vision,'" 4; Blum, *Love Canal Revisited*, 106.

35. Blum, *Love Canal Revisited*, 28.

36. Ibid., 28.

37. Ibid., 109.

38. "CERCLA Overview," *United States Environmental Protection Agency*. Available: http://www.epa.gov/superfund/policy/cercla.htm (accessed April 14, 2011).

39. Boston University School of Public Health, "Key Dates and Events at Love Canal," *Lessons from Love Canal: A Public Health Resource*, 2003. Available: http://www.bu.edu/lovecanal/canal/date.html (accessed April 14, 2011). This averaged about $15,000 per recipient.

40. "Occidental to Pay $129 Million in Love Canal Settlement," EPA Press Release, December 21, 1995. Available: http://www.justice.gov/opa/pr/Pre_96/December95/638.txt.html (accessed April 14, 2011).

41. Blum, *Love Canal Revisited*, 29.

42. Gibbs, "The Start of a Movement." Gibbs is the author of "Citizen Activism for Environmental Health: The Growth of a Powerful New Grassroots Health Movement," *Annals of the American Academy of Political and Social Science* 584 (November 2002): 97–109; *Dying from Dioxin: A Citizen's Guide to Reclaiming Our Health and Rebuilding Democracy* (Cambridge: South End, 1999); *Love Canal, My Story* (New York: Grove, 1982); *Love Canal: The Story Continues* (Gabriola Island, B.C.: New Society, 1998); and *Love Canal: And the Birth of the Environmental Health Movement* (Washington, D.C.: Island, 2010). Gibbs has been hailed as an environmental hero in works such as Patricia H. Hynes, "Ellen Swallow, Lois Gibbs, and Rachel Carson: Catalysts of the American Environmental Movement," *Women's Studies International Forum* 8 no. 4 (July 1985): 291–98.

43. See Dorceta Taylor, *Race, Gender, and American Environmentalism* (U.S. Department of Agriculture, Forest Service, Pacific Northwest Research Station, 2002); Cole and Foster, *From the Ground Up*.

44. In Stein, *New Perspectives*. See Valerie Ann Kaalund, "Witness to Truth: Black Women Heeding the Call for Environmental Justice," 78–92, and Julie Sze, "Gender, Asthma Politics, and Urban Environmental Justice Activism," 177–90. See also Bullard, *The Quest for Environmental Justice*; Simpson, "'Who Hears Their Cry?'

45. See Marie Bolton and Nancy C. Unger, "Pollution, Refineries, and People," in *The Modern Demon*, ed. Christoph Bernhardt and Genevieve Massard-Guilbaud (Clermont-Ferrand, France: Presses Universitaires, Blaise-Pascal, 2002), 425–37.

46. Traciy Curry-Reyes, "Emelda West Story," *Movies Based on True Stories Database*. [Online] June 2009. Available: http://emeldaweststory.blogspot.com/2009/06/emelda-west-story.html (accessed November 15, 2011).

47. Emelda West, "Emelda West: St. James Citizens Defeat Shintech," in Bullard, *The Quest for Environmental Justice*, 69.

48. Ibid.

49. Curry-Reyes, "Emelda West Story."

50. Ibid.

51. West, "Emelda West," 70.

52. Ibid.; "Environmental Justice Case Study: Shintech PVC Plant in Convent, Louisiana." Available: http://www.umich.edu/~snre492/shin.html (accessed November 15, 2011).

53. Curry-Reyes, "Emelda West Story."

54. West, "Emelda West," 71.

55. "Environmental Case Study: Shintech."

56. West, "Emelda West," 72–73.

57. Curry-Reyes, "Emelda West Story."

58. Warren Ross, *Funding Justice: The Legacy of the Unitarian Universalist Veatch Program* (Boston: Skinner House, 2005), 130.

59. Susana R. Almanza and Sylvia Herrera, "Susana R. Almanza and Sylvia Herrera: The Color Purple and Land Use Politics in East Austin," in Bullard, *The Quest for Environmental Justice*, 73.

60. For a case study of Chicanas in environmental justice, see Frances Ortega, "Fire in the Belly: A Case Study of Chicana Activists Working toward Environmental and Social Justice in New Mexico" (Ph.D. diss., University of New Mexico, 2005).

61. Ross, *Funding Justice*, 131.

62. Ibid., 131. "Poder" translates roughly to "power."

63. Ibid., 132.

64. Ibid.

65. Almanza and Herrera, "Almanza and Herrera," 74.

66. Ibid., 74–75.

67. Ross, *Funding Justice*, 135.

68. Ibid., 137.

69. Ibid., 138.

70. For a bibliography, see Peter Bakken, Joan Engle, and J. Roland Engel, *Ecology, Justice, and Christian Faith: A Critical Guide to the Literature* (Westport, Conn.: Greenwood, 1995). A review essay is Vecsey, "American Indian Environmental Religions," 171–76. See also Victoria Finlay and Martin Palmer, *Faith in Conservation: New Approaches to Religions and the Environment* (Washington, D.C.: World Bank Publications, 2003); Roger S. Gottlieb, *A Greener Faith: Religious Environmentalism and Our Planet's Future* (New York: Oxford University Press, 2009); Thomas R. Dunlap, *Faith in Nature: Environmentalism as Religious Quest* (Seattle: University of Washington Press, 2005); James Guth et al., "Faith and the Environment: Religious Beliefs and Attitudes on Environmental Policy," *American Journal of Political Science* 95, no. 2 (May 1995): 364–83; Benjamin Webb, ed., *Fugitive Faith: Conversations on Spiritual, Environmental, and Community Renewal* (Maryknoll, N.Y.: Orbis, 1998); Patricia Waak, *Faith, Justice, and a Healthy World: A Guide on Population and Environment for People of Faith* (New York: National Audubon Society, 1994); Action Institute, *Environmental Stewardship in the Judeo-Christian Tradition: Jewish, Catholic, and Protestant Wisdom on the Environment* (Maryknoll, N.Y.: Action Institute, 2007); Donald Mitchell and William Skudlarek, *Green Monasticism: A Buddhist-Catholic Response to an Environmental Calamity* (Herndon, Va.: Lantern, 2010); Marie George, *Stewardship of Creation: What Catholics Should Know about Church Teaching on the Environment* (Indianapolis: Saint Catherine of Sienna Press, 2009); Woodeene Koenig-Bricker, *Ten Commandments for the Environment: Pope Benedict XVI Speaks Out for Creation and Justice* (Notre Dame, Ind.: Ave Maria, 2009); Angela Smith and Simone Pulver, "Ethics-Based Environmentalism in Practice: Religious-Environmental Organizations in the United States," *Worldviews: Environment Culture Religion* 13, no. 2 (2009): 145–79.

71. Blum, *Love Canal Revisited*, 93.

72. Gail Small, "Gail Small: Voices from Northern Cheyenne Indian Country," in Bullard, *Quest for Environmental Justice*, 77.

73. "The Vision of Navajo Parks and Recreation," *Navajo Nation Parks and Recreation*. Available: http://www.navajonationparks.org/index.htm (accessed January 10, 2010); WARN Report II, June–December 1979, "The Struggle against Energy Resource Exploitation in the Four Corners Area," 16, Box 7, File 14, WARN; Doug Brugge, Timothy Benally, and Esther Yazzie-Lewis, eds., *The Navajo People and Uranium Mining* (Albuquerque: University of New Mexico Press, 2007), 1.

74. Brugge, Benally, and Yazzie-Lewis, *The Navajo People*, 2, 6.

75. WARN Report II, "The Struggle against Energy Resource Exploitation in the Four Corners Area."

76. See Tony Perrottet, "Behind the Scenes in Monument Valley," *Smithsonian*, February 2010, 72–79.

77. Peter H. Eichstaedt, *If You Poison Us: Uranium and Native Americans* (Santa Fe: Red Crane, 1994), 24.

78. Ibid., 26.

79. For a detailed study of uranium mining on Navajo land, see Judy Pasternak, *Yellow Dirt: An American Story of a Poisoned Land and a People Betrayed* (New York: Free Press, 2010), based on a series of articles she wrote for the *Los Angeles Times* in 2006.

80. Brugge, Benally, and Yazzie-Lewis, *The Navajo People*, 3.

81. Ibid.; "Addressing Uranium Contamination in the Navajo Nation," *United States Environmental Protection Agency*. Available: http://www.epa.gov/region09/superfund/navajo-nation/index.html (accessed February 5, 2010).

82. Clipping from *The Progressive*, 27, Box 4, File 10, LAND.

83. Ibid.

84. *In These Times*, February 25–March 3, 1981, Box 4, File 9, LAND.

85. Ibid.

86. Eichstaedt, *If You Poison Us*, 193.

87. Ibid., 61.

88. Pasternak, *Yellow Dirt*, 65. Michael R. Edelstein and William J. Makofske, *Radon's Deadly Daughters: Science, Environmental Policy, and the Politics of Risk* (Lanham, Md.: Rowman and Littlefield, 1998), 35. See also Clark, *Radium Girls*.

89. Eichstaedt, *If You Poison Us*, 47, 51–55.

90. Ibid., 115.

91. Brugge, Benally, and Yazzie-Lewis, *The Navajo People*, 39.

92. Quoted in ibid., 111.

93. Loretta Schwartz, "Uranium Deaths at Crown Point," *Ms. Magazine*, October 1979, 81.

94. Ibid., 60.

95. Pasternak, *Yellow Dirt*, 144, 156–58.

96. U.S. House of Representatives Committee on Oversight and Government Reform, Public Hearing, October 23, 2007, in "Voices from the Earth," Southwest Research and Information Center. Available: http://www.sric.org/voices/2009/v10n2/index.html (accessed January 8, 2010).

97. See Paul Chaat Smith and Robert Allen Warrior, *Like a Hurricane: The Indian Movement from Alcatraz to Wounded Knee* (New York: New Press, 1997); Dennis Banks and Richard Erdoes, *Ojibwa Warrior: Dennis Banks and the Rise of the American Indian Movement* (Norman: University

of Oklahoma Press, 2005); and Daniel M. Cobb and Loretta Fowler, *Beyond Red Power: American Indian Politics and Activism since 1900* (Santa Fe: School for Advanced Research Press, 2007).

98. Alvin M. Josephy Jr., Joane Nagel, and Troy R. Johnson, eds., *Red Power: The American Indians' Fight for Freedom* (Lincoln, Neb.: Bison, 1999), 51–52.

99. Eichstaedt, *If You Poison Us*, 112.

100. See Bruce Johansen, "Reprise: Forced Sterilizations," *Native Americas* 15, no. 4 (Winter, 1998): 44–47; Sally Torpy, "Native American Women and Coerced Sterilization: On the Trail of Tears in the 1970s," *American Indian Culture and Research Journal* 24, no. 2 (2000); 1–22.

101. This assertion is true if uranium-rich lands inhabited by indigenous peoples, including Canada, Australia, and Africa, are taken into account. See "Uranium Mining and Indigenous People," *WISE Uranium Project*, September 21, 2010. Available: http://www.wise-uranium.org/uip.html (accessed April 22, 2011), and "Most Mining and Milling of Uranium Occurs on Indian Lands," *Prairie Island Coalition*. Available: http://www.no-nukes.org/prairieisland/processing.html (accessed April 21, 2011).

102. Gail Robinson, "Plundering the Powerless: Uranium Mining Threatens a Land and Its People," *Environmental Action*, June 1979, 3, Box 4, File 9, LAND; WARN Report II, "The Struggle against Energy Resource Exploitation in the Four Corners Area."

103. Marcy Gilbert, WARN Statement, Box 7, File 14, LAND. Emphasis in original.

104. From 25 to 50 percent of Indian women across the United States were sterilized during this period. Jane Lawrence, "The Indian Health Service and the Sterilization of Native American Women," *American Indian Quarterly* 24, no. 3 (Summer 2000): 410; see also Torpy, "Native American Women," and Andrea Smith, *Conquest: Sexual Violence and American Indian Genocide* (Cambridge, Mass.: South End, 2005).

105. Torpy, "Native American Women."

106. WARN, "Sterilization," Box 7, File 14, LAND.

107. Torpy, "Native American Women."

108. WARN, "Sterlization."

109. WARN Report II, "The Struggle against Energy Resource Exploitation in the Four Corners Area."

110. Quoted in Lawrence, "Indian Health Service," 412.

111. Ibid., 410–12, 400.

112. Clipping from *Minneapolis Daily*, 1 June 1981, Box 4, File 9, LAND.

113. Christopher McLeod, "Birth Defects Linked to Uranium Mining," *Synapse* 25, no. 23 (April 16, 1981). Available: http://synapse.library.ucsf.edu/cgi-bin/ucsf?a=d&d=ucsf19810416-01.2.14&cl=CL2.1981.04&srpos=0&dliv=none&st=1&;e=-------en-logical-20--1------- (accessed November 27, 2011).

114. WARN Report II, "The Struggle against Energy Resource Exploitation in the Four Corners Area," 2–3.

115. Ibid., 21.

116. Torpy, "Native American Women"; Bruce E. Johnson, "Women of All Red Nations," *ABC-CLIO History and the Headlines*. Available: http://www.historyandtheheadlines.abc-clio.com/ContentPages/ContentPage.aspx?entryId=1172002&currentSection=1161468&productid=5 (accessed March 24, 2010).

117. WARN Report II, "The Struggle against Energy Resource Exploitation in the Four Corners Area," 22.

118. "Lawmaker Hears Community Concerns and Industry Promises," *Southwest Research and Information Center*. Available: http://www.sric.org/voices/2009/v10n2/index.html (January 8, 2010).

119. *In These Times*, undated, Box 4, File 9, LAND.

120. Pasternak, *Yellow Dirt*, 156. See L. S. Gottlieb and L. A. Husen, "Lung Cancer among Navajo Uranium Miners, *Chest* 81, no. 4 (April 1982): 449–52.

121. Judy Pasternak, "A Peril That Dwelt among the Navajos," LAT, November 19, 2006. Available: http://articles.latimes.com/2006/nov/19/nation/na-navajo19/13 (accessed April 21, 2011), and Pasternak, *Yellow Dirt*, 143.

122. Gail Small, "Gail Small," 77.

123. Eichstaedt, *If You Poison Us*, 145–46.

124. Pasternak, *Yellow Dirt*, 165–66.

125. Keith Schneider, "A Valley of Death for the Navajo Uranium Miners," NYT, May 3, 1993. Available: http://www.nytimes.com/1993/05/03/us/a-valley-of-death-for-the-navajo-uranium-miners.html?src=pm (accessed April 21, 2011).

126. Robert Roscoe et al., "Mortality among Navajo Uranium Miners," *American Journal of Public Health* 85, no. 4 (April 1995): 535.

127. Pasternak, *Yellow Dirt*, 143.

128. Ibid., 172.

129. Ibid., 192.

130. Kathy Helms, "'Navajo Neuropathy' Haunts Blue Gap Family," *Gallup Independent*, August 6, 2009. Available: http://nativeunity.blogspot.com/2009/08/navajo-neuropathy-sen-reid-opposes.html (accessed April 21, 2011); Charalampos L. Karadimas et al., "Navajo Neurohepatopathy Is Caused by a Mutation in the *MPV17* Gene," *American Journal of Human Genetics*, 2006. Available http://www.ncbi.nlm.nih.gov/pmc/articles/PMC1559552/ (accessed April 21, 2011).

131. U.S. House of Representatives Committee on Oversight and Government Reform, "Voices from the Earth."

132. Marjorie Childress, "Uranium Mining at Mt. Taylor Threatened; New Colo. Law Requires Cleanup," *New Mexico Independent*, June 9, 2010. Available: http://newmexicoindependent.com/56734/uranium-mining-at-mt-taylor-threatened-new-colo-law-requires-cleanup (accessed April 21, 2011).

133. See James Longhurst, *Citizen Environmentalist* (Hanover: University Press of New England, 2010).

134. Manuel Pastor and Rachel Morello-Frosch, "Assumption Is Wrong—Latinos Care Deeply about the Environment," *San Jose Mercury News*, July 8, 2002, 6B. See also Brodkin, *Power Politics*, 53–55; Margaret Rose, "'Woman Power Will Stop Those Grapes': Chicana Organizers and Middle-Class Female Supporters in the Farm Workers' Grape Boycott in Philadelphia, 1969–1970," *Journal of Women's History* 7, no. 4 (1995): 6–36; Rose, "From the Fields to the Picket Line: Huelga Women and the Boycott, 1965–1975," *California Labor History* 31 (1990): 271–93; and Rose, "Traditional and Nontraditional Patterns of Female Activism in the United Farm Workers of American, 1962–1980," *Frontiers* 11, no. 1 (1990): 26–32.

135. Rose, "Traditional and Nontraditional Patterns of Female Activism."

136. See Richard A. Garcia, "Dolores Huerta: Woman, Organizer, and Symbol," *California History* 72, no. 1 (1993): 56–71; Mario T. García, ed., *A Dolores Huerta Reader* (Albuquerque: University of New Mexico Press, 2008).

137. Garcia, *Dolores Huerta Reader*, 124, 220, 117, 297–98, 233–34; José B. Cuellar, "Labor, Farm," *Pollution Issues*. Available: http://www.pollutionissues.com/Ho-Li/Labor-Farm.html (accessed April 23, 2011). For pesticide drift, see Adrianna Quintero-Somaini and Myra Quirindongo, *Hidden Dangers: Environmental Health Threats in the Latino Community* (New York: Natural Recourses Defense Council, 2004), 40–44. Available: http://www.pollutionissues.com/Ho-Li/Labor-Farm.html (accessed April 23, 2011).

138. See Marshall Ganz, *Why David Sometimes Wins: Leadership, Organization, and Strategy in the California Farm Worker Movement* (New York: Oxford University Press, 2010).

EPILOGUE

1. Matthew Wald, "U.S. Atomic Tests in 50's Exposed Millions to Risk," NYT, July 29, 1997, A10.

2. Ruth Rosen, "Next Time Listen to Mother; a 1997 Report Vindicates the Women Who Warned in the '60s of Radioactive Fallout Hazards," LAT, August 7, 1997, 9.

3. An obituary is Margaret Ramirez, "Hazel M. Johnson, 1935–2011: South Side Activist Known as Mother of Environmental Justice Movement," *Chicago Tribune*, January 16, 2011. Available: http://articles.chicagotribune.com/2011-01-16/features/ct-met-johnson-obit-20110116_1_cancer-alley-asbestos-removal-environmental-justice (accessed May 14, 2011).

4. "Mission," *Center for Health, Environment, and Justice*. Available: http://chej.org/about/mission/ (accessed May 13, 2011).

5. "Nuclear Power Threatens Us All," *Center for Health, Environment, and Justice*, May 2011. Available: http://chej.org/2011/05/nuclear-power-threatens-us-all/ (accessed May 14, 2011).

6. Gary Langer, "Japan's Nuclear Crisis May Resonate in the U.S.," *ABCNews.com*, March 12, 2011. Available: http://blogs.abcnews.com/thenumbers/2011/03/japans-nuclear-crisis-may-resonate-in-the-us.html (accessed May 14, 2011).

7. Christopher Weber, "Dirty Secrets under the Schoolyard," *E—The Environmental Magazine* 22, no. 1 (January/February 2001): 27.

8. Torpy, "Native American Women and Coerced Sterilization," 8.

9. "Lorelei DeCora," *Robert Wood Johnson Foundation*, March 1, 2002. Available: http://www.rwjf.org/pr/product.jsp?id=52196 (accessed May 15, 2011).

10. Mark Clayton, "Women Lead a Farming Revolution in Iowa," *Christian Science Monitor*, February 25, 2009. Available: http://www.csmonitor.com/Environment/Living-Green/2009/0225/women-lead-a-farming-revolution-in-iowa (accessed February 27, 2009). See also Corry Bregendahl and Matthew Hoffman, "Women, Land and Legacy: Change Agents and Agency Change in Iowa Evaluation Results, November 2010." Available: http://journals.cambridge.org/action/displayAbstract;jsessionid=B5E5E451CCE551560C468C2EB0F7B85E.journals?fromPage=online&;aid=115029 (accessed December 26, 2011).

11. "About WFAN," Women, Food and Agriculture Network. Available: http://www.wfan.org/About_Us.html (accessed December 19, 2011).

12. Lisa Jervis, "The Bitch Holiday Gift Guide," *Bitch: Feminist Response to Popular Culture*, no. 22 (Fall 2003): 94.

13. See Miliann Kang, *The Managed Hand: Race, Gender, and the Body in Beauty Service Work* (Berkeley: University of California Press, 2010).

14. "The California Healthy Nail Salon Collaborative," Asian Health Services. Available: http://asianhealthservices.org/nailsalon/ (accessed April 23, 2011).

15. "3rd Annual Town Hall Meeting," Zerobreastcancer.org. Available: www.zerobreastcancer.org/research/th08_liou.pdf (accessed April 21, 2011).

16. Marti Kheel provides an overview in *Nature Ethics: An Ecofeminist Perspective* (Lanham, Md.: Rowman and Littlefield, 2007). Subfields continue to develop. See Josephine Donovan and Carol Adams, eds., *Animals and Women: Feminist Theoretical Explorations* (Durham: Duke University Press, 1995), *Beyond Animal Rights: A Feminist Caring Ethic for the Treatment of Animals* (New York: Continuum International, 2000), *The Feminist Care Tradition in Animal Ethics* (New York: Columbia University Press, 2007); and Lisa Kemmerer, ed., *Sister Species: Women, Animals, and Social Justice* (Champaign: University of Illinois Press, 2011). See also A. Breeze Harper, ed., *Sistah Vegan: Food, Identity, Health, and Society: Black Female Vegans Speak* (Herndon: Va.: Lantern Books, 2010).

17. See Lorraine Anderson and Thomas S. Edwards, eds., *At Home on This Earth*; Barbara T. Gates, ed., *In Nature's Name: An Anthology of Women's Writing and Illustration, 1780–1930* (Chicago: University of Chicago Press, 2002); John Murray, ed., *American Nature Writing 2000: A Celebration of Women Writers and American Nature Writing* (Corvallis: Oregon State University Press, 2000), and *American Nature Writing 2003: Celebrating Emerging Women Nature Writers* (Golden, Colo.: Fulcrom, 2003). See also Barbara J. Cook, *Women Writing Nature: A Feminist View* (Lanham, Md.: Lexington, 2007); Linda Hogan, *Intimate Nature: The Bond between Women and Animals* (New York: Ballantine, 1999), and with Brenda Peterson, *The Sweet Breathing of Plants: Women Writing on the Green World* (New York: North Point, 2002). For science fiction, see Justine Larbalestier, ed., *Daughters of Earth: Feminist Science Fiction in the Twentieth Century* (Middletown, Conn.: Wesleyan, 2006).

18. For the Chipko movement, see Vandana Shiva, *Staying Alive: Women, Ecology and Survival in India* (New Delhi: Kali for Women, 1988); Willow Ann Sirch, *Eco-Women: Protectors of the Earth* (Golden, Colo.: Fulcrum, 1996); and Shobita Jain, "Standing Up for Trees: Women's Role in the Chipko Movement," *Unasylva* 36, no. 4 (1984): 12–20. Additional studies of India include Arun Agrawal and K. Sivaramakrishnan, eds., *Agrarian Environments: Resources, Representation, and Rule In India* (Durham, N.C.: Duke University Press, 2000). Studies of other world areas include William Beinart and Joann McGregor, *Social History and African Environments* (Athens: Ohio University Press, 2003); Jean Davidson, ed., *Agriculture, Women, and Land: The African Experience* (Boulder, Colo.: Westview, 1988); Anoja Wickramasinghe, *Deforestation, Women and Forestry: The Case of Sri Lanka* (Amsterdam: International Books, 1994); Richard A. Schroeder, *Shady Practices: Agroforestry and Gender Politics in the Gambia* (Berkeley: University of California Press, 1999); Allison Goebel, *Gender and Land Reform: The Zimbabwe Experience* (Montreal: McGill-Queen's University Press, 2006); Antonia Finnane, "Water, Love, and Labor: Aspects of a Gendered Environment," in *Sediments of Time: Environment and Society in Chinese History*, ed. Mark Elvin and Ts'ui-jung Lui (Boston, Mass.: Cambridge University Press, 1998): 657–90; Patricia Crawford and Philippa Maddern, eds., *Women as Australian Citizens: Underlying Histories* (Melbourne: Melbourne University Publishing, 2001); Melody Hessing, Rebecca Raglon, Catriona Sandilands, eds., *This Elusive Land: Women and the Canadian Environment* (Vancouver: UBC Press, 2005); Jens Ivo Engels, "Gender Roles and German Anti-Nuclear Protest: The Women of Wyhl," *Le Démon Moderne: La Pollution dans les Societies Urbaines et Industrielles d'Europe* [*The Modern Demon: Pollution in Urban and Industrial European Societies*], ed. Christoph Bernhardt and Genevieve Massard-Guilbaud (Clermont-Ferrand, France: Presses Universitaires Blaise-Pascal, 2002): 407–24; Constance E. Campbell, "On the Front Lines but Struggling for Voice: Women

in the Rubber Tappers' Defence of the Amazon Forest," *Ecologist* 27 (March/April 1997): 46–54; Thomas Klubok, *Contested Communities: Class, Gender, and Politics in Chile's El Teniente Copper Mine, 1094–1951* (Durham, N.C.: Duke University Press, 1998). Comparative studies that incorporate two or more world areas include Carol MacCormack and Marilyn Strathern, eds., *Nature, Culture, and Gender* (Boston, Mass.: Cambridge University Press, 1980); Louis Fortmann and Dianne Rocheleau, "Women and Agroforestry: Four Myths and Three Case Studies," *Women in Natural Resources* 9, no. 2 (1987): 35–44, 46, 51; Carolyn E. Sachs, *Gendered Fields: Rural Women, Agriculture, and Environment* (Boulder, Colo.: Westview, 1996); Glenda Riley, *Taking Land, Breaking Land: Women Colonizing in the American West and Kenya, 1840–1940* (Albuquerque: University of New Mexico Press, 2003). The intersection of religion, ecofeminism, and globalism is examined in Rosemary Ruether, ed., *Women Healing Earth: Third World Women on Ecology, Feminism, and Religion* (Maryknoll, N.Y.: Orbis, 1996); Ruether, *Integrating Ecofeminism, Globalization, and World Religions* (Lanham, Md.: Rowman and Littlefield, 2005); and Heather Eaton and Lois Ann Lorentzen, *Ecofeminism and Globalization: Exploring Culture, Context, and Religion* (Lanham, Md.: Rowman and Littlefield, 2003).

Bibliography

BOOKS AND ESSAYS

Action Institute. *Environmental Stewardship in the Judeo-Christian Tradition: Jewish, Catholic, and Protestant Wisdom on the Environment.* Maryknoll, N.Y.: Action Institute, 2007.

Adams, Carol. *The Sexual Politics of Meat: A Feminist-Vegetarian Critical Theory.* New York: Continuum International, 2010.

Addams, Jane. *20 Years at Hull-House.* 1910. Reprint. Charleston, S.C.: BookSurge Classics, 2004.

Agnew, Eleanor. *Back from the Land.* Chicago: Ivan R. Dee, 2004.

Agrawal, Arun, and K. Sivaramakrishnan, eds. *Agrarian Environments: Resources, Representation, and Rule in India.* Durham, N.C.: Duke University Press, 2000.

Altman, Irwin, and Azra Churchman, eds. *Women and the Environment.* New York: Plenum, 1994.

Anderson, Karen. *Chain Her by One Foot: The Subjugation of Women in Seventeenth Century New France.* New York: Routledge, 1991.

Anderson, Lorraine. *Sisters of the Earth: Women's Prose and Poetry about Nature.* New York: Vintage, 2003.

Anderson, Lorraine, and Thomas S. Edwards, eds. *At Home on This Earth: Two Centuries of U.S. Women's Nature Writing.* Lebanon, N.J.: University Press of New England, 2002.

Apple, Rima D. *The Challenge of Constantly Changing Times: From Home Economics to Human Ecology at the University of Wisconsin-Madison, 1903–2003.* Madison: Parallel, 2003.

Armitage, Kevin. *The Nature Study Movement: The Forgotten Populizer of America's Conservation Ethic.* Lawrence: University Press of Kansas, 2009.

Armor, John, and Peter Wright. *Manzanar.* New York: Times Books, 1988.

269

Aron, Cindy. *Working at Play: A History of Vacations in the United States.* New York: Oxford University Press, 1999.

Atkins, Gary. *Gay Seattle.* Seattle: University of Washington Press, 2003.

Austin, Mary. *Land of Little Rain.* 1903. Reprint. Albuquerque: University of New Mexico Press, 1974.

Backus, Harriet Fish. *Tomboy Bride: A Woman's Personal Account of Life in Mining Camps of the West.* Boulder, Colo.: Pruett, 1969.

Bailes, Kendall E., ed. *Environmental History: Critical Issues in Comparative Perspective.* Lanham, Md.: Rowman and Littlefield, 1985.

Baker, Jean H. *Margaret Sanger: A Life of Passion.* New York: Hill and Wang, 2011.

Baker, Terri, and Connie Henshaw, eds. *Women Who Pioneered Oklahoma: Stories from the WPA Narratives.* Norman: University of Oklahoma Press, 2007.

Bakken, Peter, Joan Engle, and J. Roland Engel. *Ecology, Justice, and Christian Faith: A Critical Guide to the Literature.* Westport, Conn.: Greenwood, 1995.

Banks, Dennis, and Richard Erdoes. *Ojibwa Warrior: Dennis Banks and the Rise of the American Indian Movement.* Norman: University of Oklahoma Press, 2005.

Barnhart, Jacqueline. *The Fair but Frail: Prostitution in San Francisco, 1849–1900.* Reno: University of Nevada Press, 1986.

Barreiro, Jose. "Indigenous Peoples Are the 'Miners' Canary' of the Human Family." In *Learning to Listen to the Land*, ed. William Willers, 199–201. Washington, D.C.: Island, 1991.

Bederman, Gail. *Manliness and Civilization: A Cultural History of Gender and Race in the United States, 1880–1917.* Chicago: University of Chicago Press, 1996.

Beebe, Rose Marie, and Robert M. Senkewicz, eds. *Lands of Promise and Despair.* Berkeley: Santa Clara University and Heyday Books, 2001.

Beinart, William, and Joann McGregor. *Social History and African Environments.* Athens: Ohio University Press, 2003.

Benson, Maxine. *Martha Maxwell, Rocky Mountain Naturalist.* Winnipeg: Bison Books, 1999.

Berlin, Ira. *Generations of Captivity: A History of African-American Slaves.* Cambridge: Belknap Harvard, 2003.

Bernhardt, Christoph, and Genevieve Massard-Guilbaud eds. *Le Démon Moderne: La Pollution dans les Societies Urbaines et Industrielles d'Europe* [*The Modern Demon: Pollution in Urban and Industrial European Societies*]. Clermont-Ferrand, France: Presses Universitaires, Blaise-Pascal, 2002.

Berry, Daina Ramey. *Swing the Sickle for the Harvest is Ripe.* Chicago: University of Illinois Press, 2007.

Bieder, Robert. *Native American Communities in Wisconsin, 1600–1960.* Madison: University of Wisconsin Press, 1995.

Blackmon, Douglas. *Slavery by Another Name: The Re-Enslavement of Black Americans from the Civil War to World War II.* New York: Doubleday, 2008.

Blum, Deborah. *The Poisoner's Handbook.* New York: Penguin, 2010.

Blum, Elizabeth D. *Love Canal Revisited: Race, Class, and Gender in Environmental Activism.* Lawrence: University Press of Kansas, 2008.

———."Women, Environmental Rationale, and Activism during the Progressive Era." In *"To Love the Wind and the Rain": African Americans and Environmental History*, ed. Dianne D. Glave and Mark Stoll, 77–92. Pittsburgh: University of Pittsburgh, 2006.

Bonta, Marcia Myers. *American Women Afield: Writings by Pioneering Women Naturalists*. College Station: Texas A&M University Press, 1995.

———. *Women in the Field: America's Pioneering Women Naturalists*. College Station: Texas A&M University Press, 1991.

Bouvier, Virginia. *Women and the Conquest of California, 1542–1840: Codes of Silence*. Tucson: University of Arizona Press, 2001.

Bowden, Henry. *American Indians and Christian Missions: Studies in Cultural Conflict*. Chicago: University of Chicago Press, 1981.

Bowerbank, Sylvia Lorraine. *Speaking for Nature: Women and Ecologies of Early Modern England*. Baltimore: John Hopkins University Press, 2004.

Boylan, Anne M. *"Patterns of Organizations": The Origins of Women's Activism in New York and Boston, 1797–1840*. Chapel Hill: University of North Carolina Press, 2002.

Breton, Mary Joy. *Women Pioneers for the Environment*. Boston: Northeastern University Press, 1998, 2000.

Brodkin, Karen. *Power Politics: Environmental Activism in South Los Angeles*. Piscataway, N.J.: Rutgers University Press, 2009.

Brooks, Paul. *The House of Life: Rachel Carson at Work*. Boston: Houghton Mifflin, 1972.

———. *Rachel Carson: The Writer at Work*. 2nd ed. San Francisco: Sierra Club Books, 1989.

———. *Speaking for Nature: How Literary Naturalists from Henry Thoreau to Rachel Carson Have Shaped America*. Boston: Houghton Mifflin, 1980.

Brown, Dona. *Back to the Land: The Enduring Dream of Self-Sufficiency in Modern America*. Madison: University of Wisconsin Press, 2011.

Brown, Kathleen. "The Anglo-Algonquian Gender Frontier." In *Negotiators of Change: Historical Perspectives on Native American Women*, ed. Nancy Shoemaker, 24–48. New York: Routledge, 1995.

Brugge, Doug, Timothy Benally, and Esther Yazzie-Lewis, eds. *The Navajo People and Uranium Mining*. Albuquerque: University of New Mexico Press, 2007.

Bryson, Bill. *A Walk in the Woods: Rediscovering America on the Appalachian Trail*. New York: Broadway, 1998.

Bullard, Robert D., ed. *Confronting Environmental Racism: Voices from the Grassroots*. Cambridge: South End Press, 1999.

———. *The Quest for Environmental Justice: Human Rights and the Politics of Pollution*. San Francisco: Sierra Club Books, 2005.

Bullard, Robert D., and Damu Smith. "Women Warriors of Color on the Front Line." In *The Quest for Environmental Justice: Human Rights and the Politics of Pollution*, ed. Robert Bullard, 62–84. San Francisco: Sierra Club Books, 2005.

Butler, Elizabeth Beardsley. *Women and the Trades: Pittsburgh, 1907–1908*. New York: Russell Sage Foundation, 1909.

Calloway, Colin G. *First Peoples: A Documentary Survey of American Indian History*. Boston: Bedford/St. Martin's, 1999.

Camp, Stephanie M. H. "'I Could Not Stay There': Enslaved Women, Truancy, and the Geography of Everyday Forms of Resistance in the Antebellum Plantation South." In *Women, Families, and Communities*, ed. Nancy Hewitt and Kirsten Delegard, 196–209. Vol. 1. New York: Pearson Longman, 2008.

Campbell, T. N., ed. "Coahuiltecans and Their Neighbors." In *Handbook of North American Indians (Southwest)*, ed. Alonso Ortiz, 352. Washington D.C.: Smithsonian Institution, 1983.

Camp Fire Girls. 1912. Reprint, Carlisle, Mass.: Applewood, n.d.

Campisi, Jack, and Laurence M. Hauptman, eds. *The Oneida Indian Experience: Two Perspectives*. Syracuse, N.Y.: Syracuse University Press, 1988.

Carnes, Marc C., and Clyde Griffen, eds. *Meanings of Manhood*. Chicago: University of Chicago Press, 1990.

Carney, Judith. *Black Rice*. Cambridge, Mass.: Harvard University Press, 2001.

Carpenter, Stephanie A. *On the Farm Front: The Women's Land Army in World War II*. DeKalb: Northern Illinois University Press, 2003.

Carson, Rachel. *Silent Spring*. 25th Anniversary Edition. New York: Houghton Mifflin, 1987.

Carter, Patricia Anne. *"Everybody's Paid but the Teacher": The Teaching Profession and the Women's Movement*. New York: Teachers College Press, 2002.

Cartledge, Bryan, ed. *Population and the Environment: The Linacre Lectures 1993–4*. Oxford: Oxford University Press, 1995.

Caughfield, Adrienne. *True Women and Western Expansion*. College Station: Texas A&M University Press, 2005.

Chauncey, George. *Gay New York*. New York: Basic Books, 1994.

Chavez-Garcia, Miroslava. *Negotiating Conquest: Gender and Power in California, 1770s–1880s*. Tucson: University of Arizona Press, 2004.

Checker, Melissa. *Polluted Promises: Environmental Racism and the Search for Justice in a Southern Town*. New York: New York University Press, 2005.

Cheney, Joyce, ed. *Lesbian Land*. Minneapolis: Word Weavers, 1985.

Child, Lydia Marie. *The American Frugal Housewife*. Reissue ed. Carlisle, Mass.: Applewood, 1989.

Christy, David. *Cotton Is King; or, The Culture of Cotton, and Its Relation to Agriculture, Manufacturers, and Commerce, to the Free Colored People, and to Those Who Hold That Slavery Is in Itself Sinful*. 2nd ed. Cincinnati: Moore, Wilstach, Keys & Co., 1855.

Clark, Claudia. *Radium Girls: Women and Industrial Health Reform, 1910–1935*. Chapel Hill: University of North Carolina Press, 1997.

Clarke, Edward. *Sex in Education*. Boston: J.R. Osgood and Company, 1874.

Clarke, Robert. *Ellen Swallow: The Woman Who Founded Ecology*. New York: Follett, 1973.

Cobb, Daniel M., and Loretta Fowler. *Beyond Red Power: American Indian Politics and Activism since 1900*. Santa Fe: School for Advanced Research Press, 2007.

Cochran, Mary Leffler. *Fulfilling the Dream: The Story of National Garden Clubs, Inc., 1929–2004*. St. Louis: National Garden Clubs, Inc., 2004.

Cole, Luke, and Sheila Foster. *From the Ground Up: Environmental Racism and the Rise of the Environmental Justice Movement*. New York: New York University Press, 2000.

Collin, Robert W. *The Environmental Protection Agency: Cleaning Up America's Act*. New York: Greenwood, 2005.

Colten, Craig E., and Peter N. Skinner. *The Road to Love Canal: Managing Industrial Waste before EPA*. Austin: University of Texas Press, 1995.

Conard, Rebecca. *Places of Quiet Beauty: Parks, Preserves, and Environmentalism*. Iowa City: University of Iowa Press, 1997.

Conversation between Susan Wiseheart et al. "Michigan." In *Lesbian Land*, ed. Joyce Cheney, 96–98. Minneapolis: Word Weavers, 1985.

Cook, Barbara J. *Women Writing Nature: A Feminist View*. Lanham, Md.: Lexington, 2007.

Cook, Blanche Wiesen. *Eleanor Roosevelt, Vol. 2, 1933–1938*. New York: Viking, 1999.

Cook, S. F. *The Indian Population of New England in the Seventeenth Century*. Berkeley: University of California Press, 1976.

Cooper, Susan Fenimore. *Rural Hours*. New York: Gregory P. Putnam, 1850.

Corinne, Tee. *The Little Houses on Women's Lands: Feminism, Photography, and Vernacular Architecture*. Wolf Creek, Ore.: Pearlchild, 2002.

Cott, Nancy F. *The Bonds of Womanhood: "Women's Sphere" in New England, 1780–1835*. New Haven, Conn.: Yale University Press, 1977.

Cowles, Julia. *The Diaries of Julia Cowles: A Connecticut Record, 1797–1803*, ed. Laura Moseley. New Haven, Conn.: Yale University Press, 1931.

Crawford, Patricia, and Philippa Maddern, eds. *Women as Australian Citizens: Underlying Histories*. Melbourne: Melbourne University Publishing, 2001.

Cronon, William. *Changes in the Land: Indians, Colonists, and the Ecology of New England*. New York: Hill and Wang, 1983, 2003.

———. *Nature's Metropolis: Chicago and the Great West*. New York: Norton, 1992.

———. "The Trouble with Wilderness; or, Getting Back to the Wrong Wilderness." In *Out of the Woods: Essays in Environmental History*, ed. Char Miller and Hal Rothman, 28–50. Pittsburgh: University of Pittsburgh Press, 1997.

Cutter, Barbara. *Domestic Devils, Battlefield Angels: The Radicalism of American Womanhood, 1830–1865*. DeKalb: Northern Illinois University Press, 2003.

Dasmann, Richard. *California's Changing Environment*. San Francisco: Boyd and Fraser, 1981.

Davidson, Jean, ed. *Agriculture, Women, and Land: The African Experience*. Boulder, Colo.: Westview, 1988.

Davis, Angela Y. *Blues Legacies and Black Feminism*. New York: Pantheon, 1998.

Davis, Jack E. *An Everglades Providence: Marjory Stoneman Douglas and the American Environmental Century*. Athens: University of Georgia Press, 2009.

Davis, Rebecca Harding. *Life in the Iron Mills and Other Stories*. New York: Feminist Press at CUNY, 1985.

Demos, John. *The Tried and the True: Native American Women Confronting Colonization*. New York: Oxford University Press, 1995.

Devens, Carol. *Countering Colonization: Native American Women and Great Lakes Missions, 1630–1900*. Berkeley: University of California Press, 1992.

Deverell, William, and Greg Hise. *Land of Sunshine: An Environmental History of Metropolitan Los Angeles*. Pittsburgh: University of Pittsburg Press, 2006.

Diamond, Irene. *Fertile Ground: Women, Earth, and the Limits of Control*. Boston: Beacon, 1994.

Di Chiro, Giovanna. "Environmental Justice from the Grassroots: Reflections on History, Gender, and Expertise." In *The Struggle for Ecological Democracy: Environmental Justice Movements in the United States*, ed. Daniel Faber, 104–36. New York: Guilford, 1998.

Dixon, Gertrude. "How We Have Organized and Acted." In *A Primer on Nuclear Power*, ed. Jack Miller, 44. Melville, Minn.: Anvil, 1981.

Dobyns, Henry F. *Their Number Become Thinned: Native American Population Dynamics in Eastern North America*. Knoxville: University of Tennessee Press, 1983.

Donovan, Josephine, and Carol Adams, eds. *Animals and Women: Feminist Theoretical Explorations*. Durham, N.C.: Duke University Press, 1995.

———. *Beyond Animal Rights: A Feminist Caring Ethic for the Treatment of Animals*. New York: Continuum International, 2000.

———. *The Feminist Care Tradition in Animal Ethics*. New York: Columbia University Press, 2007.

Dorr, Rheta Childe. *What Eight Million Women Want*. Boston: Small, Maynard, and Co., 1910.

Dorsey, Bruce. *Reforming Men and Women: Gender in the Antebellum City*. New York: Cornell University Press, 2006.

Doughty, Robin W. *Feather and Bird Preservation: A Study in Nature Protection*. Berkeley: University of California Press, 1975.

Doyle, Susan Badger. "Women's Experiences on the Westward Emigrant Trails." In *Encyclopedia of Women in the West*, ed. Gordon Bakken and Brenda Farrington, 308–14. Thousand Oaks, Calif.: Sage, 2003.

Drumm, Stella M., ed. *Down the Santa Fe Trail and into Mexico*. Lincoln: University of Nebraska Press, 1982.

Dugan, Kathleen M. "At the Beginning Was Woman: Women in Native American Religious Traditions." In *Religion and Women*, ed. Arvind Sharma, 39–60. Albany: State University of New York Press, 1994.

Dunlap, Thomas. *DDT: Scientists, Citizens, and Public Policy*. Princeton, N.J.: Princeton University Press, 1981.

———. *Faith in Nature: Environmentalism as Religious Quest*. Seattle: University of Washington Press, 2005.

———. *In the Field, Among the Feathered: A History of Birders and Their Guides*. New York: Oxford University Press, 2011.

Dunlap, Thomas, ed. *DDT, Silent Spring, and the Rise of Environmentalism: Classic Texts*. Seattle: University of Washington Press, 2008.

Durning, Alan. *Worldwatch Paper #112: Guardians of the Land: Indigenous Peoples and the Health of the Earth*. Washington, D.C.: Worldwatch Institute, 1992.

Easlea, Brian. *Science and Sexual Oppression: Patriarchy's Confrontation with Woman and Nature*. London: Weidenfeld and Nicholson, 1981.

Eaton, Heather, and Lois Ann Lorentzen. *Ecofeminism and Globalization: Exploring Culture, Context, and Religion*. Lanham, Md.: Rowman and Littlefield, 2003.

Edelstein, Michael R., and William J. Makofske. *Radon's Deadly Daughters: Science, Environmental Policy, and the Politics of Risk*. Lanham, Md.: Rowman and Littlefield, 1998.

Egan, Timothy. *The Worst Hard Time: The Untold Story of Those Who Survived the Great American Dust Bowl*. New York: Mariner, 2006.

Eichstaedt, Peter H. *If You Poison Us: Uranium and Native Americans*. Santa Fe: Red Crane, 1994.

Eighmey, Rae. *Food Will Win the War: Minnesota Crops, Cooks, and Conservation during World War I*. St. Paul: Minnesota Historical Society Press, 2010.

Elias, Megan J. *Stir It Up: Home Economics in American Culture*. Philadelphia: University of Pennsylvania Press, 2008.

Elliott, E. N., ed. *Cotton Is King, and Pro-Slavery Arguments; Comprising the Writings of Hammond, Harper, Christy, Stringfellow, Hodge, Bledsoe, and Cartwright, on This Important Subject*. 1860. Reprint. New York: Negro Universities Press, 1969.

Engels, Jens Ivo. "Gender Roles and German Anti-Nuclear Protest: The Women of Wyhl." In *Le Démon Moderne: La Pollution dans les Societies Urbaines et Industrielles d'Europe* [*The Modern Demon: Pollution in Urban and Industrial European Societies*], ed. Christoph Bernhardt and

Genevieve Massard-Guilbaud, 407–24. Clermont-Ferrand, France: Presses Universitaires Blaise-Pascal, 2002.

English, Deirdre, and Barbara Ehrenreich. *"For Her Own Good": Two Centuries of the Experts' Advice to Women*. New York: Anchor, 2005.

Epstein, Barbara. "Grassroots Environmental Activism: The Toxics Movement and Directions for Social Change." In *Earth, Air, Fire, Water: Humanistic Studies of the Environment*, ed. Jill Kerr Conway, Kenneth Kiniston, and Leo Marx, 170–83. Amherst: University of Massachusetts Press, 1999.

Etulain, Richard W. *Does the Frontier Experience Make America Exceptional?* Boston: Bedford/St. Martins, 1999.

Faderman, Lillian. *Odd Girls and Twilight Lovers*. New York: Columbia University Press, 1991.

Fairbanks, Carol, and Sara Brooks Sundberg. *Farm Women on the Prairie Frontier: A Sourcebook for Canada and the United States*. Meutchen, N.J.: Scarecrow, 1983.

Faragher, John Mack. *Women and Men on the Overland Trail*. New Haven, Conn.: Yale University Press, 2001.

Farnham, Christie. "Sapphire? The Issue of Dominance in the Slave Family, 1830–1865." In *To Toil the Livelong Day*, ed. Carol Groneman and Mary Beth Norton, 68–86. Ithaca, N.Y.: Cornell University Press, 1987.

Fett, Sharla M. *Working Cures: Healing, Health, and Power on Southern Slave Plantations*. Chapel Hill: University of North Carolina Press, 2002.

Fink, Augusta. *I-Mary: A Biography of Mary Austin*. Tucson: University of Arizona Press, 1983.

Fink, Leon, ed. *Major Problems in the Gilded Age and Progressive Era*. 2nd ed. Boston: Houghton Mifflin, 2001.

Finlay, Victoria, and Martin Palmer. *Faith in Conservation: New Approaches to Religions and the Environment*. Washington, D.C.: World Bank Publications, 2003.

Finnane, Antonia. "Water, Love and Labor: Aspects of a Gendered Environment." In *Sediments of Time: Environment and Society in Chinese History*, ed. Mark Elvin and Ts'ui-jung Lui, 657–90. Boston: Cambridge University Press, 1998.

Fisher, Colin. "African Americans, Outdoor Recreation, and the 1919 Chicago Race Riot." In *"To Love the Wind and the Rain": African Americans and Environmental History*, ed. Dianne D. Glave and Mark Stoll, 63–76. Pittsburgh: University of Pittsburgh, 2006.

Fitzgerald, Susan. "Women in the Home." In *One Half the People: The Fight for Woman Suffrage*, ed. Anne Firor Scott and Andrew MacKay Scott, 114–115. Champaign: University of Illinois Press, 1982.

Flanagan, Maureen. *America, Reformed: Progressives and Progressivisms, 1890s–1920s*. New York: Oxford University Press, 2007.

———. *Seeing with Their Hearts: Chicago Women and the Vision of the Good City, 1871–1933*. Princeton, N.J.: Princeton University Press, 2002.

Fletcher, Thomas H. *From Love Canal to Environmental Justice: The Politics of Hazardous Waste on the Canada-U.S. Border*. Peterborough, Ont.: Broadview, 2003.

Freeman, Martha, ed. *Always, Rachel: The Letters of Rachel Carson and Dorothy Freeman, 1952–1964*. Boston: Beacon, 1995.

Frink, Margaret. *Journal of the Adventures of a Party of California Gold Seekers*. Fairfield, Wash.: Ye Galleon Press, 1987.

Furmansky, Dyana. *Rosalie Edge, Hawk of Mercy: The Activist Who Saved Nature from the Conservationists*. Athens: University of Georgia Press, 2010.

Ganz, Marshall. *Why David Sometimes Wins: Leadership, Organization, and Strategy in the California Farm Workers Movement*. New York: Oxford University Press, 2010.

García, Mario T., ed. *A Dolores Huerta Reader*. Albuquerque: University of New Mexico Press, 2008.

Garrison, Dee. *Bracing for Armageddon: Why Civil Defense Never Worked*. New York: Oxford University Press, 2006.

Gates, Barbara T. *Kindred Nature: Victorian and Edwardian Women Embrace the Living World*. Chicago: University of Chicago Press, 1999.

————. *In Nature's Name: An Anthology of Women's Writing and Illustration, 1780–1930*. Chicago: University of Chicago Press, 2002.

Gates, Barbara T., and Ann B. Shteir, eds. *Natural Eloquence: Women Reinscribe Science*. Madison: University of Wisconsin Press, 1997.

George, Marie. *Stewardship of Creation: What Catholics Should Know about Church Teachings on the Environment*. Indianapolis: Saint Catherine of Sienna Press, 2009.

Gianquitto, Tina. *Good Observers of Nature: American Women and the Scientific Study of the Natural World, 1820–1885*. Athens: University of Georgia Press, 2007.

Gibbs, Lois. *Dying from Dioxin: A Citizen's Guide to Reclaiming Our Health and Rebuilding Democracy*. Cambridge: South End, 1999.

————. *Love Canal: And the Birth of the Environmental Health Movement*. Washington, D.C.: Island, 2010.

————. *Love Canal: My Story*. New York: Grove, 1982.

————. *Love Canal: The Story Continues*. Gabriola Island, B.C.: New Society, 1998.

Giesberg, Judith Ann. *Civil War Sisterhood: The U.S. Sanitary Commission and Women's Politics in Transition*. Boston: Northeastern University Press, 2006.

Gitlin, Todd. *The Sixties*. New York: Bantam, 1993.

Glave, Dianne, and Mark Stoll, eds. *"To Love the Wind and the Rain": African Americans and Environmental History*. Pittsburgh: University of Pittsburgh Press, 2006.

Glave, Dianne. *Rooted in the Earth: Reclaiming the African American Environmental Heritage*. Chicago: Lawrence Hill, 2010.

Goebel, Allison. *Gender and Land Reform: The Zimbabwe Experience*. Montreal: McGill-Queen's University Press, 2006.

Gonzalez, Deena. *Refusing the Favor: The Spanish-Mexican Women of Santa Fe, 1820–1880*. New York: Oxford University Press, 1999.

Goodall, Heather, Paul C. Rosier, and Sylvia Washington, eds. *Echoes from the Poisoned Well: Global Memories of Environmental Injustice*. Lanham, Md.: Lexington Books, 2006.

Goodman, Susan, and Carl Dawson. *Mary Austin and the American West*. Berkeley: University of California Press, 2009.

Gottlieb, Roger S. *A Greener Faith: Religious Environmentalism and Our Planet's Future*. New York: Oxford University Press, 2009.

Gould, Lewis. *Lady Bird Johnson and the Environment*. Lawrence: University Press of Kansas, 1988.

Gowans, Fred R. *Rocky Mountain Rendezvous: A History of the Fur Trade Rendezvous 1825–1840*. Reprint. Provo, Utah: Brigham Young University Press, 1978.

Graham, Frank. *Since* Silent Spring. Boston: Houghton Mifflin, 1970.

Griffin, Susan. *Woman and Nature: The Roaring inside Her.* 2nd ed. San Francisco: Sierra Club, 1978.

Gudis, Catherine. *Buyways: Billboards, Automobiles, and the American Landscape.* New York: Routledge, 2005.

Hafen, LeRoy R., and Ann W. Hafen, eds. *Journals of Forty-Niners, Salt Lake to Los Angeles.* Winnipeg: Bison Books, 1998.

Haigh, Jane, and Claire Murphy. *Gold Rush Women.* Anchorage: Alaska Northwest Books, 1997.

Halberstam, David. *The Fifties.* New York: Ballantine, 1984.

Hale, Sarah. *Early American Cookery: "The Good Housekeeper" 1841.* Reprint. New York: Dover, 1996.

Hamilton, Alice. *Exploring the Dangerous Trades.* 1943. Reprint. Miller, S.D.: Miller Press, 2008.

Haney, Wava, and Jane Knowles, eds. *Women and Farming: Changing Roles, Changing Structures.* Boulder, Colo.: Westview, 1988.

Hansen, Debra Gold. *Strained Sisterhood: Gender and Class in the Boston Female Anti-Slavery Society.* Amherst: University of Massachusetts Press, 2009.

Harkin, Michael E., and David Lewis, eds. *Native Americans and the Environment: Perspectives on the Ecological Indian.* Lincoln: University of Nebraska Press, 2007.

Harper, A. Breeze, ed. *Sistah Vegan: Food, Identity, Health, and Society: Black Female Vegans Speak.* Herndon, Va.: Lantern, 2010.

Hepler, Allison. *Women in Labor: Mothers, Medicine, and Occupational Health in the United States, 1890–1980.* Columbus: Ohio State University Press, 2000.

Henige, David. *Numbers from Nowhere: The American Indian Population Debate.* Norman: University of Oklahoma Press, 1998.

Herman, Daniel Justin. *Hunting and the American Imagination.* Washington, D.C.: Smithsonian Institution Press, 2001.

Herring, Scott. *Another Country: Queer Anti-Urbanism.* New York: New York University Press, 2010.

Herron, John, and Andrew Kirk, eds. *Human/Nature: Biology, Culture, and Environmental History.* Albuquerque: University of New Mexico Press, 1999.

Hessing, Melody, Rebecca Raglon, and Catriona Sandilands, eds. *This Elusive Land: Women and the Canadian Environment.* Vancouver: UBC Press, 2005.

Hine, Robert V., and John Mack Faragher. *Frontiers: A Short History of the American West.* New Haven, Conn.: Yale University Press, 2000.

Hochschild, Adam. *King Leopold's Ghost.* Boston: Mariner Books, 1999.

Hodgdon, Tim. *Manhood in the Age of Aquarius: Masculinity in Two Countercultural Communities, 1965–83.* New York: Columbia University Press, 2008.

Hogan, Linda. *Intimate Nature: The Bond between Women and Animals.* New York: Ballantine, 1999.

Hogan, Linda, and Brenda Peterson. *The Sweet Breathing of Plants: Women Writing on the Green World.* New York: North Point, 2002.

Holmes, Kenneth, ed. *Covered Wagon Women: Diaries and Letters from the Western Trails.* Vols. 1–11. Glendale, Calif.: Arthur C. Clark, 1983.

Holmes, Madelyn. *American Women Conservationists: Twelve Profiles.* Jefferson, N.C.: McFarland and Company, 2004.

Houston, Jeanne Wakatsuki, and James D. Houston. *Farewell to Manzanar*. New York: Bantam, 1973.

Hoxie, W. J. *How Girls Can Help Their Country: Handbook for Girl Scouts*. 1913. Reprint, Carlisle, Mass.: Applewood, n.d.

Hoy, Suellen. *Chasing Dirt: The American Pursuit of Cleanliness*. New York: Oxford University Press, 1995.

Hughes, J. Donald. *An Environmental History of the World: Humankind's Changing Role in the Community Life*. New York: Routledge, 2002.

Hurtado, Albert. *Indian Survival on the California Frontier*. New Haven, Conn.: Yale University Press, 1990.

Isenberg, Andrew. *The Destruction of the Bison: An Environmental History, 1850–1920*. New York: Cambridge University Press, 2001.

Isenberg, Andrew, ed. *Oxford Handbook of Environmental History*. New York: Oxford University Press, forthcoming 2013.

Jacob, Jeffrey. *New Pioneers: The Back-to-the-Land Movement and the Search for a Sustainable Future*. Philadelphia: Pennsylvania University Press, 1997.

Jacobs, Wilbur R. *The Fatal Confrontation: Historical Studies of American Indians, Environment, and Historians*. Albuquerque: University of New Mexico Press, 1996.

Jarvis, Kimberley. *Franconia Notch and the Women Who Saved It*. Durham: University of New Hampshire Press, 2007.

Jefferson, Thomas. *Notes on the State of Virginia*. New York: Norton, 1954.

Jeffrey, Julie Roy. *Frontier Women: "Civilizing" the West? 1840–1880*. New York: Hill and Wang, 1998.

Jensen, Joan M. *With These Hands: Women Working on the Land*. New York: Feminist Press, 1981.

Jewett, Sarah Orne. *A White Heron*. Memphis, Tenn.: General Books, 2009.

Jiggins, Janice. *Changing the Boundaries: Women-Centered Perspectives on Population and the Environment*. Washington D.C.: Island, 1994.

Jo, Bev. "Lesbian Community: From Sisterhood to Segregation." In *Lesbian Communities: Festivals, RVs, and the Internet*, ed. Esther Rothblum and Penny Sablove, 135–43. New York: Harrington Park, 2005.

Johnson, Allan. *The Gender Knot: Unraveling Our Patriarchal Legacy*. Philadelphia: Temple University Press, 1997.

Johnson, Fran Holman. *The Gift of the Wild Things: The Life of Caroline Dormon*. Lafayette: University of Southern Louisiana Press, 1990.

Johnson, Lady Bird, and Carlton B. Lees. *Wildflowers across America*. New York: Abbeville, 1988.

Jones, Lu Ann. *Mama Learned Us to Work: Farm Women in the New South*. Chapel Hill: University of North Carolina Press, 2002.

Josephy, Alvin M., Jr., Joane Nagel, and Troy R. Johnson, eds. *Red Power: The American Indians' Fight for Freedom*. Lincoln, Neb.: Bison, 1999.

Judd, Richard W. *Common Lands, Common People: The Origins of Conservation in Northern New England*. Cambridge, Mass.: Harvard University Press, 1997.

Kang, Miliann. *The Managed Hand: Race, Gender, and the Body in Beauty Service Work*. Berkeley: University of California Press, 2010.

Kaufman, Polly Welts. *National Parks and the Woman's Voice: A History*. 1996. Updated. Albuquerque: University of New Mexico Press, 2006.

Keesing, Felix M. *The Menomini Indians of Wisconsin*. New York: Johnson Reprint Corporation, 1971.

Kemble, Frances Anne. *Journal of a Residence on a Georgian Plantation*. 1863. Reprint, New York: Knopf, 1961.

Kemmerer, Lisa, ed. *Sister Species: Women, Animals, and Social Justice*. Champaign: University of Illinois Press, 2011.

Kennedy, Elizabeth. *Boots of Leather, Slippers of Gold: The History of a Lesbian Community*. New York: Routledge, 1993.

Kerber, Linda. "The Republican Mother: Women and the Enlightenment—An American Perspective (1976)." In *Toward an Intellectual History of Women*, ed. Linda Kerber, 41–62. Chapel Hill: University of North Carolina Press, 1997.

Kersey, Harry A. Jr. *Pelts, Plumes and Hides: White Traders among the Seminole Indians 1870–1930*. Gainesville: University Presses of Florida, 1975.

Kessler-Harris, Alice. *In Pursuit of Equity: Women, Men, and the Quest for Economic Citizenship in 20th-Century America*. New York: Oxford University Press, 2003.

Kheel, Marti. *Nature Ethics: An Ecofeminist Perspective*. Lanham, Md.: Rowman and Littlefield, 2007.

Kimmel, Michael S. *Manhood in America: A Cultural History*. New York: Oxford University Press, 2005.

Kirkland, Carolyn. *A New Home—Who'll Follow?* New York: C. S. Frances, 1839.

Kline, Benjamin. *First along the River: A Brief History of the U.S. Environmental Movement*. 3rd ed. Lanham, Md.: Rowman and Littlefield, 2007.

Klubok, Thomas. *Contested Communities: Class, Gender, and Politics in Chile's El Teniente Copper Mine, 1094–1951*. Durham, N.C.: Duke University Press, 1998.

Koenig-Bricker, Woodeene. *Ten Commandments for the Environment: Pope Benedict XVI Speaks Out for Creation and Justice*. Notre Dame, Ind.: Ave Maria, 2009.

Kofalk, Harriet. *No Woman Tenderfoot: Florence Merriam Bailey, Pioneer Naturalist*. College Station: Texas A&M University Press, 1989.

Kolchin, Peter. *American Slavery, 1619–1877*. New York: Hill and Wang, 1993.

Kolodny, Annette. *The Land Before Her: Fantasy and Experience of the American Frontiers, 1630–1860*. Chapel Hill: University of North Carolina Press, 1984.

Koslow, Jennifer. *Cultivating Health, Los Angeles Women and Public Health Reform*. Piscataway, N.J.: Rutgers University Press, 2009.

Kovarik, Bill, and Mark Neuzil. "The Radium Girls." In Neuzil and Kovarik, *Mass Media and Environmental Conflict: America's Green Crusades*, 33–52. Thousand Oaks, Calif.: Sage, 1996.

Krech, Shepard. *The Ecological Indian: History and Myth*. New York: Norton, 2000.

Landes, Ruth. *Ojibwa Religion and the Midewiwin*. Madison: University of Wisconsin Press, 1968.

Langston, Nancy. *Toxic Bodies: Hormone Disruptors and the Legacy of DES*. New Haven, Conn.: Yale University Press, 2011.

Lanigan, Esther F. *Mary Austin: Song of a Maverick*. Tucson: University of Arizona Press, 1997.

Larbalestier, Justine, ed. *Daughters of the Earth: Feminist Science Fiction in the Twentieth Century*. Middletown, Conn.: Wesleyan, 2006.

Laskin, David. *The Children's Blizzard*. New York: HarperCollins, 2004.

Lear, Linda. *Rachel Carson: Witness for Nature*. New York: Owl Books, 1997.

Lemke-Santangelo, Gretchen. *Daughters of Aquarius: Women of the Sixties Counterculture*. Lawrence: University Press of Kansas, 2009.

Levine, Adeline. *Love Canal: Science, Politics, and People*. Lexington, Mass.: Lexington, 1982.

Levine, Suzanne Braun, and Mary Thom. *Bella Abzug*. New York: Farrar, Straus and Giroux, 2007.

Levy, JoAnn. *They Saw the Elephant: Women in the California Gold Rush*. Norman: University of Oklahoma Press, 1992.

Lewis, Oscar. *The Effects of White Contact upon Blackfoot Culture with Special Reference to the Role of the Fur Trade*. Seattle: University of Washington Press, 1942. Reprint, 1966.

Lindquist, Mark A., and Martin Zanger, eds. *Buried Roots and Indestructible Seeds: The Survival of American Indian Life in Story, History and Spirit*. Madison: University of Wisconsin Press, 1994.

List, Peter, ed. *Radical Environmentalism: Philosophy and Tactics*. Belmont, Calif.: Wadsworth, 1993.

Little, Jo, Linda Peake, and Pat Richardson, eds. *Women in Cities: Gender and the Urban Environment*. New York: New York University Press, 1988.

Littlefield, Daniel. *Rice and Slaves: Ethnicity and the Slave Trade in Colonial South Carolina*. Urbana: University of Illinois Press, 1991.

Long, Judith Reick. *Gene Stratton-Porter: Novelist and Naturalist*. Indianapolis: Indiana Historical Society, 1990.

Longhurst, James. *Citizen Environmentalist*. Hanover: University Press of New England, 2010.

Low, Ann Marie. *Dust Bowl Diary*. Lincoln: University of Nebraska Press, 1984.

Luchetti, Cathy, and Carol Olwell. *Women of the West*. New York: Orion, 1982.

Lytle, Mark Hamilton. *The Gentle Subversive: Rachel Carson,* Silent Spring, *and the Rise of the Environmental Movement*. New York: Oxford University Press, 2007.

MacCormack, Carol, and Marilyn Strathern, eds. *Nature, Culture, and Gender*. Boston, Mass.: Cambridge University Press, 1980.

MacDonald, Betty. *Anybody Can Do Anything*. Philadelphia: Lippincott, 1950.

———. *The Egg and I*. Philadelphia: J.B. Lippincott, 1945.

MacKinnon, Mary Heather, and Moni McIntyre, eds. *Readings in Ecology and Feminist Theology*. Kansas City: Sheed and Ward, 1995.

Magoc, Chris J. *So Glorious a Landscape: Nature and the Environment in American History and Culture*. Wilmington, Del.: Scholarly Resources, 2002.

Marco, Gino, Robert Hollingsworth, and William Durham, eds. *Silent Spring Revisited*. Washington, D.C.: American Chemical Society, 1987.

Martin, Brenda, and Penny Sparke, eds. *Women's Places: Architecture and Design, 1860–1960*. New York: Routledge, 2003.

Martin, Justin. *Genius of Place: The Life of Frederick Law Olmsted*. Cambridge, Mass.: Da Capo Press, 2011.

May, Elaine. *Homeward Bound: American Families in the Cold War Era*. New York: Basic Books, 1988.

McAdoo, Harriette Pipes, ed. *Black Families*. 3rd ed. Thousand Oaks, Calif.: Sage, 1997.

McBrewster, John, Frederic P. Miller, and Agnes F. Vandome, eds. *Ellen Swallow Richards*. Mauritius: Alphascript, 2010.

McBride, Genevieve G. *On Wisconsin Women: Working for Their Rights from Settlement to Suffrage*. Madison: University of Wisconsin Press, 1993.

McBride, Genevieve, ed. *Women's Wisconsin: From Native Matriarchies to the New Millennium*. Madison: Wisconsin Historical Society Press, 2005.

McCay, Mary A. *Rachel Carson*. New York: Twayne, 1993.

McClymer, John. *The Triangle Strike and Fire*. Orlando, Fla.: Harcourt Brace, 1998.

McCullough, David. *The Johnstown Flood: The Incredible Story behind One of the Most Devastating "Natural" Disasters America Has Ever Known*. New York: Touchstone, 1968.

McGurty, Eileen. *Transforming Environmentalism: Warren County; PCBs, and the Origins of Environmental Justice*. Piscataway, N.J.: Rutgers University Press, 2007.

McKay, Susan. *The Courage Our Stories Tell: The Daily Lives and Maternal Child Health Care of Japanese American Women at Heart Mountain*. Powell, Wy.: Western History Publications, 2002.

Melosi, Martin, ed. *Pollution and Reform in American Cities, 1870–1930*. Austin: University of Texas Press, 1980.

Merchant, Carolyn. *American Environmental History, an Introduction*. New York: Columbia University Press, 2007.

———, ed. *The Columbia Guide to American Environmental History*. New York: Columbia University Press, 2002.

———. *The Death of Nature: Women, Ecology, and the Scientific Revolution*. Reprint. San Francisco: HarperSanFrancisco, 1990.

———. *Earthcare: Women and the Environment*. New York: Routledge, 1995.

———. *Ecological Revolutions: Nature, Gender, and Science in New England*. Chapel Hill: University of North Carolina Press, 1989.

———, ed. *Green Versus Gold: Sources in California's Environmental History*. Washington D.C.: Island, 1998.

———, ed. *Major Problems in American Environmental History*. Boston: Houghton Mifflin, 1993.

———, ed. *Major Problems in American Environmental History*. 2nd ed. Boston: Houghton Mifflin, 2005.

———, ed. *Major Problems in American Environmental History*. 3rd ed. Boston: Wadsworth, 2011.

———. *Radical Ecology*. New York: Routledge, 1992.

———. *Reinventing Eden: The Fate of Nature in Western Culture*. New York: Routledge, 2003.

Miller, Robert J. *Native America, Discovered and Conquered: Thomas Jefferson, Lewis and Clark, and Manifest Destiny*. Lincoln, Neb.: Bison Books, 2008.

Miller, Susan A. *Growing Girls: The Natural Origins of Girls' Organizations in America*. Piscataway, N.J.: Rutgers University Press, 2007.

Mitchell, Donald, and William Skudlarek. *Green Monasticism: A Buddhist-Catholic Response to an Environmental Calamity*. Herndon, Va.: Lantern, 2010.

Montrie, Chad. *Making a Living: Work and Environment in the United States*. Chapel Hill: University of North Carolina Press, 2008.

Moore, John H. ed. *The Political Economy of North American Indians*. Norman: University of Oklahoma Press, 1993.

Morgan, Lael. *The Good Time Girls of the Alaska-Yukon Gold Rush*. Fairbanks, Alaska: Epicenter, 1998.

Morgan, Philip D. *Slave Counterpoint: Black Culture in the Eighteenth Century Chesapeake and the Low Country*. Chapel Hill: University of North Carolina Press, 1998.

Morris, Bonnie J. *Eden Built by Eves: The Culture of Women's Music Festivals*. New York: Alyson, 1999.

Mortimer-Sandilands, Catriona, and Bruce Erickson, eds. *Queer Ecologies: Sex, Nature, Biopolitics, and Desire*. Bloomington: University of Indiana Press, 2010.

Moynihan, Ruth, Susan Armitage, and Christian Fischer Dichamp, eds. *So Much to Be Done: Women Settlers on the Mining and Ranching Frontiers*. 2nd ed. Lincoln: University of Nebraska Press, 1998.

Murphy, Kevin P. *Political Manhood: Red Bloods, Mollycoddles, and the Politics of Progressive Reform*. New York: Columbia University Press, 2008.

Murphy, Priscilla Coit. *See What a Book Can Do: The Publication and Reception of* Silent Spring. Boston: University of Massachusetts Press, 2005.

Murphy, Vera Mary. *Mining Cultures: Men, Women, and Leisure in Butte, 1914–1941*. Urbana: University of Illinois Press, 1997.

Murray, John, ed. *American Nature Writing 2000: A Celebration of Women Writers and American Nature Writing*. Corvallis: Oregon State University Press, 2000.

———. *American Nature Writing 2003: Celebrating Emerging Women Nature Writers*. Golden, Colo.: Fulcrom, 2003.

Myres, Sandra. *Ho for California! Women's Overland Diaries from the Huntington Library*. San Marino, Calif.: Huntington Library Press: 2007.

———. *Westering Women and the Frontier Experience, 1800–1915*. Albuquerque: University of New Mexico Press, 1992.

Nash, Roderick. *American Environmentalism*. New York: McGraw-Hill, 1990.

———. *Wilderness and the American Mind*. New Haven, Conn.: Yale University Press, 1973, 1982.

Newton, Esther. *Cherry Grove, Fire Island: Sixty Years in America's First Gay and Lesbian Town*. Boston: Beacon Press, 1995.

———. "The 'Fun Gay Ladies': Lesbians in Cherry Grove, 1936–1960." In *Creating a Place for Ourselves: Lesbian, Gay, and Bisexual Community Histories*, ed. Brett Beemyn, 145–64. New York: Routledge, 1997.

Niethammer, Carolyn. *Daughters of the Earth: The Lives and Legends of American Indian Women*. New York: Simon and Schuster, 1977.

Norton, Mary Beth. *Liberty's Daughters: The Revolutionary Experience of American Women, 1750–1800*. New York: Harper Collins, 1980.

———. *Major Problems in American Women's History*. Lexington, Mass.: Heath, 1989.

Norton, Mary Beth, and Ruth Alexander, eds. *Major Problems in American Women's History*. 2nd ed. Lexington, Mass.: Heath, 1996.

———. *Major Problems in American Women's History*. 3rd ed. Boston: Houghton Mifflin, 2003.

———. *Major Problems in American Women's History*. 4th ed. Boston: Houghton Mifflin, 2007.

Norwood, Vera. *Made from This Earth: American Women and Nature*. Chapel Hill: University of North Carolina Press, 1993.

———. "Rachel Carson." In *The American Radical*, ed. Mari Jo Buhle. New York: Routledge, 1994.

Norwood, Vera, and Janice Monk. *The Desert Is No Lady: Southwestern Landscapes in Women's Writing and Art*. Tucson: University of Arizona Press, 1997.

O'Brien, Mary Barmeyer. *Into the Western Winds: Pioneer Boys Traveling the Overland Trails*. Helena, Mont.: TwoDot, 2002.

———. *Toward the Setting Sun: Pioneer Girls Traveling the Overland Trails*. Helena, Mont.: TwoDot, 1999.

Olmsted, Frederick Law. *The Cotton Kingdom: A Traveler's Observations on Cotton and Slavery in the American Slave States*. New York: Knopf, 1953.

O'Meara, Walter. *Daughters of the Country: The Women of the Fur Traders and the Mountain Men*. New York: Harcourt, Brace, and World, 1968.

Oreskes, Naomi, and Erik M. Conway. *Merchants of Doubt: How a Handful of Scientists Obscured the Truth on Issues from Tobacco Smoke to Global Warming*. New York: Bloomsbury, 2010.

Ostman, Ronald E., and Harry Littell. *William T. Clarke, Photographer: The Epic Transformation of North Central Pennsylvania from "Black Forest to Bleak Desert," Circa 1878–1917*. Ithaca, N.Y.: Six Mile Creek Press, 2008.

Paris, Leslie. *Children's Nature: The Rise of the American Summer Camp*. New York: New York University Press, 2008.

Pascoe, Peggy. *Relations of Rescue: The Search for Female Moral Authority in the American West, 1874–1939*. New York: Oxford University Press, 1993.

Pasternak, Judy. *Yellow Dirt: An American Story of a Poisoned Land and a People Betrayed*. New York: Free Press, 2010.

Peavy, Linda, and Ursula Smith. *Pioneer Women: The Lives of Women on the Frontier*. Norman: University of Oklahoma Press, 1998.

Perdue, Theda. *Cherokee Women: Gender and Culture Change, 1700–1835*. Winnipeg: Bison Books, 1999.

Philbrick, Nathaniel. *Mayflower*. New York: Penguin, 2006.

Phillips, George. *The Enduring Struggle: Indians in California History*. San Francisco: Boyd and Fraser, 1981.

Proulx, Annie. *Fine Just the Way It Is*. New York: Scribner, 2008.

Quimby, George Irving. *Indian Culture and European Trade Goods: The Archaeology of the Historic Period in the Western Great Lakes Region*. Madison: University of Wisconsin Press, 1966.

Randolph, Mary. *The Virginia Housewife*. 4th ed. Washington, D.C.: Thompson, 1830.

Ravetz, Alison. *Place of Home: English Domestic Environments, 1914–2000*. New York: Routledge, 1995.

Rawls, James J., and Walton Bean. *California: An Interpretive History*. 7th ed. New York: McGraw-Hill, 1998.

———. *California: An Interpretive History*. 10th ed. New York: McGraw-Hill, 2011.

Reed, T. V. *The Art of Protest: Culture and Activism from the Civil Rights Movement to the Streets of Seattle*. St. Paul: University of Minnesota Press, 2005.

Reiger, John F. *American Sportsmen and the Origins of Conservation*. 3rd rev. ed. Corvallis: Oregon State University Press, 2000.

Reisner, Mark. *Cadillac Desert: The American West and Its Disappearing Water*. New York: Penguin, 1986.

Rice, Richard, William Bullough, and Richard Orsi, eds. *The Elusive Eden: A New History of California*. 2nd edition. New York: McGraw-Hill, 1996.

Riddell, Francis. "Maidu and Konkow." In *Handbook of North American Indians (California)*, ed. Robert F. Heizer, 381. Washington D.C.: Smithsonian Institution, 1978.

Riley, Glenda. *The Female Frontier: A Comparative View of Women on the Prairie and the Plains*. Lawrence: University Press of Kansas, 1988.

———. *A Place to Grow: Women in the American West*. Arlington Heights, Ill.: Harlan Davidson, 1992.

———. *Taking Land, Breaking Land: Women Colonizing in the American West and Kenya, 1840–1940*. Albuquerque: University of New Mexico Press, 2003.

———. *Women and Nature: Saving the "Wild West."* Lincoln: University of Nebraska Press, 1999.

———. "Women on the Wisconsin Frontier, 1836–1848." In *Women's Wisconsin*, ed. Genevieve McBride. Madison, Wisconsin Historical Society Press, 2005.

Roberson, Susan, ed. *Women, America, and Movement: Narratives of Relocation*. Columbia: University of Missouri Press, 1998.

Rodda, Annabel, ed. *Women and the Environment*. London: Zed Books, 1993.

Rome, Adam. *The Bulldozer in the Countryside: Suburban Sprawl and the Rise of American Environmentalism*. Boston, Mass.: Cambridge University Press, 2001.

Ross, Warren. *Funding Justice: The Legacy of the Unitarian Universalist Veatch Program*. Boston: Skinner House, 2005.

Rothblatt, Donald, Daniel Garr, and Jo Sprague. *The Suburban Environment and Women*. New York: Praeger, 1979.

Rothblum, Esther, and Penny Sablove, eds. *Lesbian Communities: Festivals, RVs, and the Internet*. New York: Harrington Park, 2005.

Rotundo, E. Anthony. *American Manhood*. New York: Basic Books, 1994.

Royce, Sarah. *A Frontier Lady: Recollections of the Gold Rush and Early California*. Lincoln: University of Nebraska Press, 1960.

Rudolph, Frederick, ed. *Essays on Education in the Early Republic*. Cambridge, Mass.: Belknap Press of Harvard University Press, 1965.

Ruether, Rosemary, ed. *Integrating Ecofeminism, Globalization, and World Religions*. Lanham, Md.: Rowman and Littlefield, 2005.

———. *New Woman, New Earth: Sexist Ideologies and Human Liberation*. New York: Seabury Press, 1975.

———. *Women Healing Earth: Third World Women on Ecology, Feminism, and Religion*. Maryknoll, N.Y.: Orbis, 1996.

Russell, Dale R. *Eighteenth Century Western Cree and their Neighbors*. Gatineau, Quebec: Canadian Museum of Civilization, 1991.

Ryan, Mary P. "Mothers of Civilization: The Common Woman, 1830–1860." In Ryan, *Womanhood in America*. New York: New Viewpoints, 1975.

———. *Mysteries of Sex: Tracing Women and Men through American History*. Chapel Hill: University of North Carolina Press, 2006.

Sachs, Carolyn E. *Gendered Fields: Rural Women, Agriculture, and Environment*. Boulder, Colo.: Westview, 1996.

Sanford, Mollie Dorsey. *Mollie: The Journal of Mollie Dorsey Sanford in Nebraska and Colorado Territories, 1857–1866*. Lincoln: University of Nebraska Press, 1959.

Saum, Lewis O. *The Fur Trader and the Indian*. Seattle: University of Washington Press, 1965.

Sayers, Janet. *Biological Politics: Feminist and Anti-Feminist Perspectives*. New York: Tavistock, 1982.

Schaefer, Heike. *Mary Austin's Regionalism: Reflections on Gender, Genre, and Geography*. Charlottesville: University of Virginia Press, 2004.

Scharff, Virginia J., ed. *Seeing Nature through Gender*. Lawrence: University Press of Kansas, 2003.

Scharff, Virginia, and Carolyn Brucken. *Home Lands: How Women Made the West*. Berkeley: University of California Press, 2010.

Schlissel, Lillian. *Women's Diaries of the Westward Journey*. New York: Schocken, 1982.

Schlueter, June. *Modern American Drama: The Female Canon*. Rutherford: Fairleigh Dickinson University Press, 1990.

Schofield, Ann. *'To Do & To Be': Portraits of Four Women Activists 1893–1986*. Boston: Northeastern University Press, 1997.

Schrepfer, Susan R. *Nature's Altars: Mountains, Gender, and American Environmentalism*. Lawrence: University Press of Kansas, 2005.

Schrepfer, Susan, and Douglas Cazaux Sackman. "Gender." In *A Companion to American Environmental History*, ed. Sackman, 116–45. Hoboken, N.J.: Wiley-Blackwell, 2010.

Schroeder, Richard A. *Shady Practices: Agroforestry and Gender Politics in the Gambia*. Berkeley: University of California Press, 1999.

Schwalm, Leslie A., ed. *A Hard Fight for We: Women's Transition from Slavery to Freedom in South Carolina*. Urbana: University of Illinois Press, 1997.

Sellers, Christopher. *Crabgrass Crucible: Suburban Nature and the Rise of Environmentalism in Twentieth-Century America*. Chapel Hill: University of North Carolina Press, 2012.

Settle, Raymond W., and Mary Settle, eds. *Overland Days to Montana*. Glendale, Calif.: Arthur Clark Company, 1971.

Sharpless, Rebecca. *Fertile Ground, Narrow Choices: Women on Texas Cotton Farms, 1900–1940*. Chapel Hill: University of North Carolina Press, 1999.

Shimkin, Demitri. "Eastern Shoshone." In *Handbook of North American Indians (Great Basin)*, ed. Warren L. D'Azevedo, 330. Washington D.C.: Smithsonian Institution, 1986.

Shiva, Vandana. *Staying Alive: Women, Ecology, and Development*. Trowbridge, U.K.: Redwood, 1989.

———. *Staying Alive: Women, Ecology, and Survival in India*. New Delhi: Kali for Women, 1988.

Sicherman, Barbara. *Alice Hamilton: A Life in Letters*. Champaign: University of Illinois Press, 2003.

———. "Working It Out: Gender, Profession, and Reform in the Career of Alice Hamilton." In *Gender, Class, Race, and Reform in the Progressive Era*, ed. Noralee Frankel and Nancy S. Dye, 127–47. Lexington: University Press of Kentucky, 1991.

Sideris, Lisa H., and Kathleen Dean Moore, eds. *Rachel Carson: Legacy and Challenge*. New York: Albany State University of New York Press, 2008.

Silliman, Jael, and Ynestra King, eds. *Dangerous Intersections: Feminist Perspectives on Population, Environment, and Development*. Cambridge, Mass.: South End, 1999.

Simon, Bryant. "'New Men in Body and Soul': The Civilian Conservation Corps and the Transformation of Male Bodies and the Male Politic." In *Seeing Nature through Gender*, ed. Virginia Scharff, 80–102. Lawrence: University Press of Kansas, 2003.

Simpson, Andrea. "'Who Hears Their Cry?' African American Women and the Fight for Environmental Justice in Memphis, Tennessee." In *The Environmental Justice Reader: Politics, Poetics, and Pedagogy*, ed. Joni Anderson, Mei Mei Evans, and Rachel Stein, 82–104. Tuscon: University of Arizona Press, 2002.

Sirch, Willow Ann. *Eco-Women: Protectors of the Earth*. Golden, Colo.: Fulcrum, 1996.

Sklar, Kathryn Kish. *Catherine Beecher: A Study in American Domesticity*. New Haven, Conn.: Yale University Press, 1973.

———. *Florence Kelley and the Nation's Work: The Rise of Women's Political Culture, 1830–1900*. New Haven, Conn.: Yale University Press, 1995.

———. "Two Political Cultures in the Progressive Era: The National Consumers' League and the American Association for Labor Legislation." In *U.S. History as Women's History: New Feminist Essays*, ed. Linda K. Kerber, Alice Kessler-Harris, and Kathryn Kish Sklar, 36–62. Chapel Hill: University of North Carolina Press, 1995.

Sleeper-Smith, Susan. *Indian Women and French Men: Rethinking Cultural Encounter in the Western Great Lakes*. Amherst: University of Massachusetts Press, 2001.

Small, Gail. "Gail Small: Voices from Northern Cheyenne Indian Country." In *The Quest for Environmental Justice: Human Rights and the Politics of Pollution*, ed. Robert D. Bullard, 75–79. San Francisco: Sierra Club Books, 2005.

Smith, Andrea. *Conquest: Sexual Violence and American Indian Genocide*. Cambridge, Mass.: South End, 2005.

Smith, Daniel Scott. "Family Limitation, Sexual Control, and Domestic Feminism in Victorian America." In *Clio's Consciousness Raised*, ed. Mary Hartman and Lois Banner, 119–36. New York: Harper & Row, 1974.

Smith, Kimberly K. *African American Environmental Thought Foundations*. Lawrence: University Press of Kansas, 2000.

Smith, Paul Chaat, and Robert Allen Warrior. *Like a Hurricane: The Indian Movement from Alcatraz to Wounded Knee*. New York: New Press, 1997.

Soselisa, Hermien. "The Significance of Gender in the Fishing Economy of the Goram Islands, Maluku." In *Old World Places, New World Problems: Exploring Issues of Resource Management in Eastern Indonesia*, ed. Sandra Pannell and Franz von Benda-Beckmann, 321–35. Canberra: Australian National University, Centre for Resource and Environmental Studies, 1988.

Spain, Daphne. *Gendered Spaces*. Chapel Hill: University of North Carolina Press, 1992.

Stanley, Autumn. *Mothers and Daughters of Invention: Notes for a Revised History of Technology*. Metuchen, N.J., and London: Scarecrow, 1993.

Stanwell-Fletcher, Theodora C. *Driftwood Valley: A Woman Naturalist in the Northern Wilderness*. Corvallis: Oregon State University Press, 1999.

Starita, Joe. *"I Am a Man": Chief Standing Bear's Journey for Justice*. New York: St. Martin's Press, 2007.

Steady, Filomina Chioma, ed. *Women and Children First: Environment, Poverty, and Sustainable Development*. Rochester, Vt.: Schenkman, 1993.

Stein, Rachel. *Shifting the Ground: American Women Writers' Revisions of Nature, Gender, and Race*. Charlottesville: University of Virginia Press, 1997.

Stein, Rachel, ed. *New Perspectives on Environmental Justice: Gender, Sexuality, and Activism*. New Brunswick, N.J.: Rutgers University Press, 2004.

Steinberg, Ted. *American Green: The Obsessive Quest for the Perfect Lawn*. New York: Norton, 2006.

———. *Down to Earth: Nature's Role in American History*. New York: Oxford University Press, 2002.

Stephanson, Anders. *Manifest Destiny: American Expansion and the Empire of Right*. New York: Hill and Wang, 1996.

Stewart, Elinore Pruit. *Letters of a Woman Homesteader*. 1914. Reprint, Mineola: Dover, 2006.

Stille, Darlene. *Extraordinary Women Scientists*. Chicago: Children's Press, 1995.

Stout, David B., Erminie Wheeler-Voegelin, and Emily J. Blasingham. *Sac, Fox, and Iowa Indians II: Indians of E. Missouri, W. Illinois, and S. Wisconsin from the Proto-Historic Period to 1804.* New York and London: Garland, 1974.

Stradling, David, ed. *Conservation in the Progressive Era: Classic Texts.* Seattle: University of Washington Press, 2004.

Stratton, Joanna. *Pioneer Women: Voices from the Kansas Frontier.* New York: Simon and Schuster, 1981.

Sturgeon, Noel. *Ecofeminist Natures: Race, Gender, Feminist Theory, and Political Action.* New York: Routledge, 1997.

———. *Environmentalism in Popular Culture: Gender, Race, Sexuality, and the Politics of the Natural.* Tucson: University of Arizona Press, 2008.

Swerdlow, Amy. *Women Strike for Peace: Traditional Motherhood and Radical Politics in the 1960s.* Chicago: University of Chicago Press, 1993.

Szczygiel, Bonj, Josephine Carubia, and Lorraine Dowler, eds. *Gendered Landscapes: An Interdisciplinary Exploration of Past Place and Space.* University Park: Pennsylvania State University, 2000.

Sze, Julie. *Noxious New York: The Racial Politics of Urban Health and Environmental Justice.* Cambridge, Mass.: MIT Press, 2006.

Taylor, Dorceta. *The Environment and the People in American Cities, 1600s to 1900s: Disorder, Inequality, and Social Change.* Durham, N.C.: Duke University Press, 2009.

———. *Race, Gender, and American Environmentalism.* U.S. Department of Agriculture, Forest Service, Pacific Northwest Research Station. 2002. Online. Available: www.fs.fed.us/pnw/pubs/gtr534.pdf (accessed May 3, 2011).

Taylor, Ethel Barol. *We Made a Difference: My Personal Journey with Women Strike for Peace.* Philadelphia: Camino, 1998.

Thornton, Russell. *American Indian Holocaust and Survival: A Population History since 1492.* Norman: University of Oklahoma Press, 1987.

Thwaites, Reuben Gold, ed. *The Jesuit Relations: and Allied Documents Travels and Explorations of the Jesuit Missionaries in New France 1610–1791,* vol. 5. Cleveland: Burrows Brothers, 1896–1901. Available http://puffin.creighton.edu/jesuit/relations/relations_05.html (accessed September 1, 2011).

Tompkins, Jane. *West of Everything: The Inner Life of Westerns.* New York: Oxford University Press, 1993.

Turpin, Jennifer, and Lois Ann Lorentzen, eds. *The Gendered New World Order: Militarism, Development, and the Environment.* New York: Routledge, 1996.

Unger, Nancy C. *Fighting Bob La Follette: The Righteous Reformer.* Chapel Hill: University of North Carolina Press, 2000. Rev. ed. Madison: Wisconsin Historical Society, 2008.

———. "Gender: A Useful Category of Analysis in Environmental History." In *Oxford Handbook of Environmental History,* ed. Andrew Isenberg. New York: Oxford University Press, forthcoming 2013.

———. "Gendered Approaches to Environmental Justice: An Historical Sampling." In *Echoes from the Poisoned Well: Global Memories of Environmental Injustice,* ed. Heather Goodall, Paul C. Rosier, and Sylvia Washington, 17–34. Lanham, Md.: Littlefield/Lexington, 2006.

———. "The Role of Nature in Lesbian Alternative Environments in the United States: From Jook Joints to Sisterspace." In *Queer Ecologies: Sex, Nature, Biopolitics, and Desire*, ed. Catriona Mortimer-Sandilands and Bruce Erickson, 173–98. Bloomington: University of Indiana Press, 2010.

———. "Women, Sexuality, and Environmental Justice in American History." In *New Perspectives on Environmental Justice: Gender, Sexuality, and Activism*, ed. Rachel Stein, 45–60. New Brunswick, N.J.: Rutgers University Press, 2004.

Unger, Nancy C., and Marie Bolton. "The Case for Cautious Optimism: California Environmental Propositions in the Late Twentieth Century." In *La Californie: Périphérie ou laboratoire?*, ed. Annick Foucrier and Antoine Coppolani, 81–102. Paris: L'Harmattan, 2004.

Unger, Nancy C., and Marie Bolton. "Pollution, Refineries, and People: Environmental Justice in Contra Costa County, California, 1980." In *Le Démon Moderne: La Pollution dans les Societies Urbaines et Industrielles d'Europe* [*The Modern Demon: Pollution in Urban and Industrial European Societies*], ed. Christoph Bernhardt and Genevieve Massard-Guilbaud, 425–37. Clermont-Ferrand, France: Presses Universitaires Blaise-Pascal, 2002.

Van Kirk, Sylvia. *Many Tender Ties: Women in Fur-Trade Society, 1670–1870*. Norman: University of Oklahoma Press, 1983; Winnipeg: Watson and Dwyer, 1996.

Van Wormer, Heather. "A New Deal for Gender: The Landscapes of the 1930s." In *Shared Spaces and Divided Places: Material Dimensions of Gender Relations and the American Historical Landscape*, ed. Deborah L. Rotman and Ellen-Rose Savulis, 190–224. Knoxville: University of Tennessee Press, 2003.

Vecsey, Christopher. "American Indian Environmental Religions." In *American Indian Environments*, ed. Vecsey and Robert Venables, 1–37. Syracuse: Syracuse University Press, 1980.

Vecsey, Christopher, and Robert W. Venables, eds. *American Indian Environments: Ecological Issues in Native American History*. Syracuse, N.Y.: Syracuse University Press, 1980.

Vermaas, Lori. *Sequoia: The Heralded Tree in American Art and Culture*. Washington, D.C.: Smithsonian, 2003.

Vo, Linda Trinh, and Marian Sciachitano, eds. *Asian American Women: The Frontiers Reader*. Lincoln: University of Nebraska Press, 2003.

Waak, Patricia. *Faith, Justice, and a Healthy World: A Guide on Population and Environment for People of Faith*. New York: National Audubon Society, 1994.

Wagner, Tricia Martineau. *African American Women of the Old West*. Helena, Mont.: TwoDot, 2007.

Walker, Melissa. *All We Knew Was to Farm: Rural Women in the Upcountry South, 1919–1941*. Baltimore: John Hopkins University Press, 2000.

Warren, Karen, ed. *Ecofeminism: Women, Culture, and Nature*. Bloomington: Indiana University Press, 1997.

Washington, Sylvia Hood. *Packing Them In: An Archeology of Environmental Racism in Chicago, 1865–1954*. Lanham, Md.: Lexington, 2005.

Watts, Sarah. *Rough Rider in the White House: Theodore Roosevelt and the Politics of Desire*. Chicago: University of Chicago Press, 2003.

Weaver, Jace, ed. *Defending Mother Earth: Native American Perspectives on Environmental Justice*. Maryknoll, N.Y.: Orbis, 2001.

Webb, Benjamin, ed. *Fugitive Faith: Conversations on Spiritual, Environmental, and Community Renewal*. Maryknoll, N.Y.: Orbis, 1998.

Webster, Juliet. *Shaping Women's Work: Gender, Employment and Information Technology*. New York: Longman, 1996.

Weinstein, Laurie, ed. *Enduring Traditions: The Native Peoples of New England*. Westport, Conn.: Bergin and Garvey, 1994.

Weisman, Leslie Kanes. *Discrimination by Design: A Feminist Critique of the Man-Made Environment*. Urbana and Chicago: University of Illinois Press, 1992.

Weiss, Elaine F. *Fruits of Victory: The Woman's Land Army of America in the Great War*. Washington, D.C.: Potomac, 2008.

Westmacott, Richard. *African-American Gardens and Yards in the Rural South*. Knoxville: University of Tennessee Press, 1992.

Westra, Laura, and Peter S. Wenz, eds. *Faces of Environmental Racism: Confronting Issues of Global Justice*. Lanham, Md.: Rowman and Littlefield Publishers, Inc., 1995.

"What Is Meant by Conservation?" *Ladies' Home Journal* 28 (November 1911). In *Conservation in the Progressive Era: Classic Texts*, ed. David Stradling, 32–34. Seattle: University of Washington Press, 2004.

Wheeler-Voegelin, Erminie, and J. A. Jones. *Indians of Western Illinois and Southern Wisconsin*. New York: Garland, 1974.

White, Deborah Gray. *"Ar'n't I a Woman?" Female Slaves in the Plantation South*. 1985. Reprint, New York: Norton, 1999.

White, Richard. *The Roots of Dependency: Subsistence, Environment, and Social Change among the Choctaws, Pawnees, and Navajos*. Lincoln: University of Nebraska Press, 1983.

Wickramasinghe, Anoja. *Deforestation, Women, and Forestry: The Case of Sri Lanka*. Amsterdam: International Books, 1994.

Wilbur, Charles Dana. *The Great Valleys and Prairies of Nebraska and the Northwest*. Omaha, Neb.: Daily Republican Print, 1881.

Wilder, Laura Ingalls. *Little House on the Prairie*. New York: Harper and Row, 1971.

Williams, Robert Chadwell. *Horace Greeley: Champion of American Freedom*. New York: New York University Press, 2006.

Willoughby, Robert J, ed. *The Great Western Migration to the Gold Fields of California, 1849–1850*. Jefferson, N.C.: McFarland and Company, 2003.

Wollstonecraft, Mary. "Unfortunate Situation of Females, Fashionably Educated, and Left without a Fortune." 1787. Reprinted in *Thoughts on the Education of Daughters: with Reflections on Female Conduct, in the Most Important Duties in Life*, 71. Bristol: Thoemmes, 1995.

Woloch, Nancy. *Women and the American Experience*. New York: McGraw-Hill, 1984.

Wood, Mary I. *The History of the General Federation of Women's Clubs for the First Twenty-Two Years of Its Organization*. Norwood, Mass.: Norwood, 1912.

Woodward, Mary Dodge. "The Diary of Mary Dodge Woodward." In *The Checkered Years*, ed. Mary Cowdrey. Caldwell, Idaho: Caxton, 1937.

Worster, Donald. *Dust Bowl: The Southern Plains in the 1930s*. New York: Oxford University Press, 1979.

———. *Rivers of Empire: Water, Aridity, and the Growth of the American West*. New York: Oxford University Press, 1985.

Yung, Judy. *Unbound Feet: A Social History of Chinese Women in San Francisco*. Berkeley: University of California Press, 1995.

———. *Unbound Voices: A Documentary History of Chinese Women in San Francisco*. Berkeley: University of California Press, 1999.

ARTICLES

"3rd Annual Town Hall Meeting." Zerobreastcancer.org. Online. Available: http://zerobreast-cancer.org/research/th08_liou.pdf (accessed April 21, 2011).

Adams-Williams, Lydia. "A Million Women for Conservation." *Conservation: Official Organ of the American Forestry Association* 15 (1909): 346–47.

Addams, Jane. *Why Women Should Vote, 1915. Modern History Sourcebook.* 1999. Online. Available: http://www.fordham.edu/halsall/mod/1915janeadams-vote.html (accessed June 23, 2010).

"Addressing Uranium Contamination in the Navajo Nation." *United States Environmental Protection Agency.* Online. Available: http://www.epa.gov/region09/superfund/navajo-nation/index.html (accessed February 5, 2010).

"About WFAN." Women, Food and Agriculture Network. Online. Available: http://www.wfan.org/About_Us.html (accessed December 19, 2011).

"Alapine Community Association, Inc." Online. Available: http://alapine.com (accessed February 3, 2009).

Albertine, Susan. "The Life Writings of Harriet Strong." *Biography* 17, no. 2 (Spring 1994): 161–86.

Alexander, Ruth. "'We Are Engaged as a Band of Sisters': Class and Domesticity in the Washingtonian Temperance Movement, 1840–1850." *Journal of American History* 75, no. 3 (December 1988): 763–85.

American Farmer. "Kitchen Garden, for March." *Illinois Gazette.* March 28, 1829. D.

Armitage, Kevin C. "Bird Day for Kids: Progressive Conservation in Theory and Practice." *Environmental History* 13, no. 3 (July 2007): 528–51.

Bailey, Ronald. "Earth Day, Then and Now." *Reason.* May 2000. Online. Available: http://www.reason.com/issues/show/346.html (available May 10, 2009).

Bauman, Paula Mae. "Single Women Homesteaders in Wyoming, 1880–1930." *Annals of Wyoming* 58, no. 1 (1986): 39–53.

Bayers, Peter. "Fredrick Cook, Mountaineering in the Alaskan Wilderness and the Regeneration of Progressive Era Masculinity." *Western American Literature* 38, no. 2 (Summer 2003): 170–93.

Becker, Paula. "Betty MacDonald's *The Egg and I* Is Published on October 3, 1945." *HistoryLink: The Free Online Encyclopedia of Washington State History.* August 14, 2007. Online. Available: http://www.historylink.org/index.cfm?DisplayPage=pf_output.cfm&file_id=8261 (accessed May 17, 2010).

———. "Libel Trial against Betty MacDonald of *Egg and I* Fame Opens in Seattle on February 5, 1951." *HistoryLink: The Free Online Encyclopedia of Washington State History.* August 31, 2007. Online. Available: http://www.historylink.org/index.cfm?DisplayPage=pf_output.cfm&file_id=8270 (accessed May 17, 2010).

———. "Seattle Jury Finds for the Defendants in Libel Suit against *Egg and I* Author Betty MacDonald on February 20, 1951." *HistoryLink: The Free Online Encyclopedia of Washington State History.* September 5, 2007. Online. Available: http://www.historylink.org/index.cfm?DisplayPage=output.cfm&file_id=8271 (accessed May 30, 2011).

Benton-Cohen, Katherine. "Common Purposes, Worlds Apart: Mexican-American, Mormon, and Midwestern Women Homesteaders in Cochise County, Arizona." *Western Historical Quarterly* 36 (Winter 2005): 429–52.

Binkley, Cameron. "'No Better Heritage Than Living Trees': Women's Clubs and Early Conservation in Humboldt County." *Western Historical Quarterly* 33 (Summer 2002): 179–203.

Blauvelt, Martha Tomhave. "The Work of the Heart: Emotion in the 1805–35 Diary of Sarah Connell." *Journal of Social History* 35 (2002): 577–92.

Block, Daniel. "Saving Milk through Masculinity: Public Health Officers and Pure Milk, 1880–1930." *Food and Foodways: History and Culture of Human Nourishment* 15, no. 1/2 (January–June 2005): 115–34.

Blum, Elizabeth. "Linking American Women's History and Environmental History: A Preliminary Historiography." Online. Available: http://www.h-net.org/~environ/historiography/uswomen.htm (accessed January 15, 2009).

———. "Power, Danger, and Control: Slave Women's Perception of Wilderness in the Nineteenth Century." *Women's Studies* 31, no. 2 (2002): 247–66.

Boris, Eileen. "The Power of Motherhood: Black and White Activist Women Redefine the 'Political.'" *Yale Journal of Law and Feminism* 2, no. 1 (Fall 1989): 25–49.

Boston University School of Public Health. "Key Dates and Events at Love Canal." *Lessons from Love Canal: A Public Health Resource*. 2003. Online. Available: http://www.bu.edu/lovecanal/canal/date.html (accessed April 14, 2011).

Bregendahl, Corry, and Matthew Hoffman. "Women, Land and Legacy: Change Agents and Agency Change in Iowa Evaluation Results, November 2010." Online. Available: http://journals.cambridge.org/action/displayAbstract;jsessionid=B5E5E451CCE551560C468C2EB0F7B85E.journals?fromPage=online&;aid=115029 (accessed December 26, 2011).

Bruchac, Joseph. "Otstungo: A Mohawk Village in 1491." *National Geographic* 180 (October 1991): 68–83.

Bullard, Robert. "Environmental Justice Movement Loses Southside Chicago Icon Hazel Johnson." *OpEdNews.com*. January 14, 2011. Online. Available: http://www.opednews.com/articles/Environmental-Justice-Move-By-Robert-Bullard-110114323.html (accessed February 7, 2011).

Bullard, Robert, and Beverly Hendris. "The Politics of Pollution: Implications for the Black Community." *Phylon* 47, no. 1 (1986): 71–78.

"The California Healthy Nail Salon Collaborative." *Asian Health Services*. Online. Available: http://asianhealthservices.org/nailsalon/ (accessed April 23, 2011).

Campbell, Constance E. "On the Front Lines but Struggling for Voice: Women in the Rubber Tappers' Defence of the Amazon Forest." *Ecologist* 27 (March/April 1997): 46–54.

Cashin, Joan E. "'Decidedly Opposed to "The Union"': Women's Culture, Marriage, and Politics in Antebellum South Carolina." *Georgia Historical Quarterly* 78 (1994): 735–59.

"CERCLA Overview." *United States Environmental Protection Agency*. Online. Available: http://www.epa.gov/superfund/policy/cercla.htm (accessed April 14, 2011).

Chiang, Connie. "Imprisoned Nature: Toward an Environmental History of the World War II Japanese American Incarceration." *Environmental History* 15 (April 2010): 237–67.

Childress, Marjorie. "Uranium Mining at Mt. Taylor Threatened; New Colo. Law Requires Cleanup." *New Mexico Independent*. June 9, 2010. Online. Available: http://newmexicoindependent.com/56734/uranium-mining-at-mt-taylor-threatened-new-colo-law-requires-cleanup (accessed April 21, 2011).

Ciasullo, Ann. "Making Her (In)Visible: Cultural Representations of Lesbianism and the Lesbian Body in the 1990s." *Feminist Studies* 27, no. 3 (2001): 577–608.

Clavers, Mary. "Spring in the Woodlands." *Lady's Book* 31 (June 1845): 22.

Clayton, Mark. "Women Lead a Farming Revolution in Iowa." *Christian Science Monitor*, February 25, 2009. Online. Available: http://www.csmonitor.com/Environment/Living-Green/2009/0225/women-lead-a-farming-revolution-in-iowa (accessed February 27, 2009).

Clinton, Katherine. "Pioneer Women in Chicago, 1833–1837." *Journal of the West* 12, no. 2 (April 1973): 317–24.

Comment, Kristin M. "Charles Brockden Brown's 'Ormond' and Lesbian Possibility in the Early Republic." *Early American Literature* 40, no. 1 (2005): 57–78.

Corinne, Tee. "Little Houses on Women's Lands." 1998. Online. No longer available: http://www.lib.usc.edu/~retter/teehouse.html (accessed June 1, 2002). See instead Corinne, Tee. *The Little Houses on Women's Lands: Feminism, Photography, and Vernacular Architecture*. Wolf Creek, Ore.: Pearlchild, 2002.

Corbett, Julia. "Women, Scientists, Agitators: Magazine Portrayal of Rachel Carson and Theo Colborn." *Journal of Communication* 51 (2001): 720–49.

Cronin, Mary. "Redefining Woman's Sphere." *Journalism History* 25, no. 1 (1999): 13–25.

Cuellar, José B. "Labor, Farm." *Pollution Issues*. Online. Available: http://www.pollutionissues.com/Ho-Li/Labor-Farm.html (accessed April 23, 2011).

Cumbler, John T. "Conflict, Accommodation, and Compromise: Connecticut's Attempt to Control Industrial Wastes in the Progressive Era." *Environmental History* 5 (July 2000): 314–35.

Curry-Reyes, Traciy. "Emelda West Story." *Movies Based on True Stories Database*. June 2009. Online. Available: http://emeldaweststory.blogspot.com/2009/06/emelda-west-story.html (accessed November 15, 2011).

Dawson, Jan C. "'Lady Lookouts' in a 'Man's World' during World War II: A Reconsideration of American Women and Nature." *Journal of Women's History* 8 (Fall 1996): 99–113.

Devens, Carol. "'If We Get the Girls, We Get the Race': Missionary Education of Native American Girls." *Journal of World History* 3, no. 2 (1992): 219–37.

———. "Separate Confrontations: Gender as a Factor in Indian Adaptation to European Colonization in New France." *American Quarterly* 38, no. 3 (1986): 461–80.

deWit, Cary W. "Women's Sense of Place on the American High Plains." *Great Plains Quarterly* 21 (2001): 29–44.

Dobrzynski, Judith H. "The Grand Women Artists of the Hudson River School." Smithsonian.com. July 21, 2010. Online. Available: http://www.smithsonianmag.com/arts-culture/The-Grand-Women-Artists-of-the-Hudson-River-School.html?c=y&;page=2# (accessed April 28, 2011).

Edgington, Ryan H. "'Be Receptive to the Good Earth': Health, Nature, and Labor in Countercultural Back-to-the-Land Settlements." *Agricultural History* 82, no. 3 (Summer 2008): 279–308.

Edwards, Rebecca, John R. Richard, and Richard Bensel. "Should We Abolish the 'Gilded Age'?" *Journal of the Gilded Age and Progressive Era* 8, no. 4 (October 2009): 461–85.

Eliassen, Meredith. "Got Pure Milk? Dr. Adelaide Brown's Crusade for San Francisco's Safe Milk Supply." *Argonaut* 18, no. 1 (January 2007): 36–51.

Ellett, Mrs. E. F. "The Pioneer Mothers of Michigan." *Godey's Lady's Book* 44 (April 1852): 266.

Eltis, David, Philip Morgan, and David Richardson. "Agency and Diaspora in Atlantic History: Reassessing the African Contribution to Rice Cultivation in the Americas." *American Historical Review* 112, no. 5 (December 2007): 1328–58.

Eltis, David et al. "*AHR* Exchange: The Question of 'Black Rice.'" *American Historical Review* 115, no. 1 (February 2010): 123–71.

"Environmental Justice Case Study: Shintech PVC Plant in Convent, Louisiana." Online. Available: http://www.umich.edu/~snre492/shin.html (accessed November 15, 2011).

Farmer's Cabinet. "Kitchen Garden." *New-Hampshire Statesman and State Journal.* April 6, 1839. E.

Fish, Virginia Kemp. "Widening the Spectrum: The Oral History Technique and Its Use with LAND, a Grass-Roots Group." *Sociological Imagination* 31, no. 2 (1994): 101–10.

Flanagan, Maureen A. "The City Profitable, the City Livable: Environmental Policy, Gender, and Power in Chicago in the 1910s." *Journal of Urban History* 22 (January 1996): 163–90.

Forest History Society. *Environmental History Bibliography.* Online. Available: http://www.foresthistory.org/Research/biblio.html (accessed March 27, 2012).

Fortmann, Louis, and Dianne Rocheleau. "Women and Agroforestry: Four Myths and Three Case Studies." *Women in Natural Resources* 9, no. 2 (1987): 35–51.

Fox, Warwick. "The Deep Ecology-Ecofeminism Debate and Its Parallels." *Environmental Ethics* 11 (Spring 1989): 5–26.

"Free Soil for Free Men." *National Era* 11, no. 549 (July 9, 1857): 112.

Fujita-Rony, Thomas Y. "Remaking the 'Home Front' in World War II: Japanese American Women's Work and the Colorado River Relocation Center." *Southern California Quarterly* 88, no. 2 (June 2006): 161–204.

Gabrial, Brian. "A Woman's Place." *American Journalism* 25, no. 1 (2008): 7–29.

Garber, Eric. "A Spectacle in Color: The Lesbian and Gay Subculture of Jazz Age Harlem." Online. Available: http://xroads.virginia.edu/~UG97/blues/garber.html (accessed January 10, 2009).

Garcia, Richard A. "Dolores Huerta: Woman, Organizer, and Symbol." *California History* 72, no. 1 (1993): 56–71.

"Garden Borders." *Godey's Lady's Book* 50 (May 1855): 477.

Gearheart, Dona G. "Coal Mining Women in the West: The Realities of Difference in an Extreme Environment." *Journal of the West* 37, 1 (1988): 60–68.

Geiser, Ken. "Toxic Times and Class Politics." *Radical America* 17, no. 2 (1983): 39–50.

"General Festival Information." *Michigan Womyn's Music Festival.* Online. Available: http://www.michfest.com/festival/index.htm (accessed April 7, 2007).

Genovese, Eugene D. "Cotton, Slavery, and Soil Exhaustion in the Old South." *History Review* 2, no. 1 (1961): 3–17.

Gibbs, Lois. "Citizen Activism for Environmental Health: The Growth of a Powerful New Grassroots Health Movement." *Annals of the American Academy of Political and Social Science* 584 (November 2002): 97–109.

———. "The Start of a Movement." In *Lessons from Love Canal: A Public Health Resource.* Ed. Boston University School of Public Health. 2003. Online. Available: http://www.bu.edu/lovecanal/canal/date.html (accessed April 14, 2011).

Gibbs, Rafe. "What Became of the Shelter Belt?," *Popular Mechanics* 99, no. 5 (May 1953): 102–6, 238, 240, 242.

Giesen, James C. "'The Truth about the Boll Weevil': The Nature of Planter Power in the Mississippi Delta." *Environmental History* 14, no. 4 (October 2009): 683–704.

Glave, Dianne. "'A Garden So Brilliant with Colors, So Original in Its Design': Rural African American Women, Gardening, Progressive Reform, and the Foundation of an African American Environmental Perspective." *Environmental History* 8, no. 3 (2003): 395–411.

Gonzalez, Gilbert G. "Women, Work, and Community in the Mexican *Colonias* of the Southern California Citrus Belt." *California History* 74, no. 1 (Spring 1995): 58–70.

Gottlieb, L. S. and L. A. Husen. "Lung Cancer among Navajo Uranium Miners." *Chest* 81, no. 4 (April 1982): 449–52.

Grant, Edwin H. "To the Unemployed of Our Eastern Cities" (Washington, D.C.) *National Era* 11, no. 572 (December 17, 1857): 201.

Greenberg, Dolores. "Reconstructing Race and Protest: Environmental Justice in New York City." *Environmental History* 5 (April 2000): 223–50.

Gross, Joan. "Capitalism and Its Discontents: Back-to-the-Lander and Freegan Foodways in Rural Oregon." *Food and Foodways: History and Culture of Human Nourishment* 17, no. 2 (April–June 2009): 57–79.

Gugliotta, Angela. "Class, Gender, and Coal Smoke: Gender Ideology and Environmental Injustice in Pittsburgh, 1868–1914." *Environmental History* 5 (April 2000): 165–93.

Gugliotta, Guy. "Indians Hunted Carelessly, Study Says." *Seattle Times*, February 21, 2006.

Guth, James, et al. "Faith and the Environment: Religious Beliefs and Attitudes on Environmental Policy." *American Journal of Political Science* 95, no. 2 (May 1995): 364–83.

Hain, Danielle. *Stopping Stereotypes: Problem Drinking and Alcoholism in the LGBT Community.* Tempe, Ariz.: Do It Now Foundation, 2007. Online. Available: http://doitnow.org/pages/110.html (accessed May 11, 2009).

Harris, Emma. "Emma the Welder." *San Jose Mercury News.* February 24, 2000, 1E; 9E.

Hay, Amy M. "A New Earthly Vision: Religious Community Activism in the Love Canal Chemical Disaster." *Environmental History* 14, no. 3 (2009): 502–27.

———. "Recipe for Disaster: Motherhood and Citizenship at Love Canal." *Journal of Women's History* 21, no. 1 (Spring 2009): 111–34.

Hayes, Sarah Hepburn. "Peter Allan's Panther Chase." *Godey's Lady's Book* 40 (February 1850): 134.

Helms, Kathy. "'Navajo Neuropathy' Haunts Blue Gap Family." *Gallup Independent.* August 6, 2009. Online. Available: http://native unity.blogspot.com/2009/08/Navajo-neuropathy-sen-reid-opposes.html (accessed April 21, 2011).

Hewitt, Nancy A. "Taking the True Woman Hostage." *Journal of Women's History* 14, no. 1 (2002): 156–62.

Hicks, Cheryl D. "'Bright and Good Looking Colored Girl': Black Women's Sexuality and 'Harmful Intimacy' in Early-Twentieth-Century New York." *Journal of History of Sexuality* 18 (September 2009): 418–56.

"History." *Campfire USA.* Available: http://www.campfireusaia.org/about-us-history.php (accessed November 1, 2011).

History—Girl Scouts Timeline 1912–1919; 1920s, *Girl Scouts.* Available: http://www.girlscouts.org/who_we_are/history/timeline/1920s.asp (accessed November 1, 2011).

Huftalen, Sarah. "The Use of the Handbook in Rural Schools." *Midland Schools: A Journal of Education* 24, no. 10 (July 1910): 300.

Hurtado, Albert. "Sexuality in California's Franciscan Missions: Cultural Perceptions and Sad Realities." *California History* 72 (Fall 1992): 370–85.

Hynes, Patricia H. "Ellen Swallow, Lois Gibbs, and Rachel Carson: Catalysts of the American Environmental Movement." *Women's Studies International Forum* 8, no. 4 (July 1985): 291–98.

"Ill Advised Sarcasm," *La Follette's Magazine* 5, no. 35 (August 30, 1913): 3.

Irving, Mary. "A Peep at the Prairie." *National Era* 5, no. 212 (January 23, 1851): 13.

———. "The Romance of Society." *National Era* 5, no. 213 (January 30, 1851): 17.

Jabour, Anya. "Albums of Affection: Female Friendship and Coming of Age in Antebellum Virginia." *Virginia Magazine of History and Biography* 107 (1999): 125–58.

Jackson, Cecile. "Women/Nature or Gender/History? A Critique of Ecofeminist 'Development.'" *Journal of Peasant Studies* 20 (April 1993): 389–419.

Jain, Shobita. "Standing Up for Trees: Women's Role in the Chipko Movement." *Unasylva* 36, no. 4 (1984): 12–20.

James, William. "The Moral Equivalent of War." *McClure's Magazine* 35 (August 1910): 463–68.

Jason, Sonya. "From Gunpowder Girl to Working Woman." *Newsweek.* February 23, 2004, 20.

Jervis, Lisa. "The Bitch Holiday Gift Guide." *Bitch: Feminist Response to Popular Culture*, no. 22 (Fall 2003): 94.

Jewett, Sarah Orne, and Marjorie Pryse. "Outgrown Friends." *New England Quarterly* 69 (1996): 461–72.

Johansen, Bruce. "Reprise: Forced Sterilizations." *Native Americas* 15, no. 4 (Winter 1998): 44–47.

Johnson, Bruce E. "Women of All Red Nations." *ABC-CLIO History and the Headlines.* Online. Available: http://www.historyandtheheadlines.abc-clio.com/ContentPages/ContentPage. aspx?entryId=1172002&current Section=1161468&productid=5 (accessed March 24, 2010).

Jordan, Ben. "'Conservation of Boyhood': Boy Scouting's Modest Manliness and Natural Resource Conservation, 1910–1930." *Environmental History* 15, no. 4 (October 2010): 612–42.

Kalof, Linda, et al. "Race, Gender, and Environmentalism: The Atypical Values and Beliefs of White Men." *Race, Gender, and Class* 9, no. 2 (2002): 1–19.

Kaminer, Wendy. "Crashing the Locker Room." *Atlantic.* July 1992. Online. Available: http://theatlantic.com/issues/92jul/kaminer.htm (accessed June 5, 2009).

Kant, Curtis. "Warning Signs." *Reminisce* (May/June 2006): 66.

Kanter, Deborah. "Native Female Land Tenure and Its Decline in Mexico, 1750–1900." *Ethnohistory* 42 (Fall 1995): 607–16.

Karadimas, Charalampos L., et al. "Navajo Neurohepatopathy Is Caused by a Mutation in the *MPV17* Gene." *American Journal of Genetics.* 2006. Online. Available: http://www.ncbi.nlh. nih.gov/pmc/articles/PMC1559552/ (accessed April 21, 2011).

Karaim, Reed. "Not So Fast with the DDT: Rachel Carson's Warnings Still Apply." *American Scholar* 74, no. 3 (Summer 2005): 53–59.

Kastner, Joseph. "Long before Furs, It Was Feathers That Stirred Reformist Ire." *Smithsonian* 25, no. 4 (July 1994): 96–105.

Katz, Esther. "Sanger, Margaret." *American National Biography Online.* February 2000. Online. Available: http://www.anb.articles//15-00598.html (accessed September 28, 2002).

Kelley, Mary. "Beyond the Boundaries." *Journal of the Early Republic* 21, no. 1 (2001): 73–78.

Kershaw, Sarah. "My Sister's Keeper." *New York Times.* February 1, 2009. Online. Available: http://www.nytimes.com/2009/02/01/fashion/01womyn.html?_r=2&scp=2&scp=sarah%20 kershaw&st=cse (accessed February 3, 2009).

Kleinberg, S. J. "Race, Region, and Gender in American History." *Journal of American Studies* 33, no. 1 (1999): 85–88.

Kolbert, Elizabeth. "Turf War." *New Yorker*. July 21, 2008, 82.

Koslow, Jennifer. "Putting It to a Vote: The Provision of Pure Milk in Progressive Era Los Angeles." *Journal of the Gilded Age and Progressive Era* 3, no. 2 (April 2004): 111–44.

Kraig, Beth. "Betty and the Bishops: Was *The Egg and I* Libelous?" *Columbia Magazine* 12, no. 1 (Spring 1998): 17–22.

L., Susan A. "Another Letter from the Western Wilds." *Godey's Lady's Book* 50 (February 1855): 130.

"Landscape Gardening. Operating with Wood." *Godey's Lady's Book* 46 (March 1853): 252.

Langer, Gary. "Japan's Nuclear Crisis May Resonate in U.S." *ABCNews.com*. March 12, 2011. Online. Available: http://www.blogs.abcnews.com/thenumbers/2011/03/japans-nuclear-crisis-may-resonate-in-the-us.html (accessed May 14, 2011).

Larson, B. L., and K. E. Ebner. "Significance of Strontium-90 in Milk: A Review." *Journal of Dairy Science* 41, no. 12 (1995): 1647–62.

Lasser, Carol. "Beyond Separate Spheres: The Power of Public Opinion." *Journal of the Early Republic* 21, no. 1 (2001): 115–23.

"Lawmaker Hears Community Concerns and Industry Promises." *Southwest Research and Information Center*. Online. Available: http://www.sric.org/voices/2009/v10n2/index.html (accessed January 8, 2010).

Lawrence, Jane. "The Indian Health Service and the Sterilization of Native American Women." *American Indian Quarterly* 24, no. 3 (Summer 2000): 400–420.

Leach, Melissa, and Cathy Green. "Gender and Environmental History: From Representation of Women and Nature to Gender Analysis of Ecology and Politics." *Environment and History* 3 (1997): 343–70.

"Learning Housekeeping Continued." (San Francisco) *Daily Evening Bulletin*. May 9, 1856, issue 28, col. A.

Lerner, Gerda. "The Lady and the Mill Girl: Changes in the Status of Women in the Age of Jackson." *Mid-Continental American Studies Journal* 10 (Spring 1969): 5–15.

Lo, Malinda. "Behind the Scenes at the Michigan Womyn's Festival." April 20, 2005. Online. Available: http://www.afterellen.com/archive/ellen/Music/2005/4/michigan3.html (accessed May 17, 2007).

London Magazine. "Garden Operations for Ladies." *Lady's Book*. July 1831. 10.

Loo, Tina. "Of Moose and Men: Hunting for Masculinities in British Columbia, 1880–1939." *Western Historical Quarterly* 32, no. 3 (Autumn 2001): 296–320.

"Lorelei DeCora." *Robert Wood Johnson Foundation*. March 1, 2002. Online. Available: http://www.rwjf.org/pr/product.jsp?id=52196 (accessed May 15, 2011).

"Lorrie Otto: 'Godmother of Natural Landscaping.'" Milwaukee Public Television. Online. Available: http://www.mptv.org/garden_paths/guests.php (accessed January 26, 2009).

"Lou Henry Hoover: A Biographical Sketch." *The Herbert Hoover Presidential Library and Museum*. Online. Available: http://hoover.archives.gov/education/louhenrybio.html. Accessed September 12, 2009.

M. C. P. "Letter to the Publisher." *Godey's Lady's Book* 49 (December 1854): 538.

Mackenzie, Fiona. "Political Economy of the Environment, Gender and Resistance under Colonialism: Murang'a District, Kenya, 1910–1950." *Canadian Journal of African Studies* 25, no. 2 (1991): 226–56.

Matsumoto, Valerie. "Japanese Women during World War II." *Frontiers: A Journal of Women Studies* 8, no. 1 (1984): 6–14.

May, Jessie. "My Little Neighbor." *Godey's Lady's Book* 64 (April 1862): 356.

McBane, Margo. "The Role of Gender in Citrus Employment: A Case Study of Recruitment, Labor, and Housing Patterns at the Limoneira Company, 1893 to 1940." *California History: The Magazine of the California Historical Society* 74, no. 1 (Spring 1995): 68–87.

McLeod, Christopher. "Birth Defects Linked to Uranium Mining." *Synapse* 25, no. 23 (April 16, 1981). Online. Available: http://synapse.library.ucsf.edu/cgi-bin/ucsf?a=d&d=ucsf19810416-01.2.14&cl=CL2.1981.04&srpos=0&dliv=none&st=1&e=-------en-logical-20--1-------- (accessed November 27, 2011).

McNeill, J. R. "Observations on the Nature and Culture of Environmental History." *History and Theory* 42, no. 4 (December 2003): 5–43.

Melosi, Martin. "Environmental Justice, Political Agenda Setting, and the Myths of History." *Journal of Policy History* 12, no. 1 (2000): 43–71.

Merchant, Carolyn. "Gender and Environmental History." *Journal of American History* 76 (1990): 1117–21.

———. "George Bird Grinnell's Audubon Society: Bridging the Gender Divide in Conservation." *Environmental History* 15 (January 2010): 3–30.

———. "Women of the Progressive Conservation Movement, 1900–1916." *Environmental Review* 8, no. 1 (Spring 1984): 55–85.

Mintz, S. "Slavery Fact Sheet. " *Digital History* (2007). Online. Available: http://www.digitalhistory.uh.edu/historyonline/slav_fact.cfm (accessed February 16, 2008).

Mishler, Paul C. Review of *Children's Nature*, by Paris. *Journal of American History* 96, no. 2 (September 2009): 590.

"Mission." *Center for Health, Environment, and Justice*. May 2011. Online. Available: http://chej.org/about/mission (accessed May 13, 2011).

Mock, Michele L. "'A Message to Be Given': The Spiritual Activism of Rebecca Harding Davis." *NWSA Journal* 12, no. 1 (Spring 2000): 44–67.

Montrie, Chad. "Expedient Environmentalism: Opposition to Coal Surface Mining in Appalachia and the United Mine Workers of America, 1945–1975." *Environmental History* 6 (October 2001): 75–98.

———. "'Men Alone Cannot Settle a Country': Domesticating Nature in the Kansas-Nebraska Grasslands." *Great Plains Quarterly* 25, no. 4 (2005): 245–58.

Moore, Ted. "'Democratizing the Air': The Salt Lake Women's Chamber of Commerce and Air Pollution, 1936–1945." *Environmental History* 12 (January 2007): 80–106.

Morris, Bonnie J. "Valuing Women-Only Spaces." *Feminist Studies* 31, no. 3 (Fall 2005): 619–30.

"Most Mining and Milling of Uranium Occurs on Indian Lands." *Prairie Island Coalition*. Online. Available: http://www.no-nukes.org/prairieisland/processing.html (accessed April 21, 2011).

Nash, Linda. "The Fruits of Ill-Health: Pesticides and Workers' Bodies in Post-World War II California." *Osiris* 19 (2004): 203–19.

Nash, Margaret. "Rethinking Republican Motherhood: Benjamin Rush and the Young Ladies Academy of Philadelphia." *Journal of the Early Republic* 17 (Summer 1997): 171–92.

Neal, Alice B. "Life at a Post." *Godey's Lady's Book* 46 (January 1853): 58.

Nelson, Paula. "Women and the American West: A Review Essay." *Annals of Iowa* 50, nos. 2–3 (1989): 269–73.

Nestle, Joan. "Excerpts from the Oral History of Mabel Hampton." *Signs* 18, no. 4 (Summer 1993): 925–35.

Neushul, Peter. "Love Canal: A Historical Review." *Mid America* 69, no. 3 (1987): 125–38.

Newman, Richard. "'Where There Is No Vision, the People Perish,' Comparative Religious Responses to Hurricane Katrina and Love Canal." *Journal of Southern Religion* 12 (2010): 3. Online. Available: http://jsr.fsu.edu (accessed April 12, 2011).

Norwood, Vera. "Disturbed Landscape/Disturbing Process: Environmental History for the Twenty-First Century." *Pacific Historical Review* (February 2001): 77–90.

———. "Western Women and the Environment: A Review Essay." *New Mexico Historical Review* 65 (April 1990): 267–76.

"Nuclear Power Threatens Us All." *Center for Health, Environment, and Justice.* May 2011. Online. Available: http://chej.org/2011/05/nuclear-power-threatens-us-all (accessed May 14, 2011).

"Occidental to Pay $129 Million in Love Canal Settlement." *EPA Press Release.* December 21, 1995. Online. Available: http://www.justice.gov/opa/pr/Pre_96/December95/638.txt.html (accessed April 14, 2011).

"Our Eleventh Birthday." *Everygirls' Magazine* 10, no. 6 (February 1923).

Palone, Roxane S. "Women in Forestry—Past and Present." *Pennsylvania Forests* 92 (Spring 2001): 7–10.

Pasternak, Judy. "Oases in Navajo Desert Contained 'A Witch's Brew'" *Los Angeles Times.* November 20, 2006. Online. Available: http://articles.latimes.com/2006/nov/20/nation/na-navajo20/1-11 (accessed April 14, 2011).

———. "The Peril That Dwelt among the Navajo." *Los Angeles Times.* November 19, 2006. Online. Available: http://articles.latimes.com/2006/nov/19/nation/na-navajo19 (accessed January 8, 2010).

Pastor, Manuel, and Rachel Morello-Frosch. "Assumption Is Wrong—Latinos Care Deeply about the Environment." *San Jose Mercury News*, July 8, 2002, 6B.

Pattison, Fawn. "Examining the Evidence on Pesticide Exposure and Birth Defects in Farmworkers: An Annotated Bibliography with Resources for Lay Readers." Raleigh, N.C.: Agricultural Resources Center and Pesticide Education Project, 2006. Online. Available: http://www.beyondpesticides.org/documents/Evidence_May06.pdf (accessed June 8, 2009).

Perrin, Liese M. "Resisting Reproduction: Reconsidering Slave Contraception in the Old South." *Journal of American Studies* 35, no. 2 (2001): 258–59; 263; 266.

Perrottet, Tony. "Behind the Scenes in Monument Valley." *Smithsonian*, February 2010, 72–79.

Peterson, Sarah Jo. "Voting for Play: The Democratic Potential of Progressive Era Playgrounds." *Journal of the Gilded Age and Progressive Era* 3 (July 2007): 145–75.

Pollitt, Katha. "Phallic Balloons against the War." *The Nation.* March 24, 2003, 9.

Prescott, Cynthia. "'Why She Didn't Marry Him': Love, Power, and Marital Choice on the Far Western Frontier." *Western Historical Quarterly* 38, no. 1 (2007): 25–45.

Quiggin, John, and Tim Lambert. "Rehabilitating Carson." *Prospect Magazine.* May 24, 2008. Online. Available: http://www.prospectmagazine.co.uk/2008/05/rehabilitatingcarson (accessed November 23, 2008).

Quintero-Somaini, Adrianna, and Myra Quirindongo. *Hidden Dangers: Environmental Health Threats in the Latino Community.* New York: Natural Resource Defense Council, 2004. Online. Available: http://www.pollutionissues.com/Ho-Li/Labor-Farm.html (accessed April 23, 2011).

Rabin, Joan S., and Barbara Slater. "Lesbian Communities across the United States: Pockets of Resistance and Resilience." *Journal of Lesbian Studies* 9, no. 1/2 (2005): 169–82.

"Rachel Carson Dies of Cancer, *Silent Spring* Author Was 56." *New York Times*, April 15, 1964, 1, 25.

Ramey, Daina. "'She Do a Heap of Work': Female Slave Labor on Glynn County Rice and Cotton Plantations." *Georgia Historical Quarterly* 82, no. 4 (Winter 1998): 707–34.

Ramirez, Margaret. "Hazel M. Johnson, 1935–2011: South Side Activist Known as Mother of Environmental Justice Movement." *Chicago Tribune.* January 16, 2011. Online. Available: http://articles.chicagotribune.com/2011-01-16/features/ct-met-johnson-obit-20110116_1_cancer-alley-asbestos-removal-environmental-justice (accessed May 14, 2011).

Rawson, Donald M. "Caroline Dormon: A Renaissance Spirit of Twentieth Century Louisiana." *Louisiana History* 24, no. 2 (1983): 121–39.

Regalado, Samuel O. "Incarcerated Sport: Nisei Women's Softball and Athletics during Japanese American Internment." *Journal of Sport History* 27, no. 3 (Fall 2000): 431–44.

Riley, Glenda. "'Wimmin Is Everywhere': Conserving and Feminizing Western Landscapes, 1870–1940." *Western Historical Quarterly* 29, no. 1 (Spring 1988): 5–23.

Roberts, Gregory. "Environmental Justice and Community Empowerment: Learning from the Civil Rights Movement." *American University Law Review* 48, no. 1 (October 1998): 229–67.

Roberts, Mary Louise. "True Womanhood Revisited." *Journal of Women's History* 14, no. 1 (2002): 150–55.

Rollings-Magnusson, Sandra. "Canada's Most Wanted: Pioneer Women on the Western Prairies." *Canadian Review of Sociology and Anthropology* 37 (2000): 223–38.

Rome, Adam. "Building on the Land: Toward an Environmental History of Residential Development in American Cities and Suburbs, 1870–1990." *Journal of Urban History* 20, no. 3 (May 1994): 407–34.

———. "The Genius of Earth Day." *Environmental History* 15 (April 2010): 194–205.

———. "'Give Earth a Chance': The Environmental Movement and the Sixties." *Journal of American History* 90, no. 2 (September 2003): 525–54.

———. "'Political Hermaphrodites': Gender and Environmental Reform in Progressive America." *Environmental History* 11, no. 3 (July 2006): 440–63.

Roscoe, Robert, et al. "Mortality among Navajo Uranium Miners." *American Journal of Public Health* 85, no. 4 (April 1995): 535–40.

Rose, Margaret. "From the Fields to the Picket Line: Huelga Women and the Boycott, 1965–1975." *California Labor History* 31 (1990): 271–93.

———. "Traditional and Nontraditional Patterns of Female Activism in the United Farm Workers of America, 1962–1989." *Frontiers: A Journal of Women Studies* 11, no. 1 (1990): 26–32.

———. "'Woman Power Will Stop Those Grapes': Chicana Organizers and Middle-Class Female Supporters in the Farm Workers' Grape Boycott in Philadelphia, 1969–1970." *Journal of Women's History* 7, no. 4 (1995): 6–36.

Rosen, Ruth. "Next Time Listen to Mother; a 1997 Report Vindicates the Women Who Warned in the '60s of Radioactive Fallout Hazards." *Los Angeles Times*, August 7, 1997, 9.

Rowsome, Frank Jr. "The Verse by the Side of the Road." *American Heritage* 17 (December 1965): 102–5.

Ruiz, Vicki. "Shaping Public Space/Enunciating Gender: A Multiracial Historiography of the Women's West, 1995–2000." *Frontiers* 22, no. 3 (2001): 22–25.

Sandilands, Catriona. "Lesbian Separatist Communities and the Experience of Nature." *Organization and Environment* 15, no. 2 (June 2002): 131–61.

Scauzillo, Retts. "Women-Only Festivals: Is There a Future?" *About: Lesbian Life*. Online. Available: http://lesbianlife.about.com/od/musicreviews/a/WomenOnlyFest.htm (accessed April 7, 2007).

Schneider, Keith. "A Valley of Death for the Navajo Uranium Miners." *New York Times*. May 3, 1993. Online. Available: http://www.nytimes.com/1993/05/03/us/a valley of death-for-the-navajo-uranium-miners.html?src=pm (accessed April 21, 2011).

Schwartz, Loretta. "Uranium Deaths at Crown Point." *Ms. Magazine* (October 1979): 53–54, 59, 81–82.

Schweitzer, Ivy. "Foster's 'Coquette': Resurrecting Friendship from the Tomb of Marriage." *Arizona Quarterly* 61, no. 2 (2005): 1–32.

Scott, Joan W. "Gender: A Useful Category of Historical Analysis." *American Historical Review* 91 (December 1986): 1053–75.

Searle, Newell. "Minnesota National Forest: The Politics of Compromise, 1898–1908." *Minnesota History* 42, no. 7 (1971): 243–57.

Smalley, Andrea. "'Our Lady Sportsmen': Gender, Class, and Conservation in Sport Hunting Magazines, 1873–1920." *Journal of the Gilded Age and Progressive Era* 4, no. 4 (October 2005): 355–80.

Smith, Angela, and Simone Pulver. "Ethics-Based Environmentalism in Practice: Religious-Environmental Organizations in the United States." *Worldviews: Environmental Culture and Religion* 13, no. 2 (2009): 145–79.

Smith, Michael B. "'The Ego Ideal of the Good Camper' and the Nature of Summer Camp." *Environmental History* 11 (January 2006): 70–101.

———. "'Silence, Miss Carson!' Science, Gender, and the Reception of *Silent Spring*." *Feminist Studies* 27, no. 3 (Fall 2001): 733–52.

Smith, Sarah. "They Sawed up a Storm." *Northern Logger and Timber Processor* 48 (April 2000): 8–10, 61–62.

Smith-Rosenberg, Carroll. "The Female World of Love and Ritual: Relations between Women in the Nineteenth-Century America." *Signs: Journal of Women in Culture and Society* 1 (1975): 1–30.

Speltz, Mark. "'An Interest in Health and Happiness as Yet Untold': The Woman's Club of Madison, 1893–1917." *Wisconsin Magazine of History* 89 (Spring 2006): 2–15.

Taugher, Mike. "Her Spoiled View Inspired a Clearer Vision for the Bay." *San Jose Mercury News*, November 7, 2011, A1, A9.

Thomas, William G. III and Edward L. Ayers. "The Differences Slavery Made: A Close Analysis of Two American Communities." *American Historical Review* 108 (December 2003): 1299–1307.

Torpy, Sally. "Native American Women and Coerced Sterilization: On the Trail of Tears in the 1970s." *American Indian Culture and Research Journal* 24, no. 2 (2000): 1–22.

Trice, Dawn Turner. "Far South Side Environmental Activist Hazel Johnson and Her Daughter 'Decided to Stay Here and Fight.'" *Chicago Tribune*. March 1, 2010. Online. Available: http://articles.chicagotribune.com/2010-03-01/news/ct-met-trice-altgeld-0229-20100228_1_sewer-line-hazel-johnson-cancer (accessed April 4, 2011).

Tydeman, William E. "No Passive Relationship: Idaho Native Americans in the Environment." *Idaho Yesterdays* 39, no. 2 (Summer 1995): 23–28.

Udall, Stewart. "How the Wilderness Was Won." *American Heritage*. February–March 2000, 98–105.

Underwood, June. "Western Women and True Womanhood: Culture and Symbol in History and Literature." *Great Plains Quarterly* 5 (Spring 1985): 93–106.

Unger, Nancy C. "The Role of Gender in Environmental History." *Environmental Justice* 1, no. 3 (September 2008): 115–20.

———."The Two Worlds of Belle La Follette." *Wisconsin Magazine of History* 83 (Winter 1999–2000): 82–110.

———. "The We Say What We Think Club." *Wisconsin Magazine of History* 90, no. 1 (Autumn 2006): 16–27.

———. "Wisconsin's League Against Nuclear Dangers: The Power of Informed Citizenship." *Wisconsin Magazine of History*. Forthcoming 2013.

———. "Women for a Peaceful Christmas: Wisconsin Homemakers Seek to Remake American Culture." *Wisconsin Magazine of History* 93, no. 2 (Winter 2009–2010): 2–15.

Unger, Nancy C., and Marie Bolton. "Housing Reconstruction after the Catastrophe: The Failed Promise of San Francisco's 1906 'Earthquake Cottages.'" In *Annales de Demographie Historique*, no. 2 (2010): 217–40.

"Uranium Mining and Indigenous People." *WISE Uranium Project*. September 21, 2010. Online. Available: http://wise-uranium.org/uip.html (accessed April 22, 2011).

U.S. House of Representatives Committee on Oversight and Government Reform, Public Hearing. "Voices from the Earth." Southwest Research and Information Center. October 23, 2007. Online. Available: http://www.sric.org/voices/2009/v10n2/index.html (accessed January 8, 2010).

Vance, Linda. "May Mann Jennings and Royal Palm State Park." *Florida Historical Quarterly* 65 (1976): 1–17.

Van Kirk, Sylvia. "The Role of Native Women in the Creation of Fur Trade Society in Western Canada, 1670–1830." *Frontiers* 7, no. 3 (1984): 9–13.

Van Wienen, Mark. "Poetics of the Frugal Housewife: A Modernist Narrative of the Great War and America." *American Literary History* 7 (Spring 1995): 55–91.

Vecsey, Christopher. "American Indian Religion, Christianity, and the Environment." *Environmental Review* 9 (Summer 1985): 171–76.

Verhovek, Sam Howe. "After 10 Years, the Trauma of Love Canal Continues." *New York Times*. August 5, 1988, B1.

"The Vision of Navajo Parks and Recreation." *Navajo Nation Parks and Recreation*. Online. Available: http://www.navajonationparks.org/index.htm (accessed January 10, 2010).

Wald, Matthew. "U.S. Atomic Tests in 50's Exposed Millions to Risk." *New York Times*. July 29, 1997, A10.

Watkins, Frances. "Southwestern Athapascan Women." *Southwestern Lore* 10, no. 33 (December 1944): 32–35.

Webber, David J. "Senator Gaylord Nelson, Founder of Earth Day." January 1996. Online. Available: http://web.missouri.edu/~polidjw/Nelson.html (accessed December 29, 2005).

Weber, Christopher. "Dirty Secrets under the Schoolyard." *E—The Environmental Magazine* 22, no. 1 (January/February 2001): 22–28.

Weiner, Douglas. "A Death-Defying Attempt to Articulate a Coherent Definition of Environmental History." *Environmental History* 10, no. 3 (July 2005): 404–20.

Welter, Barbara. "The Cult of True Womanhood: 1820–1860." *American Quarterly* 18, no. 2, pt. 1 (Summer 1966): 151–74.

White, Bruce. "The Woman Who Married a Beaver: Trade Patterns and Gender Roles in the Ojibwa Fur Trade." *Ethnohistory* 46, no. 1 (1999): 109–47.

White, Richard. "Environmental History: Watching a Historical Field Mature." *Pacific Historical Review* 70, no. 1 (February 2001): 103–11.

Wisconsin Development Authority. "Wisconsin Gets a Power Program." Madison, Wisc.: WDA, 1938. *Wisconsin Historical Society*. Online. Available: http://www.wisconsinhistory.org/turningpoints/search.asp?id=1007 (accessed January 20, 2006).

"The Women's Pentagon Action." *Women and Life on Earth*. Online. Available: http://www.wloe.org/women-s-pentagon-action.77.0.html (accessed April 27, 2009).

Wong, Victoria. "Square and Circle Club: Women in the Public Sphere." *Chinese America: History and Perspectives* (1995): 127–53.

Wyat, W. K. "50,000 Baby Teeth." *The Nation*. June 13, 1959, 535–37.

Zaleski, Rob. "Soaring Triumph: Banning DDT Brought Eagles Back, and It Started in Wisconsin." *Capital Times*. July 14, 2007, A1.

Zimmerman, Jonathon. "*Brown*-ing the American Textbook: History, Psychology, and the Origins of Modern Multiculturalism." *History of Education Quarterly* 44 (Spring 2004): 46–69.

Zschoche, Sue. "Dr. Clarke Revisited: Science, True Womanhood, and Female Collegiate Education." *History of Education Quarterly* 29 (1989): 545–69.

THESES AND DISSERTATIONS

Almquist, Jennifer Marie. "Incredible Lives: An Ethnography of Southern Oregon's Women's Lands." M.A. thesis, Oregon State University, 2004.

Battle, Whitney. "A Yard to Sweep: Race, Gender, and the Enslaved Landscape." Ph.D. diss., University of Texas at Austin, 2004.

Dillon, Clarissa. "'A Large, an [sic] Useful, and a Grateful Field': Eighteenth-Century Kitchen Gardens in Southeastern Pennsylvania, the Uses of Plants, and Their Place in Women's Work." Ph.D. diss., Bryn Mawr, 1986.

Hay, Amy. "Recipe for Disaster: Chemical Wastes, Community Activists, and Public Health at Love Canal, 1945–2000." Ph.D. diss., Michigan State University, 2005.

Keefover-Ring, Wendy. "Municipal Housekeeping, Domestic Science, Animal Protection, and Conservation: Women's Political and Environmental Activism in Denver, Colorado, 1894–1912." M.A. thesis, University of Colorado, 2002.

Kleiner, Catherine B. "'Doin' It for Themselves': Lesbian Land Communities in Southern Oregon, 1970–1995." Ph.D. diss., University of New Mexico, 2003.

Massmann, Priscilla G. "A Neglected Partnership: The General Federation of Women's Clubs and the Conservation Movement, 1890–1920." Ph.D. diss., University of Connecticut, 1997.

Miller, Tamara. "'Seeking to Strengthen the Ties of Friendship': Women and Community in Southeastern Ohio, 1788–1850." Ph.D. diss., Brandeis University, 1995.

Miyamoto, Melody. "No Home for Domesticity? Gender and Society on the Overland Trails." Ph.D. diss., Arizona State University, 2006.

Ortega, Frances. "Fire in the Belly: A Case Study of Chicana Activists Working toward Environmental and Social Justice in New Mexico." Ph.D. diss., University of New Mexico, 2005.

Perrin, Liese. "Slave Women and Work in the American South." Ph.D. diss., University of Birmingham, 1999.

Ploughman, Penelope Denise. "The Creation of Newsworthy Events: An Analysis of Newspaper Coverage of the Man-Made Disaster at Love Canal." Ph.D. diss., State University of New York, Buffalo, 1984.

Slaight, Wilma Ruth. "Alice Hamilton: First Lady of Industrial Medicine." Ph.D. diss., Case Western Reserve University, 1974.

Warner, Nancy J. "Taking to the Field: Women Naturalists in the Nineteenth-Century West." M.A. thesis, Utah State University, 1995.

Young, Angela Nugent. "Interpreting the Dangerous Trades: Workers' Health in America and the Career of Alice Hamilton, 1910–1935." Ph.D. diss., Brown University, 1982.

FILMS

After Stonewall. Produced by John Scagliotti, Janet Baus, and Dan Hunt. Directed by John Scagliotti. First Run Features, 1999.

Before Stonewall. Produced by Robert Rosenberg, John Scagliotti, and Greta Schiller. Directed by Greta Schiller. First Run Features, 1985.

Index